Proceedings of the Conference on Finite Groups

ACADEMIC PRESS RAPID MANUSCRIPT REPRODUCTION

Proceedings of the Conference on Finite Groups

Edited by
William R. Scott
Fletcher Gross
Department of Mathematics
University of Utah
Salt Lake City, Utah

Academic Press, Inc. New York San Francisco London 1976

A Subsidiary of Harcourt Brace Jovanovich, Publishers

ACADEMIC PRESS, INC.
111 Fifth Avenue, New York, New York 10003

United Kingdom Edition published by
ACADEMIC PRESS, INC. (LONDON) LTD.
24/28 Oval Road, London NW1

Library of Congress Cataloging in Publication Data

Conference on Finite Groups, Park City, Utah, 1975.
Proceedings of the Conference on Finite Groups.

Bibliography: p.
Includes index.
1. Finite groups–Congresses. I. Scott, William Raymond, (date) II. Gross, Fletcher.
QA171.C679 1975 512'.2 75-13083
ISBN 0-12-633650-4

PRINTED IN THE UNITED STATES OF AMERICA

Contents

PART III. REPRESENTATIONS

PART IV. PERMUTATION GROUPS

---

*Not presented at Conference, but included because of relevance.

List of Contributors

Leo Alex
Department of Mathematics
SUNY at Oneonta
Oneonta, New York 13820

Jonathan Alperin
Department of Mathematics
University of Chicago
Chicago, Illinois 60637

Michael Aschbacher
Department of Mathematics
California Institute of Technology
Pasadena, California 91109

Clark Benson
Department of Mathematics
University of Arizona
Tuscon, Arizona 85721

Thomas R. Berger
Department of Mathematics
University of Minnesota
Minneapolis, Minnesota
55455

Gerhard Betsch
University of Tubingen
Math Inst. (74)
Tubingen, Fed. Rep. Germany

Ben Brewster
Department of Mathematics
SUNY at Binghamton
Binghamton, New York 13901

Nick Burgoyne
Department of Mathematics
University of California
Santa Cruz, California 95060

Roger Carter
Department of Mathematics
University of Warwick
Coventry, Warwickshire
ENGLAND CV4 7AL

James R. Clay
Department of Mathematics
University of Arizona
Tucson, Arizona 85721

Edward Cline
Department of Mathematics
University of Virginia
Charlottesville, Virginia 22901

Michael Collins
Department of Mathematics
University College
Oxford, ENGLAND OX1 4BH

Bruce N. Cooperstein
Department of Mathematics
University of Michigan
Ann Arbor, Michigan 48104

Katherine Bolling Farmer
Department of Mathematics
University of Florida
Gainesville, Florida 32611

Walter Feit
Department of Mathematics
Yale University
New Haven, Connecticut 06520

Arnold Feldman
Department of Mathematics
Rutgers University
New Brunswick, New Jersey 08903

Charles Ford
Department of Mathematics
Washington University
St. Louis, Missouri 63130

LIST OF CONTRIBUTORS

Joseph A. Gallian
Department of Mathematics
University of Minnesota
Duluth, Minnesota 55812

John Gillam
Department of Mathematics
Ohio University
Athens, Ohio 45701

George Glauberman
Department of Mathematics
University of Chicago
Chicago, Illinois 60637

David Goldschmidt
Department of Mathematics
University of California
Berkeley, California 94720

Daniel Gorenstein
Department of Mathematics
Rutgers University
New Brunswick, New Jersey 08903

Robert Griess, Jr.
Department of Mathematics
University of Michigan
Ann Arbor, Michigan 48104

Larry Grove
Department of Mathematics
University of Arizona
Tucson, Arizona 85721

Mark Hale, Jr.
Department of Mathematics
University of Florida
Gainesville, Florida 32601

Marshall Hall
Department of Mathematics
California Institute of Technology
Pasadena, California 91109

Koichiro Harada
Department of Mathematics
Ohio State University
Columbus, Ohio 43210

Morton E. Harris
Department of Mathematics
University of Minnesota
Minneapolis, Minnesota 55455

Trevor Hawkes
Department of Mathematics
University of Warwick
Coventry, Warwickshire
ENGLAND CV4 7AL

John Hayden
Department of Mathematics
Bowling Green State University
Bowling Green, Ohio 43403

Donald Higman
Department of Mathematics
University of Michigan
Ann Arbor, Michigan 48104

William Cary Huffman
Department of Mathematics
Dartmouth College
Hanover, New Hampshire 03755

Anthony Hughes
Department of Mathematics
University of Illinois
at Chicago Circle
P.O. Box 3438
Chicago, Illinois 60680

Elayne A. Idowu
Department of Mathematics
University of Pittsburgh
Pittsburgh Pennsylvania 15260

Wayne Jones
Department of Mathematics
University of Virginia
Charlottesville, Virginia 22901

Peter Kornya
Department of Mathematics
University of Oregon
Eugene, Oregon 97403

Jeffrey Leon
Department of Mathematics
University of Illinois
at Chicago Circle
P.O. Box 3438
Chicago, Illinois 60680

Henry S. Leonard
Department of Mathematics
Northern Illinois University
DeKalb, Illinois 60115

Richard N. Lyons
Department of Mathematics
Rutgers University
New Brunswick, New Jersey 08903

Spyros Magliveras
Department of Mathematics
SUNY at Oswego
Oswego, New York 13126

Geoffrey Mason
Department of Mathematics
University of California
Santa Cruz, California 95060

John McKay
Computing Department
Concordia University
1455 de Maisonneuve Blvd.
West Montreal, Quebec H36 1M8
CANADA

Michael O.Nan
Department of Mathematics
Rutgers University
New Brunswick, New Jersey 08903

Brian Parshall
Department of Mathematics
University of Virginia
Charlottesville, Virginia 22903

Leonard Scott
Department of Mathematics
University of Virginia
Charlottesville, Virginia 22903

David Sibley
Department of Mathematics
Pennsylvania State University
University Park, Pennsylvania 16802

Stephen Smith
Department of Mathematics
California Institute of Technology
Pasadena, California 91109

Ronald Solomon
Department of Mathematics
Ohio State University
Columbus, Ohio 43210

David B. Surowski
Department of Mathematics
University of Arizona
Tucson, Arizona 85721

David Wales
Department of Mathematics
California Institute of Technology
Pasadena, California 91109

Jay Yellen
Department of Mathematics
Colorado State University
Fort Collins, Colorado 80523

Preface

This volume consists of the papers presented at a conference on finite groups, which took place in Park City, Utah, 10-13 February, 1975. The conference was sponsored by the University of Utah and received financial support from both the University of Utah and the National Science Foundation.

As is indicated in the table of contents, the subjects discussed at the conference were in one of five main areas of finite group theory. In Part I, the problem considered is that of characterizing simple groups in terms of the local structure of a group. The structure and representations of specific simple groups is treated in Part II. Part III is concerned with the general theory of representations and characters of finite groups. Permutation groups and the connection between group theory and geometry are discussed in Part IV. Finally, Part V deals with finite solvable groups and the theory of formations. In addition to the papers presented at the conference, two other papers have been included because of their relevance to the subjects discussed at the conference.

We wish to thank Professor Jonathan Alperin who served on the organizing committee of the conference. The smooth running of the conference from day to day was due in large part to the efforts of Mrs. Ann Reed. In the editing of these proceedings, we were greatly assisted by Dr. Kenneth Klinger. This volume would not exist without the hard work and dedication of Mrs. Sandy Everett, who refused to let mere hospitalization and surgery interfere with her typing of these papers.

Finally, the editors wish to thank all of those who were present at Park City for making the conference such an enjoyable experience for both of us.

William R. Scott
Fletcher Gross

PART I

LOCAL STRUCTURE

A CHARACTERIZATION OF CERTAIN CHEVALLEY GROUPS AND ITS APPLICATION TO COMPONENT TYPE GROUPS

BY

MICHAEL ASCHBACHER

It is the purpose of this note to announce a characterization of the Chevalley groups over fields of odd order and to indicate the role of this characterization in the theory of component type groups.

THEOREM 1. Let G be a finite group with $F^*(G)$ simple. Let z be an involution in G and K a subnormal subgroup of $C_G(z)$ such that K has nonabelian Sylow 2-groups and z is the unique involution in K. Assume for each 2-element $k \in K - \langle z \rangle$ that $k^G \cap C(z) \subseteq N(K)$ and for each $g \in C(z) - N(K)$ that $[K, K^g] \leqslant O(C(z))$. Then $F^*(G)$ is a Chevalley group of odd characteristic, M_{11}, M_{12}, or $Sp_6(2)$.

COROLLARY 2. Let G be a finite group with $F^*(G)$

Partial support supplied by the Alfred P. Sloan Foundation and by NSF GP - 35678.

simple and K tightly embedded in G with quaternion Sylow 2-subgroups. Then $F^*(G)$ is a Chevalley group of odd characteristic, M_{11}, or M_{12}.

COROLLARY 3. Let G be a finite group with $F^*(G)$ simple. Let z be an involution in G and K a 2-component or solvable 2-component of $C_G(z)$ of 2-rank 1, containing z. Then $F^*(G)$ is a Chevalley group of odd characteristic or M_{11}.

The restriction on $F^*(G)$ can be removed, somewhat enlarging the class of examples. All Chevalley groups of odd characteristic, with the exception of $L_2(q)$ and ${}^2G_2(q)$, satisfy the hypothesis. The embedding of K is essentially uniquely determined. Corollary 2 follows directly from Theorem 1. Corollary 3 is not immediate but follows from [3]. The proof of Theorem 1 will appear in [4].

The possibility of such a theorem was first suggested by J. G. Thompson in January, 1974, during his lectures at the winter meeting of the American Mathematical Society in San Francisco. At that time Thompson also pointed out the significance of a certain section of the group, which is crucial to the proof.

The theorem finds its motivation in the study of component type groups. It seems best to begin the discussion by recalling some of the notation and terminology particular to this area. More basic notation can be found in [10].

A group G is *quasisimple* if $G = [G, G]$ and $G/Z(G)$ is simple. A *component* of G is a subnormal quasisimple subgroup of G. $E(G)$ is the subgroup generated by all components of G. $F(G)$ is the Fitting subgroup of G, and $F^*(G) = F(G)\,E(G)$ is the *generalized Fitting subgroup* of G. The generalized Fitting subgroup has the property that

$$C_G(F^*(G)) \leqslant F^*(G) .$$

In particular $F^*(G)$ is simple exactly when G is contained in the automorphism group of the simple group $F^*(G)$.

A 2-*component* of G is a subnormal subgroup L such that $L = [L, L]$ and $L/Z(L)$ is quasisimple. A *solvable* 2-*component* is a subnormal subgroup K such that $O(G) = O(K)$ and $K/O(K)$ is isomorphic to $L_2(3)$ or $SL_2(3)$. $B(G)$ is the subgroup of G generated by all 2-components of G which are *not* quasisimple. $L(G) = E(G)\,B(G)$. Notice that G is 2-constrained exactly when $L(G) = 1$.

A finite group G is said to be of *component type* if $L(C_G(t)) \neq 1$ for some involution t in G. Hence G is of component type when the centralizer of some involution is not 2-constrained. The main problem of component type groups is easily stated:

MAIN PROBLEM: Find all component type groups G with $F^*(G)$ simple such that the simple composition factor of some 2-component of the centralizer of some involution is of known isomorphism type.

The following conjecture and theorem are basic to this problem.

Thompson B-conjecture. $B(C_G(t)) \leqslant B(G)$, for each finite group G and each involution t in G.

Component Theorem (Aschbacher, Foote, [2], [9]). Let G be of component type with $F^*(G)$ simple and G satisfying the B-conjecture. Then, with known exceptions, G possesses a standard subgroup.

A *standard subgroup* of G is a quasisimple subgroup A of G such that $K = C_G(A)$ is tightly embedded in G, $N_G(K) = N_G(A)$, and A commutes with none of its conjugates. A subgroup K of G is *tightly embedded* in G if K has even order while K intersects its distinct conjugates in subgroups of odd order.

Once the B-conjecture is established, the Component Theorem reduces the Main Problem to the solution of *standard form problems* for the known quasisimple groups. That is, find all groups G with a standard subgroup A such that $A/Z(A)$ is isomorphic to some fixed known simple group. Hence the

Main Problem divides into two smaller problems: Establish the B-conjecture; solve the standard form problems.

We consider the second subdivision first.

THEOREM 4. (Aschbacher-Seitz [5]) Let A be a standard subgroup of G of known isomorphism type such that $m(C(A)) > 1$. Then $< A^G >$ is of known isomorphism type or of Conway type.

Hence in solving standard form problems one may assume C(A) has 2-rank 1. If C(A) has quaternion Sylow 2-subgroups, then Corollary 2 implies $< A^G >$ is a known group. This is the first application of Theorem 1. It reduces standard form problems to the case where C(A) has cyclic Sylow 2-subgroups.

A number of standard form problems have already been solved. Work of J. Walter, now in progress and discussed below, would solve all standard form problems where A is a Chevalley group of odd characteristic distinct from ${}^2G_2(q)$ or $L_2(q)$. M. Harris and R. Solomon [14] are near a solution in the case $A/Z(A) \cong A_6$ or A_7. Other solutions known to the author are: A/Z(A) of type Janko-Ree, L. Finkelstein [7], $A/Z(A) \cong M_{23}$, L. Finkelstein [8], A/Z(A) a Bender group, R. Griess and G. Seitz [13], $A/Z(A) \cong L_3(4)$, Chang, Kai Nah [6], $A/Z(A) \cong A_n$, $n \geqslant 8$, R. Solomon [15].

We now turn to the B-conjecture. Rather than establish this conjecture directly it seems best to attempt a somewhat more general problem. A group G is said to be *balanced* if $O(C_G(t)) \leq O(G)$ for each involution t in G. Notice that if G is balanced then G satisfies the B-conjecture. Hence it is sufficient to determine the unbalanced groups.

The following is a list of the known unbalanced groups with $F^*(G)$ simple.

I. Chevalley groups of odd characteristic distinct from $L_2(q)$ and ${}^2G_2(q)$.

II. $L_2(q)$, q odd.

III. A_n, n odd.

IV. $L_3(4)$ and Held's group He.

To get a start on the unbalanced group problem one appeals to the following theorem.

THEOREM 5. (Aschbacher, Gorenstein, Harada, Walter, [1], [11], [12]).

Let G be unbalanced. Then, with known exceptions, G contains an involution t and a 2-component L of $C_G(t)$, such that $Aut_G(L/Z^*(L))$ is unbalanced.

Hence if G is an unbalanced group, minimal with respect to not appearing on the list above, then G contains an involution t and a 2-component L of $C_G(t)$ such that $L/Z^*(L)$ is on the list. R. Solomon has shown that, in such a minimal counter example, $L/Z^*(L)$ is not of type III. Work of J. Walter, now in progress, is aimed at showing that if G is a group with an involution t and a 2-component L of $C_G(t)$ with $L/Z^*(L)$ of type I, then G possesses an involution z and a 2-component or solvable 2-component K of $C_G(z)$ with $z \in K$ and $m(K) = 1$. At this point Corollary 3 is applicable and shows G to be of type I. This is the second application of Theorem 1. J. G. Thompson [16] has already established this result in the case where G is a minimal counter example to the unbalanced group problem, and L exhibits the unbalance.

The cases where $L/Z^*(L)$ is of type II or IV are, to the author's knowledge, still open.

REFERENCES

1. M. Aschbacher, *Finite groups with a proper 2-generated core*, Trans. A.M.S. 197 (1974), 87-112.

2. ____________, *On finite groups of component type*, (unpublished).

3. ____________, *2-components in finite groups*, (unpublished).

4. ____________, *A characterization of the Chevalley groups over fields of odd order*, (unpublished).

5. M. Aschbacher and G. Seitz, *On groups with a standard component of known type*, (unpublished).

6. Chang, Kai Nah, (unpublished).

7. L. Finkelstein, *Finite groups with a standard component of type Janko-Ree*, (unpublished).

8. ____________, *Finite groups with a standard component isomorphic to* M_{23}, (unpublished).

9. R. Foote, *Finite groups with components of 2-rank* , I, II, (unpublished).

10. D. Gorenstein, *Finite Groups*, Harper and Row, New York, 1968.

11. D. Gorenstein and K. Harada, *Finite groups whose Sylow 2-subgroups are generated by at most* 4 *elements*, Memoirs A.M.S. 147 (1974), 1-464.

12. D. Gorenstein and J. Walter, *Centralizers of involutions in balanced groups*, J. Alg. 20 (1972), 284-319.

13. R. Griess and G. Seitz, (unpublished).

14. M. Harris and R. Solomon, *Finite groups with a standard component isomorphic to* A_6 *or* A_7, (unpublished).

15. R. Solomon, *Finite groups with* 2-*components of alternating type*, (unpublished).

16. J. Thompson, *Notes on the* B-*conjecture*, (unpublished).

California Institute of Technology
Pasadena, California

FINITE GROUPS OF ALTERNATING TYPE

BY

RONALD SOLOMON

In this article we discuss certain results in the direction of a classification of finite groups with 2-components of alternating type and some contributions to the unbalanced group problem.

By a 2-component of a group, G , we mean a perfect subnormal subgroup, K , of G with K/O(K) quasi-simple. The product of all 2-components of G is a characteristic subgroup, L(G) , called the 2-layer of G. A group, G, has trivial 2-layer if and only if G is 2-constrained. The study of the 2-layer of 2-local subgroups of finite simple groups was initiated by Gorenstein and Walter. They proved in [12] the crucial L-Balance Theorem: If G is a finite group and T a 2-subgroup of G , then $L(C_G(T)) \subseteq L(G)$. A key objective of their analysis was to prove that if G is a core-free finite group and H is a 2-local subgroup of G , then

$$O(L(H)) \subseteq Z(L(H)) ;$$

that is, L(H) is a central product of quasi-simple components.

This has recently come to be known as the $B(G)$ Conjecture.

For $\mathcal{K}$ a set of isomorphism classes of finite groups and G a finite group, we define $\mathcal{K}(G)$ to be the product of all 2-components, K, of G with $K/O(K)$ isomorphic to a member of some class in $\mathcal{K}$. We refer to such a 2-component as being of type $\mathcal{K}$. We call $\mathcal{K}$ G-maximal if for every subset, $\mathcal{K}_0$, of $\mathcal{K}$, every proper section, H, of G and every 2-subgroup, T, of H, we have $\mathcal{K}_0(C_H(T)) \subseteq \mathcal{K}_0(H)$. We remark that this is very nearly equivalent to saying that 2-components of type $\mathcal{K}$ are maximal in the ordering defined by Aschbacher in [2]. Using Theorem 5 of [2] and some signalizer functor methods of Goldschmidt [6], we can prove an analogue for 2-components of Aschbacher's Component Theorem (Theorem 1 of [2]).

THEOREM [20]: Let G be a finite group. Let $\mathcal{K}$ be a G-maximal set of isomorphism classes of quasi-simple groups such that if $[K] \in \mathcal{K}$, then

(1) $m_2(K) \geqslant 3$ and K is 2-connected in the sense of [12],

(2) $K/Z(K)$ is not involved in $\mathrm{Aut}\, K/\mathrm{Inn}\, K$,

(3) If t is an involution of $\mathrm{Aut}\, K$, then

$$O(C_{\mathrm{Aut}\, K}(t))$$

is an abelian subgroup of $\mathrm{Inn}\, K$.

If H is a 2-local subgroup of G, then either $\mathcal{K}(H) \subseteq \mathcal{K}(G)$

or $K(H)/O(K(H))$ is quasi-simple.

The following corollary is suggestive of the types of conclusions one may draw from the above theorem concerning the 2-local structure of G.

COROLLARY [20]: Let G, K, N be as above. Suppose that $L(G)C_G(L(G))$ is simple. Suppose that $K(N) \neq 1$, $|K(N)|_2$ is maximal and $|N|_2$ is maximal subject to these assumptions. Let $S \in Syl_2(N)$. Let $R = S \cap K(N)$ and $Q = S \cap C_N(K(N)/O(K(N)))$. Let $S \subseteq T \in Syl_2(G)$. Then either

(1) There exists $g \in G$ with $Q^g \cap S \neq 1$ but $Q^g \cap Q = 1$,

or (2) $S = T$ and R is strongly closed in S with respect to G.

We remark that case (2) is rendered highly unlikely by results of Goldschmidt [8].

The above results and related results on 2-components of type $PSL(2,q)$ or A_7 may be used in the analysis of unbalanced groups. We call a finite group, G, unbalanced if for some involution, t, of G, $O(C_G(t)) \not\subseteq O(G)$. The main objective of current research on finite simple groups of component type is the proof of the following conjecture.

UNBALANCED GROUP CONJECTURE: Let G be a finite unbalanced group with $L(G)C_G(L(G))$ quasi-simple. Then

$L(G)/Z(L(G))$ is isomorphic to one of the following:

(1) A Chevalley group of odd characteristic,

(2) An alternating group of odd degree,

(3) $PSL(3,4)$ or Held's group.

We remark that the $B(G)$ Conjecture would follow from the "Unbalanced Group Theorem" by inspection of the list of conclusions.

By a theorem of Gorenstein and Walter [12], if G is a finite 2-connected, unbalanced group, then there exists a pair (s,t) of commuting involutions and a 2-component, L, of $C_G(t)$ such that $< O(C_G(s)) \cap C_G(t), s >$ normalizes $O(C_G(t))L$ and $[O(C_G(t))L , O(C_G(s)) \cap C_G(t)] \not\leq O(C_G(t))$. In particular, $\mathrm{Aut}\,(L/O(L))$ is a finite unbalanced group. We call such a 2-component, L, unbalancing in G . As finite groups which are not 2-connected are known by the work of Gorenstein-Harada [10] and Aschbacher [1], the existence of unbalancing 2-components gives some inductive leverage in attacking the unbalanced group conjecture. The method of attack has been to attempt to successively rule out all of the known unbalanced groups as unbalancing 2-components in a minimal counterexample, G , to the unbalanced group conjecture. A general result with applications in this context is the following theorem.

THEOREM [19]: Let G be a finite group with $L(G)C_G(L(G))$ quasi-simple. Suppose that t is an involution of G with $t \in L \lhd\lhd C_G(t)$ and $L/O(L) \cong \hat{A}_n$ for some $n \geq 8$. Then either $G' \cong \hat{A}_n$ or $G' \cong \hat{M}c$ or Mc or $G' \cong LyS$.

Here Mc is the simple group of McLaughlin in which $C_G(t) \cong \hat{A}_8$; $\hat{M}c$ is the perfect 3-fold covering group of Mc ; and LyS is the simple group of Lyons and Sims in which $C_G(t) \cong \hat{A}_{11}$. The theorem depends on strong closure and embedding theorems of Goldschmidt [7] and Aschbacher [1] and relies ultimately on the classification by Janko [15], Janko-Wong [16], Lyons [17] and Gorenstein-Harada [9] of simple groups in which

$$C_G(t)/O(C_G(t)) \cong \hat{A}_n$$

for $8 \leq n \leq 11$. As an easy corollary one deduces the following application.

COROLLARY [19]: Let G be a finite group with $L(G)C_G(L(G))$ quasi-simple. Suppose that L is an unbalancing 2-component in G with $L/O(L) \cong \hat{A}_n$ for some $n \geq 8$. Then $G' \cong \hat{A}_m$ for some $m \geq n$.

Using the above corollary and work of Aschbacher [4] and Burgoyne [5], Thompson has proved that if G is a finite group with $L(G)C_G(L(G))$ quasi-simple and L is an unbalancing 2-component in G with $L/O(L)$ a Chevalley group of odd characteristic other than $PSL(2,q)$, then $L(G)$ is a Chevalley group of odd characteristic or $L(G)$ is isomorphic to $\hat{A}_n$ for some odd n or $L(G)$ is isomorphic to a 16-fold or 48-

fold covering group of PSL(3,4) . With these results in mind we approached the problem of ruling out unbalancing components of alternating type in a minimal counterexample to the unbalanced group conjecture.

THEOREM [22]: Let G be a minimal counterexample to the Unbalanced Group Conjecture. Then G has no unbalancing 2-component, L, with $L/O(L)$ isomorphic to A_n for $n \geq 9$.

Moreover, if all unbalancing 2-components have 2-rank 2, then the only unbalancing 2-components, L, have $L/O(L) \cong PSL(2,q)$. Thus if no unbalancing 2-components are of PSL(3,4) type or of Held type, then all unbalancing 2-components are of PSL(2,q) type.

This reduction of the Unbalanced Problem depends on our standard form results for 2-components, additional signalizer arguments and, finally, on the following identification of the alternating and symmetric groups. We recall that a tightly embedded subgroup, Q, of a group G is a subgroup of even order with $|Q \cap Q^g|$ odd for all $g \in G - N_G(Q)$. A standard component, A, of G is a quasi-simple subgroup of G with $K = C_G(A)$ tightly embedded in G, $N_G(A) = N_G(K)$ and $[A,A^g] \neq 1$ for all $g \in G$.

THEOREM (Aschbacher [3], Solomon [21]): Let G be a finite group with $L(G)C_G(L(G))$ simple. Suppose that A is a standard component of G with $A \cong A_n$ for some $n \geq 8$. Then

(1) $G' \cong A_{n+2}$ or $G \cong A_{n+4}$

or (2) $n = 8$ and $G' \cong HiS$ or $G' \cong PSL(4,4)$

or (3) $n = 10$ and $G' \cong F_5$.

Here, HiS denotes the Higman-Sims sporadic simple group and F_5 denotes the simple subgroup of the "Monster" investigated by Harada.

The heart of this problem is the identification of the alternating and symmetric groups in the cases where

$$C_G(A)/O(C_G(A))$$

is isomorphic to A_4 or to Z_2. Fusion analysis forces the G-conjugates of a Sylow 2-subgroup, Q, of $C_G(A)$ to be Q together with the "root subgroups" of A^*, where A^* is either A or a subgroup of $N_G(A)$ containing A and isomorphic to S_n. If the rank of A is large enough, a full set of generators for an A_{n+4} or S_{n+2} is "visible" inside $\bigcup_g C_G(Q^g)$. Thus one can construct an A_{n+4} or S_{n+2} inside G. Then, a criterion of Aschbacher [1] may be invoked to prove that $G \cong A_{n+4}$ or $G \cong S_{n+2}$. This latter argument fails for $n = 10$ and one constructs a standard component, K in G with $K \cong HiS$ and $C_G(K)/O(C_G(K)) \cong Z_4$. A theorem of Harada [14] then identifies G as Aut F_5. The entire construction collapses for $n = 8$ or 9 and one must resort to the construction of a maximal 2-local subgroup of G and ultimately invoke McBride [18] for the identification of PSL(4,4) and Gorenstein-Harris [11] for HiS .

For the A_7 case we need the following result of Harris and Solomon [14], discussed by Harris in this book.

THEOREM: Let G be a finite group with $L(G)C_G(L(G))$ simple. Let L be a 2-component in G with $L/O(L) = A_7$ and $C_G(L/O(L))$ of 2-rank 1. Then either $G \cong S_9$ or $G \cong$ Aut Held .

We hope to obtain a classification of finite groups with 2-components of alternating type, which is independent of any hypotheses on other unbalancing 2-components. This will require a delicate blend of fusion analysis and signalizer arguments. In any case, when the smoke clears, it is anticipated that the following result will emerge: Let G be a finite group with $L(G)C_G(L(G))$ quasi-simple. Suppose that the centralizer of some involution of G has a 2-component, L , with $L/Z^*(L) \cong A_n$ for some $n \geqslant 8$. Then $L(G)/Z(L(G))$ is isomorphic to A_m for some $m \geqslant 8$ or to one of: HiS, PSL(4,4), Mc, F_5, LyS .

REFERENCES

1. M. Aschbacher, *Finite groups with a proper 2-generated core*, TAMS, 197 (1974), 87-112.

2. __________, *On finite groups of component type*, Ill. J. Math., to appear March 1975.

3. __________, *Standard components of alternating type centralized by a 4-group*, to appear.

4. __________, *A characterization of the Chevalley groups over finite fields of odd order*, to appear.

5. N. Burgoyne, to appear.

6. D. Goldschmidt, *Weakly embedded 2-local subgroups of finite groups*, J. Algebra 21 (1972), 341-351.

7. __________, *2-Fusion in finite groups*, Ann. of Math. 99 (1974), 70-117.

8. __________, *Strongly closed 2-subgroups of finite groups*, to appear.

9. D. Gorenstein and K. Harada, *On finite groups with Sylow 2-subgroups of type* $\hat{A}_n$, $n = 8$, 9 , 10 *and* 11 , J. Algebra 19 (1971), 185-227.

10. D. Gorenstein and K. Harada, *Finite groups whose 2-subgroups are generated by at most 4 elements*, Mem. Amer. Soc. 147 (1974), 1-464.

11. D. Gorenstein and M. Harris, *A characterization of the Higman-Sims simple group*, J. Algebra 24 (1973), 565-590.

12. D. Gorenstein and J. Walter, *Balance and generation in finite groups*, J. Algebra 33 (1975), 224-287.

13. M. Harris and R. Solomon, *Finite groups with a 2-component of type* A_6 , to appear.

14. K. Harada, *Some topics related to the simple group* F , to appear.

15. Z. Janko, *The nonexistence of a certain type of simple group*, J. Algebra 18 (1971), 245-253.

16. Z. Janko and S. K. Wong, *A characterization of the McLaughlin's simple group*, J. Algebra 20 (1972), 203-225.

17. R. Lyons, *Evidence for a new finite simple group*, J. Algebra 20 (1972), 540-569.

18. P. McBride, *Finite groups with a Sylow 2-subgroup of type* $PSL(n,2^m)$, Ph.D. Thesis, University of Illinois at Chicago Circle, 1975.

19. R.Solomon, *Finite groups with intrinsic* 2-*components of type* $\hat{A}_n$, J. Algebra 33 (1975), 498-522.

20. R. Solomon, *A standard form theorem for* 2-*components in finite groups*, to appear.

21. R. Solomon, *Standard components of alternating type*, to appear.

22. R. Solomon, *Finite groups with* 2-*components of alternating type*, to appear.

RUTGERS UNIVERSITY
NEW BRUNSWICK, NEW JERSEY

FINITE GROUPS OF 2-LOCAL 3-RANK AT MOST 1

BY

DANIEL GORENSTEIN* AND RICHARD LYONS

The known finite simple groups of 2-local 3-rank at most one--that is, in which all 2-local subgroups have cyclic 3-subgroups--are given by the following table:

3-RANK 0

$Sz(2^n)$, n odd

3-RANK 1

$L_2(q)$, $q > 3$, $q \neq 3^n$,

$L_3(q)$, $q \equiv -1 \pmod 3$,

$U_3(q)$, $q \equiv 1 \pmod 3$,

J_1

*Presented by Daniel Gorenstein

3-RANK 2

$L_2(9)$,

$L_3(3)$,

$U_3(3)$,

$L_3(2^n)$, n even, $n \not\equiv 0 \pmod 3$,

$U_3(2^n)$, n odd, $n > 1$,

$Sp_4(2^n)$, n odd, $n > 1$,

$G_2(2^n)$, n odd, $n > 1$,

${}^3D_4(2^n)$, n odd,

${}^2F_4(2^n)$, n odd, $n > 1$,

${}^2F_4(2)'$,

M_{11}

3-RANK AT LEAST 3

$L_2(3^n)$, $n \geqslant 3$.

As is customary, one begins the classification of all finite simple groups of 2-local 3-rank at most one with the conjecture that the above list is complete; and one then proceeds to investigate a minimal counterexample G to this proposed theorem.

The natural place to begin the analysis is in the case in which the centralizer of some involution of G is not 2-constrained--equivalently, G is of *component type*. However, the general problem of groups of component type is at present the object of intense investigation and there are very real prospects that within the next few years all simple groups of component type whose proper subgroups have composition factors of known type will have been determined. Even if this goal is not completely reached, it is even more likely that the results obtained will be strong enough to yield as a corollary the classification of simple groups of component type of 2-local 3-rank at most one.

As a result, we have decided to leave this case aside for the present and to focus our attention on the case in which the centralizer of every involution of G is 2-constrained. Using some standard theorems, one reduces quickly to the case in which G is of *characteristic 2-type*--that is, every 2-local subgroup of G is 2-constrained and has a trivial core and $SCN_3(2)$ is nonempty in G.

As Thompson's work on N-groups and 3'-groups indicates, this case itself divides into three major subcases. To describe these, we introduce the following terminology:

$m_{2,p}(G)$ = max {rank (A) | A ranging over all abelian p-subgroups which lie in a 2-local subgroup of G, p an odd prime} .

$e(G)$ = max $\{m_{2,p}(G) | p$ ranging over all odd primes$\}$.

The major subdivision corresponds to the following three possibilities:

(I.) $e(G) \geqslant 3$. (II.) $e(G) = 2$. (III.) $e(G) = 1$.

Note that when $e(G) = 1$, every 2-local subgroup of G has cyclic Sylow p-subgroups for all odd primes p and hence G is a *thin* group. In the extension of Thompson's N-group analysis to the classification of groups with solvable 2-local subgroups, Janko treated the case $e(G) = 1$, F. Smith the case $e(G) = 2$, and Lyons and I treated the case $e(G) \geqslant 3$. It was therefore natural for us to begin our work on groups of 2-local 3-rank one with the same case.

Thus our working hypothesis is as follows:

HYPOTHESIS:
1. G is simple with $m_{2,3}(G) \leqslant 1$;
2. G is of characteristic 2 type;
3. The nonsolvable composition factors of every proper subgroup of G are all isomorphic to groups in the above table;
4. $m_{2,p}(G) \geqslant 3$ for some odd prime p.

We remark that if G itself is isomorphic to one of the groups in the table, then, in fact $e(G) \leqslant 2$. Thus there exist no known finite simple groups which satisfy our hypothesis. Hence if our conjecture is indeed correct, the object of our analysis in this case must be to derive a contradiction. The global strategy to accomplish this involves a generalization of the techniques which Thompson developed to treat the corresponding cases in his classification of N-groups and of 3'-groups. These themselves divide into three major parts:

A. For each odd prime p such that $m_{2,p}(G) \geqslant 3$, prove that G possesses a so-called "uniqueness subgroup" M_p and that M_p can be taken to be a 2-local subgroup of G.

B. Restrict the possibilities for $O_2(M_p)$ by analysis of the weak closure in M_p (with respect to G) of certain elementary abelian normal 2-subgroups of M_p.

C. In these residual cases, prove that M_p is strongly embedded in G by an analysis of the weak closure in M_p (with respect to G) of certain subgroups of $O_2(M_p)$.

At the present time, Lyons and I have completed the analysis of Part A and have made some preliminary investigations of Part B. I should like now to state the results which we have proved under Part A and then to make a conjecture concerning the nature of some of the residual cases which we anticipate will occur in the analysis of Part B.

We first recall some terms which we shall need for the

statements of our results. Let X be an arbitrary finite group, p a fixed prime, and P a Sylow p-subgroup of X. Then by definition, for any positive integer k:

$$\Gamma_{P,k}(X) = \langle N_X(Q) \mid Q \leq P, \ Q \text{ has p-rank at least } k \rangle.$$

$\Gamma_{P,k}(X)$ is called the k-*generated* p-*core* of X. It is determined up to conjugation by the choice of the Sylow p-subgroup P of X. If $\Gamma_{P,1}(X)$ is a *proper* subgroup of X, we say that $\Gamma_{P,1}(X)$ is *strongly* p-*embedded* in X. This agrees with the usual definition of strong embedding in the case $p = 2$.

Next we define

$$L(X) = \text{unique largest semisimple normal subgroup of } X.$$

$L(X)$ is called the *layer* of X. (In the Bender terminology, it is denoted by $E(X)$).

Finally we define

$$L_p(X) = \text{unique normal subgroup of } X \text{ which is minimal subject to covering } L(X/O_{p'}(X)).$$

$L_p(X)$ is called the p-*layer* of X.

Now we can state our results. Here G will denote a simple group which satisfies Hypotheses (1), (2), (3), (4) listed above.

THEOREM A. If p is an odd prime such that

$$m_{2,p}(G) \geq 3$$

and if P is a Sylow p-subgroup of G, then the 2-generated p-core $\Gamma_{P,2}(G)$ of G lies in a 2-local subgroup of G.

THEOREM B. If p and P are as in Theorem A, then $\Gamma_{P,2}(G)$ lies in a unique maximal 2-local subgroup of G which we denote by $M(P)$. Moreover, the following conditions hold:

(i) $M(P)$ controls G-fusion in P. In particular, $O^p(M(P)) = M(P)$;

(ii) $M(P) \cap M(P)^g$ has cyclic Sylow p-subgroups for $g \in G - M(P)$;

(iii) Either $M(P)$ is strongly p-embedded in G or else the following conditions are satisfied:

(1) P is abelian;

(2) $M(P)$ is solvable of p-length 1;

(3) There is P_0 of order p in P such that P_0 is weakly closed in P with respect to G;

(4) If we set $C = C_G(P_0)$, then $L_p(C) = L(C) \cong L_2(p^n)$, $n \geq 2$, and $L_p(C) \not\leq M$;

(5) $C_C(L_p(C)) = O_{p',p}(C)$; and

(6) $O_{p'}(M)P_0$ is a Frobenius group with kernel $O_2(M)$ and complement $O_{p'}(C)P_0$.

Theorems A and B relate primarily to the p-local structure of G. Our last result concerns the 2-local structure of G.

THEOREM C. M(P) contains any 2-local subgroup H of G such that $H \cap P$ is noncyclic.

We remark that Theorems A, B, and C apply, in particular, if G is a 3'-group inasmuch as any such group certainly has 2-local 3-rank at most 1. Hence they provide alternative proofs of the corresponding assertions in Thompson's analysis of 3'-groups. (Note that Theorem B(iii) reduces then to the assertion that M(P) is strongly p-embedded in G).

We also remark that the proof of Theorem A relies heavily on signalizer functor methods for odd primes. It turns out that each of the known simple groups K with $m_{2,3}(G) \leqslant 1$ is *locally 2-balanced* with respect to any odd prime p--that is, if K is normal in a group H with $C_H(K) = 1$ and if B is any subgroup of H of type (p,p), then

$$\Delta_H(B) = \bigcap_{b \in B^\#} O_{p'}(C_H(b)) = 1.$$

Furthermore, they satisfy a condition which we call p-Schreier --namely, for any Sylow p-subgroup P of K, $C_{Aut(K)}(P)$ has a normal p-complement (for p = 2, this is true for any finite simple group K by a theorem of Glauberman).

As a consequence of these conditions, we can prove that G itself is 2-*balanced with respect to* any odd prime p--that is, if B is any subgroup of G of type (p,p) and x is any element of G of order p which centralizes B, then

$$\Delta_G(B) \cap C_G(x) \leqslant O_{p'}(C_G(x)).$$

This leads us to construct the standard "2-balanced signalizer functor". For any elementary abelian p-subgroup A of G of rank at least 4 and any $a \in A^\#$, we set

$$\theta(a) = \langle C_G(a) \cap \Delta_G(B) | B \leqslant A,\ B \text{ of type } (p,p) \rangle;$$

and we set

$$\theta(G;A) = \langle \theta(a) | a \in A^\# \rangle.$$

In a separate paper we give some general conditions which suffice to prove that θ is an A-signalizer functor on G and that $\theta(G;A)$ is a p'-group. (When $\theta(a)$ is nonsolvable for some $a \in A^\#$, it is an open question whether either of these statements holds in an arbitrary group G which is 2-balanced with respect to p). The results of this paper are critical for the proof of Theorem A.

These uniqueness theorems are the precise analogues of the corresponding results which Thompson uses in his analysis of N-groups and 3'-groups in the case $e(G) \geqslant 3$. If $M = M(P)$

is such a uniqueness subgroup, Thompson subjects $O_2(M)$ to a long and difficult analysis, the object of which is to prove that M possesses no elementary abelian normal 2-subgroups of order exceeding 4. The residual cases which are left at this point are then treated by separate arguments, which lead to ultimate contradictions in both the case of N-groups and of 3'-groups.

The first portion of Thompson's argument is the assertion that $V = \Omega_1(R_2(M))$ has order at most 4. Here, by definition, $R_2(M)$ is the unique abelian normal 2-subgroup of M which is maximal subject to the condition $O_2(M/C_M(R_2(M)) = 1$.

For the past several months, Lyons and I have been attempting to generalize this result to our simple group G in the 2-local 3-rank at most 1 situation. Although this analysis is not yet completed, I should like to conclude with a precise statement of the analogue we are aiming for. At one place in our argument, we make a choice of the prime p; and at the present time at least, our hoped-for result will apply only to the corresponding uniqueness subgroup $M = M(P)$. Again $V = \Omega_1(R_2(M))$ and we put $\overline{M} = M/C_M(V)$.

CONJECTURE. For some choice of p and M, one of the following holds:

(i) $|V| \leqslant 4$;

(ii) $|V| = 2^n$, $n \geqslant 3$, $\overline{M}$ is solvable, and $\overline{M}$ contains a cyclic normal subgroup $\overline{F}$ of order $2^n - 1$ acting transitively on $V^\#$;

(iii) $|V| = 2^{2n}$, $n \geqslant 2$, and $\overline{M}$ contains a normal subgroup $\overline{L} \cong SL(2,2^n)$ acting transitively on $V^\#$; or

(iv) $|V| = 8$ or 16 and $\overline{M} \cong GL(3,2)$.

RUTGERS UNIVERSITY
NEW BRUNSWICK, NEW JERSEY

FINITE SIMPLE GROUPS OF CHARACTERISTIC 2,3-TYPE

BY

GEOFFREY MASON

1. INTRODUCTION: In the Introduction to his work on N-groups J.G. Thompson [3] wrote, "The work is flawed because as yet I have been unable to axiomatize the properties of solvable groups which are "really" needed ... . If this is done, the usual benefits will undoubtedly accrue: stronger theorems, shorter proofs".

Whether or not the work *is* flawed is a matter for one's own conscience, however the validity of the second statement is undeniable. We shall discuss below some recent progress in this direction.

2. BACKGROUND: It has been apparent for a long while that the correct setting for axiomatization of the N-group paper is that in which all 2-local subgroups are 2-constrained. (Recall that if p is a prime and X a group, X

is called p-constrained if $C_X(P) \leqslant O_{p',p}(X)$ whenever P is a Sylow p-subgroup of $O_{p',p}(X)$). As a consequence of Hall and Higman's lemma 1.2.3 a solvable group is p-constrained for every prime p, so in an N-group all p-locals are p-constrained.

So the ultimate problem in this field is

PROBLEM 1: Find all simple groups G, all of whose 2-local subgroups are 2-constrained.

A number of results, many of which are themselves extensions of results in [3], have been obtained over the past ten years, and which are important for studying problem 1. We mention explicitly Bender's classification of groups with a strongly embedded subgroup and the Gorenstein-Goldschmidt-Glauberman Signalizer Functor theorem. As a consequence of these and other results, problem 1 has been completely solved in the following cases:

(a) T is a Sylow 2-subgroup of G and $SCN_3(T) = \emptyset$.

(b) T is a Sylow 2-subgroup of G and T normalizes a non-identity subgroup of odd order.

In the language of [3], these correspond to the case that $2 \notin \pi_4$. So we are faced with

PROBLEM 1′: Find all simple groups G such that

$2 \in \pi_4$ and all 2-local subgroups are 2-constrained.

If G satisfies the conditions of problem 1' we say that it is of characteristic 2-type.

One way of studying groups of characteristic 2-type is by focusing attention on the prime 3. There are a number of reasons why this is a good idea, not the least of which is that Thompson's classification of 3'-groups at least provides us with some elements of order 3! In any case, having decided on this course, we further subdivide the problem into two, according as the 3-local subgroups are, or are not, all 3-constrained. The part we are concerned with here is

PROBLEM 2: Find all simple groups of characteristic 2-type, all of whose 3-local subgroups are 3-constrained.

We observe that problem 2 is much closer in spirit to the N-group situation than problems 1 or 1'. Almost all of the difficulties of [3] lie in the {2,3}-subgroups, and in problem 2 the primes 2 and 3 are both assumed to be reasonably well-behaved.

3. STATEMENT OF MAIN THEOREMS:

One may well be inclined to believe that, as with the prime 2, the hardest part of problem 2 is the case $3 \in \pi_4$. The pessimist might even think it is then necessary to pass to consideration of the prime 5, subdivide according as the 5-locals are 5-constrained or not, and so on *ad infinitum* ... . One of the consequences of theorem 1 is that such a procedure is unnecessary.

To state our theorems we need a little more notation.

Let p be a prime, X a group. Then define

$A(p) = A_X(p) = \{V \leqslant X \mid V$ is elementary abelian of order p^2 and V is contained in an elementary abelian subgroup of order $p^3\}$.

$B(p) = B_X(p) = \{V \leqslant X \mid V$ is elementary abelian of order p^2 and $V \notin A(p)\}$.

(Goldschmidt has suggested the following appropriate mnemonic: A = All-right, B = Bad.)

In sections 7-9 of [3] Thompson studies the situation in which $\{2,3\} \subseteq \pi_4$ and in which there is a "big enough" $\{2,3\}$-subgroup. The following theorem is the appropriate extension of his results to the framework of problem 2.

THEOREM 1: Let G be a finite simple group of characteristic 2-type, all of whose 3-local subgroups are 3-constrained. Let R be a Sylow 3-subgroup of G, and suppose that the following conditions hold:

(a) $SCN_3(R) \neq \emptyset$

(b) R normalizes no non-trivial 2-groups of G.

(c) Some 2-local contains an element of $A_G(3)$.

Then G is isomorphic to one of the following groups: PSp(4,3), $G_2(3)$, $U_4(3)$.

It is in some sense unfortunate that one has to differentiate between the elements of A(p) and B(p), however the methods of handling these two sets are completely different. They also differ in the sense that elements of B(p) never appear unless they have to! This is the content of

THEOREM 2: Let G be a finite simple group of characteristic 2-type, all of whose 3-local subgroups are 3-constrained. Let R be a Sylow 3-subgroup of G, and suppose that

(a) R has rank at least 2.

(b) Some 2-local contains an elementary subgroup of order 9.

(c) No 2-local contains an element of $A_G(3)$. Then R has rank exactly 2.

4. OUTLINE OF THE PROOF:

We will here give the barest outline of the proof of theorems 1 and 2. Full details will appear in [2].

The hypotheses of theorem 1 are patently non-inductive,

so to prove theorem 1 we proceed with a direct construction of the centralizer of a central involution. We ultimately show that if t is a central involution of G and $C = C_G(t)$ then C is isomorphic to the centralizer of a central involution of either $PSp(4,3)$, $G_2(3)$ or $U_4(3)$, then quote prior characterizations due to Janko and Phan.

In these three simple groups C has a quite simple structure. In each case $O_2(C)$ is the central product of two quaternion groups, while $C/O_2(C)$ has order 18, 18 and 36 respectively.

Starting with our group G and the subgroup C, all of the difficulty lies in the determination of $O_2(C)$. The first reductions are obtained during the course of some joint-work with Ken Klinger [1]. As a result of some fairly general work we first show that $O_2(C)$ is of symplectic-type. Then a closer scrutiny yields that $O_2(C)$ can in fact be taken to be extra-special of width 2, 3 or 4.

The analysis is taken up again in [2]. By this time the 3-structure of G is quite limited, and we utilize this knowledge to ultimately show, in case $O_2(C)$ has width 3 or 4, that $\langle t \rangle = Z(O_2(C))$ is weakly closed in C. Glauberman's Z^*-theorem now yields a contradiction, so we are forced to conclude that $O_2(C)$ has width 2. The precise structure of C is now easily obtained, and the theorem is proved.

Turning to theorem 2, we are confronted with the Bad elements of B(3)! As in [3] they are handled by means of transitivity results. Indeed theorem 2 is an immediate consequence of the following result:

PROPOSITION: Assume the hypotheses of theorem 2, and let $V \in B(3)$ be such that V lies in a 2-local subgroup of G . Suppose further that R has rank at least 3. Then $O_{3'}(C(A))$ is transitive on $\mathcal{U}_G^*(A;2)$.

The only problem in proving the proposition is to identify the simple sections of G involved in the appropriate 2-local subgroups. As the situation is again non-inductive one has to do this constructively. This requires some effort and one has to quote a number of deep characterization theorems. In any case, we get our theorem.

5. CONCLUDING REMARKS:

By combining theorems 1 and 2 we obtain the following contribution to problem 2.

THEOREM 3. Let G be a finite simple group of characteristic 2-type, all of whose 3-local subgroups are 3-constrained. Let R be a Sylow 3-subgroup of G. Then exactly one of the following holds:

(a) G has 2-local 3-rank at most 1.

(b) G has 2-local 3-rank 2 and R has rank 2.

(c) R has rank at least 3 and R normalizes a non-trivial 2-subgroup of G.

The groups occurring in (a) and (b) are presently under investigation by various authors, and we may hope that they will eventually be completely determined. It will then remain to analyze groups in (c). This is closely connected with the problem of determining those groups which possess a strongly 3-embedded subgroup. Should this ever be solved, problem 2 will no longer be a problem!

REFERENCES

1. K. Klinger and G. Mason, *Centralizers of* p-*groups in groups of characteristic* 2,p-*type*, (to appear).

2. G. Mason, *Two theorems on groups of characteristic* 2-*type*, (to appear).

3. J. G. Thompson, *Non-solvable finite groups all of whose local subgroups are solvable*, Bull. Amer. Math. Soc. (1968), 383-437.

UNIVERSITY OF CALIFORNIA, SANTA CRUZ
SANTA CRUZ, CALIFORNIA

3-STRUCTURE IN FINITE SIMPLE GROUPS

BY

MICHAEL J. COLLINS

In the study of finite simple groups in which all 2-local subgroups are 2-constrained, attention passes to the prime 3. Gorenstein has discussed the situation where the 3-rank of each 2-local subgroup is at most one, and Mason that in which every 3-local subgroup is 3-constrained; we shall be concerned with situations in which neither of these hold in general. Work is presently at a very early stage; what we hope to do is to provide the motivation for a programme of attack.

First we consider the 3-structure of the known simple groups in which all 2-locals are 2-constrained. These can be conveniently divided into three categories:

(a) groups of Lie type of small rank over fields of odd characteristic, and A_7 ;

(b) Sporadic groups; and

(c) groups of Lie type over fields of characteristic 2.

In case (a), all elements of order 3 have soluble centralizers. In case (b), if t is an element of order 3 such that $C(t)$ is not 3-constrained, then $O^2(C(t))$ has one of the two forms $\langle t \rangle \times A$, where A is simple, or $\hat{A}$, a perfect central extension of a simple group A by a group of order 3; respectively we have $N(A) = N(\langle t \rangle)$ or $N(\hat{A}) = N(\langle t \rangle)$.

Burgoyne has computed the centralizers of elements of order 3 in groups of Lie type of characteristic 2. It turns out that all components of such centralizers (in the strict sense of minimal subnormal quasisimple subgroups) are themselves groups of Lie type of characteristic 2; furthermore, if t is an element of order 3 and $E(C(t))$ is the product of the components of $C(t)$, then $C(t)/E(C(t))$ is soluble. There may be as many as three components; however, we particularly note the following in contrast with the situation for sporadic groups. Let G be of Lie type over $GF(q)$, for $q = 2^m \geqslant 4$, other than $A_1(q)$, $A_2(q)$, ${}^2A_2(q)$, ${}^2B_2(q)$ or ${}^2F_4(q)$. Then G contains an element t of order 3 such that $E(C(t))$ is quasisimple and, putting $A = C_G(E(C(t)))'$, $t \in A$ and $A \cong SL(2,q)$. For $q = 2$, similar statements can be made, allowing for the solubility of $SL(2,2)$ and $SU(3,2)$.

It would seem that this property may afford an approach to distinguishing between the groups of Lie type and sporadic groups, but before pursuing this we make some observations of a general nature. We first note that if G is a known simple group in which all 2-local subgroups are 2-constrained, and if N is a 3-local subgroup of G, then $O_2(N) = 1$ unless $G \cong PSL(2,q)$ for $q \equiv \pm 1 \pmod{12}$, in which case N will be dihedral of order divisible by 12, or $G \cong A_7$ or M_{22}, both of which have centralizers of the form $Z_3 \times A_4$. So we first pose the following.

PROBLEM 1. Let G be a finite simple group in which all 2-local subgroups are 2-constrained. Let N be a 3-local subgroup of large enough 3-rank. Is $O_2(N) = 1$?

By "large enough," one would like results for "at least three." Maybe the problem is impossibly difficult in this generality; at a glance it may look like an analogue of the unbalanced group problem, though we do not expect groups to arise. However, 2-constraint does force $O_2(N_G(O_2(N)))$ to be strictly larger than $O_2(N)$ in a counterexample, though one may end up with a 2-local subgroup which is "strongly 3-embedded" or some "similar" configuration, but questions of this nature will have to be faced. Perhaps, given Mason's work, one should restrict one's attention to the case where N is assumed *not* to be 3-constrained. Our next problem definitely relates to such situations.

PROBLEM 2. Find an analogue to Aschbacher's component theorem for the prime 3 for groups in which all 2-locals are 2-constrained.

This has been left deliberately vague. What is wanted is to assume that such a group is "of component type for the prime 3," take E as a maximal component, and obtain information about $C(E)$ under suitable assumptions. Possibly one should assume a "$B_3(G)$-conjecture" (and prove that separately); alternatively, it may be possible to obtain results for "3-components" under a slightly stronger version of Problem 1, namely that $O_{3'}(N)$ have odd order. Gorenstein and Lyons have shown that under suitable hypotheses, $C(E)$ will have cyclic Sylow 3-subgroups; actually, they can do this for any odd prime in place of 3. However, we would have in mind conclusions relating to the 2-structure of $C(E)$, hopefully of a type which would allow us to deduce that if $O^{2'}(C(E)) \neq 1$, then $O^{2'}(C(E))$ was in a list of groups including $SL(2,q)$ for q even, and not too much else. For this, of course, one would need an affirmative answer to Problem 1, or even to the stronger version mentioned above. This $SL(2,q)$ will in many cases be the subgroup A of our discussion of groups of Lie type earlier, though the linear groups provide examples to show that this does not always arise from a maximal component.

However, for a conclusion of this type it is critical to take the prime 3; any SL(n,q) having an irreducible element of prime order can appear in this way for that particular prime in a linear group of higher dimension.

Suppose now that we have an affirmative answer to Problem 2 with the type of conclusion indicated above. Then there is an element t of order 3 and a component E of C(t) such that C(E) has a characteristic subgroup A isomorphic to SL(2,q). By Gorenstein-Lyons, we shall suppose that $t \in A$. Then $E \lhd C(t)$ so that, if $g \in C(t)$, g normalizes C(E) and hence also A. This argument does not involve the maximality of E, and may be modified if E is only a 3-component. This leads to the following hypothesis.

(*) G contains a subgroup A isomorphic to SL(2,q), where $q = 2^m$, such that, if t is an element of order 3 in A, then $C(t) \subseteq N(A)$.

This condition is satisfied by all groups of Lie type of characteristic 2 with the exceptions of PSL(3,q) for $q \equiv 1 \pmod 3$, PSU(3,q) for $q \equiv -1 \pmod 3$, Sz(q) and ${}^2F_4(q)$, although it does hold in SL(3,q) and SU(3,q). In ${}^2F_4(q)$ there is one class of elements of order 3, having centralizers isomorphic to SU(3,q). In the Chevalley groups, the subgroup A may be chosen to be generated by root subgroups corresponding to a long root and its negative; in $F_4(q)$ and $B_\ell(q)$, though not $G_2(q)$, one may also take short roots. By abuse of terminology, we shall refer to such subgroups as root $SL(2,q)$'s. There are no special exceptions for $q = 2$ or groups of small rank; however, the corresponding component E need not then exist.

Hypothesis (*) is the natural analogue in characteristic 2 of the classical involution in odd characteristic; if q were assumed odd and t taken as an involution, then A would be tightly embedded. So, parallelling Aschbacher's work, we should like to characterize the groups of Lie type of characteristic 2 by this property. For $q = 2$, however, any group generated by a class of 3-transpositions will have this property, but there seem to be no known exceptions for $q \geqslant 4$.

To characterize these groups, it is desirable to forget about 2-constraint and use (*) as the sole hypothesis. By doing so, it is possible to apply induction, and the following result gives rise to a natural division into three cases.

PROPOSITION. Assume Hypothesis (*). Then the following hold:

(i) if $t^g \in N(A)$ for some $g \in G$, then $t^g \in A\cdot C(A)$, and

(ii) if $t^g \in C(A)$, then $A^g \subseteq C(A)$.

PROOF. Assume (i) false. Then t^g acts on A as a field automorphism, so we may assume that $q \geqslant 8$. Put $A_1 = C_A(t^g)$; then $A_1 \cong SL(2,q_1)$ where $q_1{}^3 = q$. Without loss, we may suppose that $t \in A_1$. Then

$$A_1{}^{g^{-1}} \subseteq C(t) .$$

Since $N(A)/A{\cdot}C(A)$ is abelian, for $q > 8$ we have

$$t^{g^{-1}} \in A_1{}^{g^{-1}} = (A_1{}^{g^{-1}})^{(\infty)} \subseteq C(t)^{(\infty)} \subseteq C(A) ,$$

while for $q = 8$ we have $[N(A):A{\cdot}C(A)] = 3$ so that

$$t^{g^{-1}} \in A_1{}^{g^{-1}} = O^{2'}(A_1{}^{g^{-1}}) \subseteq O^{2'}(C(t)) \subseteq C(A) ;$$

in either case, $A^g \subseteq C(t)$ so that

$$t^g \in A^g = (A^g)^{(\infty)} \subseteq C(A) ,$$

contrary to assumption. So (i) holds.

Now suppose that $t^g \in C(A)$. Then $A^{g^{-1}} \subseteq C(t)$. If $q = 2$, $N(A) = A \times C(A)$, whence $A^{g^{-1}} \subseteq C(A)$. If $q \geqslant 4$, A is perfect and $A^{g^{-1}} \subseteq C(t)^{(\infty)} \subseteq C(A)$. In either case,

$$[A^{g^{-1}},A] = 1 \quad \text{and} \quad [A,A^g] = 1 .$$

The value of (ii) is that in an inductive situation we would immediately have information about the subgroup

$$< A^g | A^g \subseteq C(A) > .$$

So we divide into cases as follows:

I. $\langle t \rangle$ is weakly closed in $C(t)$.

II. $\langle t \rangle$ is not weakly closed, but $t^G \cap C(A) = \emptyset$.

III. $t^G \cap C(A) \neq \emptyset$.

Clearly in any full characterization the steps should be taken in this order. However, we have considered some particular configurations that occur in III and will be concerned primarily with them.

First we remark that Hypothesis (*) imposes some severe restrictions for step I. Clearly $N(A)$ contains a Sylow 3-subgroup of G, and an elementary transfer argument allows us to assume that 3 does not divide the index $[N(A):A\cdot C(A)]$. So a Sylow 3-subgroup of A is a direct factor of a Sylow 3-subgroup of G. Now the obvious approach is to attempt to determine $\langle t^G \rangle$; the goal should be that $\langle t^G \rangle / O_{3'}(\langle t^G \rangle) \cong A$ except for $q = 2$.

Step II is potentially the most difficult since the situation does arise in known groups. In $SL(3,q)$ and $SU(3,q)$ one has $C(A)$ cyclic; in $G_2(q)$ and ${}^3D_4(q)$, $C(A)$ is isomorphic to $SL(2,q)$ and $SL(2,q^3)$ respectively.

We now turn to step III and consider a special case for the remainder of the paper, though the methods almost certainly generalize. We fix A and t, and put

$$B = \langle A^g | A^g \subseteq C(A) \rangle .$$

We shall assume that $B \cong SL(n,q)$, and restrict our attention to $q \geqslant 4$. Suppose that $t^g \in B$; we must first determine its conjugacy class in B. Hypothesis (*) holds for A^g as a subgroup of B; since $t^g \in A^g$ and $A^g \subseteq C_G(C(t^g)^{(\infty)})$, consideration of rational canonical forms shows that A^g can be represented in B as a group of matrices having a fixed space of codimension 2. Thus A^g is a root $SL(2,q)$ in B. The class of t^g is now uniquely determined. Now fix g with $t^g \in B$, and put $H = \langle B, B^g \rangle$.

THEOREM. Let H be defined as above, and suppose that $n \geqslant 6$. Then either H is a homomorphic image of $SL(n+2,q)$ or H is a universal covering group of $E_6(q)$.

The principle of the proof is to use the Steinberg relations for B and B^g, and show that the restricted Steinberg relations [1] can be satisfied by a set of generators for H. In practice this is done by taking a set of root $SL(2,q)$'s corresponding to a set of fundamental roots starting with $B \cap B^g$, and is possible as Dynkin diagrams contain no loops. This will yield H as a homomorphic image of the universal covering group of the appropriate simple group (without restrictions for small q). One must now check to see which homomorphic images satisfy the original hypothesis. Any non-trivial image of $SL(n+2,q)$ does, but in the simple group $E_6(q)$ the appropriate subgroup is $SL(2,q) \times PSL(n,q)$; for $q \equiv 1 \pmod 3$ this lifts to $SL(2,q) \times SL(n,q)$ in the covering group (most easily seen inside $E_7(q)$).

We now sketch the main points of the proof. Let $L = B \cap B^g$. Then $L \cong SL(n-2,q)$. L is canonical in each of B and B^g as $C(A^g)$ and $C(A)$ respectively, so we may choose conjugates $A_2, \ldots, A_{n-2}$ of A to generate L as a set of "diagonal" root $SL(2,q)$'s. Putting $A = A_0$ and $A^g = A_n$, we may choose A_1, a diagonal root $SL(2,q)$ in B^g, so that $B^g = \langle A_0, A_1, \ldots, A_{n-2} \rangle$ and similarly A_{n-1} so that

$$B = \langle A_2, \ldots, A_{n-1}, A_n \rangle .$$

Ignoring $E_6(q)$ for the moment, suppose that we are trying to obtain Steinberg relations for $SL(n+2,q)$. Then we represent this configuration diagrammatically by Figure 1. The dotted squares represent A_1 and A_{n-1} respectively. To obtain those relations not given inside B or B^g, we have only to show that $[A_1, A_{n-1}] = 1$. This will be done inside $C(A_4)$. So, if $A_4 = A^h$, put $B_4 = B^h$.

Suppose that $n \geqslant 8$. Then

$$A_0 \times A_2 \times \langle A_6, \ldots, A_{n-2} \rangle \times A_n \subseteq B_4 .$$

B_4 can be parameterized with this subgroup "naturally filling" the diagonal; let the corresponding ordered basis for a vector space be $(e_1, \ldots, e_n)$. Since $A_1 \subseteq B_4$ and

$$A_1 \subseteq C(< A_6, \ldots, A_{n-2} > \times A_n) ,$$

A_1 fixes $\{e_5, \ldots, e_n\}$ and the subspace $< e_1, \ldots, e_4 >$. By a similar argument, A_{n-1} is "captured" by A_{n-2} and A_n. Since $n \geqslant 8$, this implies that $[A_1, A_{n-1}] = 1$. These arguments also hold for $q = 2$.

Suppose that $n = 7$. Then we no longer have the subgroup $< A_6, \ldots, A_{n-2} >$. However, $< A_0, A_1, A_2 >$ is a subgroup of B_4 isomorphic to $SL(4,q)$ and generated by transvections, and McLaughlin's theorem [2] forces this to be a canonical $SL(4,q)$ since $n < 8$, provided that $q \geqslant 4$. Since $< A_{n-1}, A_n >$ is a subgroup of B_4 isomorphic to $SL(3,q)$ and centralizing A_0 and A_2, it also centralizes A_1.

If $n = 6$, we must also obtain the covering group of $E_6(q)$. So we assume that $[A_1, A_5] \neq 1$ for all possible parameterizations. The extended Dynkin diagram for E_6 is

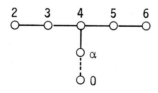

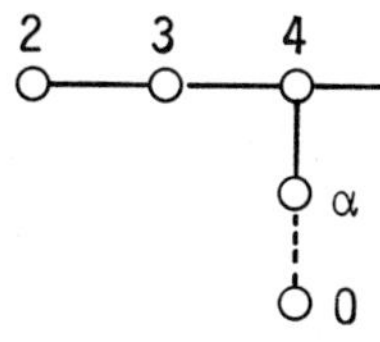

so, if we are to use a generating set for B as part of one for H, A_0 will *not* form part of the generating system. The appropriate root $SL(2,q)$ corresponding to the root α must, together with A_0 and A_4, generate $SL(4,q)$. Hence, if B is parameterized canonically corresponding to a vector space basis $(f_1,\ldots,f_6)$, it should be the subgroup A_α fixing $< f_1,f_6 >$ and $\{f_2,\ldots,f_5\}$. We need then to show that $[A_\alpha,A_5] = 1$; this involves careful choice of the parameterization, especially when $q \equiv 1 \pmod 3$, but this can be done in such a way that an assumption that $[A_1,A_5] \neq 1$, $[A_\alpha,A_5] \neq 1$ and $[A_1,A_\alpha] = 1$ leads to a contradiction inside $C(A_3)$. Thus H has the desired structure.

This completes the proof of the theorem as stated. Suppose, however, that we had assumed that B was a nonidentity homomorphic image of $SL(n,q)$. Then it is easy to see that exactly the same "proof" holds; the arguments involving linear algebra are strictly formal and are used to prove properties which still hold in homomorphic images. The only real difference is that we would not start with presentations for B and B^g, but the final step of checking which homomorphic images satisfy the original hypothesis will force B to be isomorphic to $SL(n,q)$, except for the possibility that $B \cong PSL(6,q)$ if $q \equiv 1 \pmod 3$, in which case $H \cong E_6(q)$. It seems likely that most of the arguments can be written in terms of Lie theory; then it should be possible to prove a similar theorem taking

B as an arbitrary group of Lie type provided the rank is large enough.

Returning to the particular situation we considered (though again our remarks will probably be true more generally), we note that $N(A_0) = B\cdot(N(A_0) \cap N(A_n))$ since $Aut(B)$ is known. It follows that $N(A_0)$ normalizes H; hence

$$H \lhd \langle N(A_0), N(A_n) \rangle .$$

Also, $N(H) = H\cdot C(H)\cdot N(B)$. In general, we would hope to show that $H \lhd G$. This seems very difficult, but suppose that we had also assumed that all 2-local subgroups of G were 2-constrained. Since H contains a subgroup containing t which is a Frobenius group of order 21, Hypothesis (*) forces H to centralize any 2-subgroup that it normalizes; hence 2-constraint will force $C_G(A \times B)$ to have odd order. It seems likely that one can get a grip on the 2-structure of G in such circumstances.

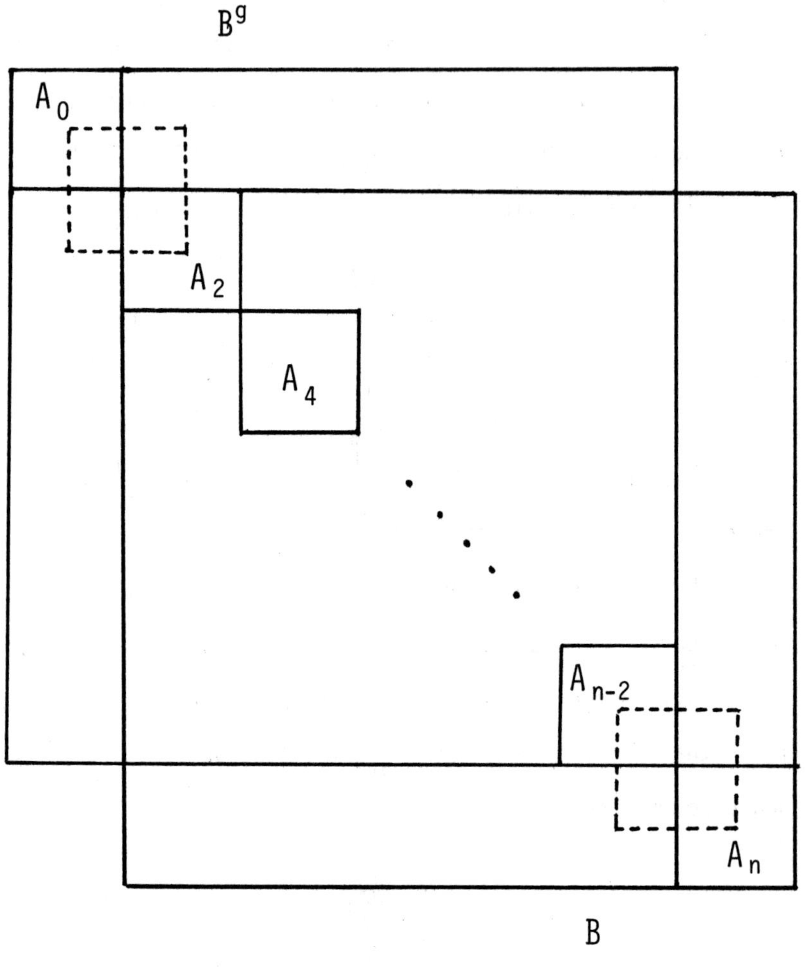

FIGURE 1.

REFERENCES

1. C. Curtis, *Central extensions of groups of Lie type*, J. Reine Angew. Math. 220 (1965), 174-185.

2. J. McLaughlin, *Some groups generated by transvections*, Arch. Math. 18 (1967), 364-368.

Institute for Advanced Study
Princeton, New Jersey

University College
Oxford, England

FINITE GROUPS OF PSL(2,q)-TYPE

A REPORT ON CURRENT JOINT WORK WITH RONALD SOLOMON

BY

MORTON E. HARRIS

We are interested in the following:

PROBLEM: Classify all finite groups G such that

1) $O(G) = 1$, and

2) there exists an involution $t \in G - Z(G)$ and a 2-component L of $C_G(t)$ such that

   (i) $L/O(L) \cong PSL(2,q)$ with q an odd prime power, and

   (ii) $C_G(L/O(L))$ has 2-rank 1.

(For the definitions of 2-component, 2-rank and $C_G(L/O(L))$, etc.; see [3, pp. 225-228]).

This problem is a "minimal configuration" that must be treated, in some fashion, in any resolution of both the B-conjecture (or balance conjecture) and the:

B(G)-CONJECTURE: $B(C_G(j)) \leqslant B(G)$ for any finite group G and any involution $j \in G$.

(See [6] for the definition of $B(G)$ and [6, Grand Conjecture] for the statement of the B-conjecture).

Note that current research seems to indicate that the B(G)-conjecture may be derivable from the B-conjecture.

Then assuming that the B(G)-conjecture has been dealt with, our problem is also a "minimal configuration" that must be treated in classifying all finite (simple) groups of component type by [1, Theorem 1].

REMARK: If a finite group G satisfies hypotheses (1) and (2) of our problem, then $O_2(G) = 1$.

PROOF: Let $Q = O_2(G) \neq 1$. Then $R = C_Q(t) \neq 1$ and hence $[L,R] \leqslant R$. But $[L,R] \leqslant O(C_G(t))$, so that

$$R \leqslant C_G(L/O(L))$$

which has 2-rank 1. Thus $\langle t \rangle = \Omega_1(R)$ char R and hence $R = Q$ and $\langle t \rangle = \Omega_1(Q) \leqslant Z(G)$, a contradiction. Hence $O_2(G) = 1$.

For our first attack on this problem, we prove:

PROPOSITION: Assume that the finite group G satisfies (1) and (2) of our problem and

3) $r_2(G) > 4$, and

4) $N_G(L)/LC_G(L/O(L))$ is cyclic.

Then the following conditions hold:

a) $\langle t \rangle \in Syl_2(C_G(L/O(L))$ and

b) if $t \in S \in Syl_2(N_G(L))$ and $D = S \cap L \in Syl_2(L)$, then $\langle t \rangle = S \cap C_G(L/O(L)) \lhd S$, $D \lhd S$, $\langle t \rangle D = \langle t \rangle \times D$ and there exists an element $f \in S\#$ such that $\langle f \rangle$ acts like "field automorphisms" on $L/O(L)$ and $S = (\langle t \rangle \times D) \langle f \rangle$ with

$$\langle f \rangle \cap (\langle t \rangle \times D) = 1 .$$

In the above, $r_2(G)$ denotes the sectional 2-rank of G; i.e., the maximal 2-rank of every section of G. Since all simple groups X with $r_2(X) \leqslant 4$ have been classified in [2, Main Theorem], it follows that all finite groups G with $r_2(G) \leqslant 4$ and satisfying conditions (1) and (2) of our problem have been classified. Thus in our investigation of the problem above, we may assume that (3) holds.

Our proof of the proposition involves a technical study of 2-fusion. I would like to present here a portion of the proof of this proposition. Thus let G be a minimal counterexample and let $Q = S \cap C_G(L/O(L))$. Then Q is cyclic or generalized quaternion, $Q \lhd S$, $D \lhd S$, $Q \cap D = 1$, $QD =$

$Q \times D \trianglelefteq S$, $S/(Q\times D)$ is cyclic, $< t > = \Omega_1(Q)$ and

$$S \in Syl_2(C_G(t)).$$

Our first step in the proof of the proposition is to show that $S \notin Syl_2(G)$. Once this is established, it follows that there exists a 2-element $x \in N_G(S) - S$ such that $x^2 \in S$. In order to impart some of the flavor of our work, I will now present our proof of the second step in our proof of the proposition:

Q is cyclic and $|Q| \geqslant 4$.

PROOF: Assume that Q is generalized quaternion. Then, since $S \in Syl_2(C_G(t))$, we have $t^x \neq t$. Hence $Q^x \cap Q = 1$ and $Q^x \trianglelefteq S$. Letting $\overline{S} = S/Q$, we have $Q^x \cong \overline{Q^x} \trianglelefteq \overline{S}$. But $\overline{S}$ is isomorphic to a subgroup of $Aut(L/O(L))$ with $\overline{D} \trianglelefteq \overline{S}$ and $\overline{D}$ corresponding to a Sylow 2-subgroup of $Inn(L/O(L))$. An easy lemma now implies that $\overline{S}$ contains a normal semi-dihedral subgroup $\overline{U}$ containing $\overline{D}$ with $|\overline{U}/\overline{D}| = 2$. But $\overline{S}/\overline{D}$ is cyclic by (4). This forces $\overline{S} = \overline{U} = \overline{D}\,\overline{Q^x}$. Hence $S = QQ^xD = QC_S(Q)$ where $C_S(Q) = (< t > \times D)Q^x$. Since $|S/(Q\times D)| = 2$, $Q^x \cap (< t > \times D)$ is a maximal subgroup of Q^x. Thus

$$< t^x > \leqslant \mho^1(Q^x \cap (< t > \times D)) \leqslant \mho^1(< t > \times D)$$

and hence $t^x = z$ where $< z > = Z(D)$. But $\Omega_1(S) = < t > \times D$

since $\Omega_1(\overline{S}) = \Omega_1(\overline{U}) = \overline{D}$. Hence $\langle z \rangle = \Omega_1(\Omega_1(S)')$char S. Since x normalizes S and $t^x = z$, we obtain a contradiction. Thus Q is cyclic. Next assume that $Q = \langle t \rangle$ and suppose that S/D is cyclic. Then $\Omega_1(S) = \langle t \rangle \times D$ and since conclusion (b) of the proposition does not hold, we have $C_S(\Omega_1(S)) = \langle y \rangle \times \langle z \rangle$ where $\langle z \rangle = Z(D)$ and

$$y^2 \in \{t, tz\}.$$

Hence $\langle y^2 \rangle$char S, $\Omega_1(\Omega_1(S)') = \langle z \rangle$char S and thus

$$\langle t \rangle \text{char } S,$$

a contradiction. We conclude that S/D is not cyclic. But $S/(\langle t \rangle \times D)$ is cyclic, whence $tD \notin \mho^1(S/D)$ and there is an element $f \in S$ such that $S/D = \langle tD \rangle \times \langle fD \rangle$. Clearly now conclusion (b) of the proposition holds, which is false.

At this point, it is not difficult to complete a proof of our proposition by showing that $r_2(G) \leqslant 4$.

Thus, in view of the proposition, we began to investigate our problem for the smallest possible value for q, namely $q = 9$; i.e., with $L/O(L) \cong PSL(2,9) \cong A_6$. We are consequently reduced to the problem of classifying finite groups G satisfying:

(*) $O_{2',2}(G) = 1$ and G contains an involution t such that $C_G(t)/O(C_G(t)) \cong Z_2 \times S_6$.

Note that a theorem of Gaschütz ([4, I, 17.4]) implies that (*) is equivalent to:

(**) $O_{2',2}(G) = 1$ and G contains an involution t and a subgroup R of $C_G(t)$ such that $C_G(t) = \langle t \rangle \times R$ and $R/O(R) \cong S_6$.

There are nine known groups G satisfying this condition. Since $PSL(2,9) \cong A_6$, we get:

1) $G \cong S_8$, $(|G|_2 = 2^7)$.

2) $G = E(G)X$ where $E(G) \cong A_6 \times A_6$, $E(G) \cap X = 1$, $t \in X$, and X is a 4-group, $(|G|_2 = 2^8)$.

3,4) Two non-isomorphic groups G (one of which is S_6 wr Z_2) with $G/E(G) \cong D_8$ and $E(G) \cong A_6 \times A_6$, $(|G|_2 = 2^9)$.

Since $S_6 \cong Sp(4,2) \cong O(5,2)$, we get:

5) G is an extension of $GL(5,2)$ by a "graph automorphism" of order 2, $(|G|_2 = 2^{11}$ and G has a maximal 2-local subgroup that is an extension of $D_8 * D_8 * D_8$ by an extension of $GL(3,2)$ by a "graph automorphism").

6) G is an extension of $PSU(5,4)$ by a "field automorphism" of order 2, $(|G|_2 = 2^{11}$ and G has a maximal 2-local subgroup that is an extension of $Q_8 * Q_8 * Q_8$ by an extension of $SU(3,2)$ by a "field

automorphism" of order 2).

7) G is an extension of $Sp(4,4)$ by a "field automorphism" of order 2, ($|G|_2 = 2^9$ and G has a maximal 2-local subgroup that is an extension of $(((E_{16} \cdot Sp(2,4)) \times E_4)$ by a "field automorphism" of order 2).

The final two examples are:

8) $G \cong SO(5,3)$, ($|G|_2 = 2^7$, a Sylow 2-subgroup of G is isomorphic to D_8 wr Z_2 and G has a maximal 2-local subgroup that is an extension of E_{16} by S_5).

9) $G \cong PSO^-(6,3)$, ($|G|_2 = 2^8$ and G has a maximal 2-local subgroup that is an extension of $Z_4 \times Z_4 \times Z_4$ by $Z_2 \times S_3$).

We continue to assume that G is a finite group satisfying (**).

As mentioned above, utilizing [2, Main Theorem], we may assume that $r_2(G) > 4$. Hence [5, Four generator theorem] implies that if $T \in Syl_2(G)$, then $SCN_3(T) \neq \phi$. Moreover, two doctoral students of Dieter Held at Mainz have recently completed the classification of all simple groups X with $|X|_2 \leqslant 2^{10}$. Thus we may also assume that $|G|_2 > 2^{10}$.

I shall now describe some consequences of hypothesis (**) that we need for our proof.

Let $H = C_G(t)$, $S \in Syl_2(H)$ and $D = S \cap R'$. Then $D \cong D_8$ and $D \lhd S$. Let $D = \langle x,y \rangle$ with $x^2 = y^2 = 1$ and $|xy| = 4$ and let $D' = Z(D) = \langle z \rangle$. Then $S = \langle t,u \rangle \times D$ where $Z(S) = \langle t,u,z \rangle \cong E_8$ and $Z(S)^{\#}$ is a set of representatives for the H-conjugacy classes of involutions of S. Hence $N_G(S)$ controls the G-fusion of $t^G \cap Z(S)$ and $t \underset{G}{\not\sim} z$ since $S' = \langle z \rangle$. Also $\langle H, N_G(S) \rangle$ controls the G-fusion of $t^G \cap S$.

Clearly

$$C_G(t) \cap N_G(S) = N_H(S) = O(N_H(S)) \times S = C_G(Z(S)) \cap N_G(S).$$

Letting $\overline{N_G(S)} = N_G(S)/(N_G(S) \cap C_G(Z(S)))$, we conclude that

$$|t^{\overline{N_G(S)}}| = |\overline{N_G(S)}| \leqslant 6$$

since $|Z(S)^{\#}| = 7$ and $t \underset{G}{\not\sim} z$. Also $2 \,\big|\, |\overline{N_G(S)}|$ since $S \notin Syl_2(G)$, so that there is a natural division of the problem into 3 cases:

(I) $|\overline{N_G(S)}| = 2$,

(II) $|\overline{N_G(S)}| = 4$, and

(III) $|\overline{N_G(S)}| = 6$.

By choice of notation, we may assume that u, z and uz are representatives of the 3 conjugacy classes of involutions in R. Let $A = \langle t,u,z,y \rangle$ and $B = \langle t,u,z,x \rangle$. Then $S = \langle A,B \rangle$, $m(S) = 4$, $I(S) = A^{\#} \cup B^{\#}$ and $\mathcal{E}_4(S) = \{A,B\}$. Moreover A and B are not conjugate in $H = C_G(t)$.

Let $\overline{H} = H/O(H)$. Then $C_{\overline{H}}(\overline{A}) = \overline{A}$, $C_{\overline{H}}(\overline{B}) = \overline{B}$ and $C_{\overline{H}}(\overline{z}) = C_{\overline{H}}(\overline{tz}) = \overline{S} = N_{\overline{H}}(\overline{S})$. Also there exist 3-elements $\rho \in N_H(A)$, $\rho_1 \in N_H(B)$ with $|\overline{\rho}| = |\overline{\rho_1}| = 3$, $\rho^x = \rho^{-1}$ and $\rho_1{}^y = \rho_1{}^{-1}$ and such that

$$C_{\overline{H}}(\overline{u}) = C_{\overline{H}}(\overline{tu}) = N_{\overline{H}}(\overline{A}) = \langle \overline{t},\overline{u} \rangle \times \langle \overline{y},\overline{z},\overline{\rho},\overline{x} \rangle$$

and

$$C_{\overline{H}}(\overline{uz}) = C_{\overline{H}}(\overline{tuz}) = N_{\overline{H}}(\overline{B}) = \langle \overline{t},\overline{uz} \rangle \times \langle \overline{z},\overline{xu},\overline{\rho_1},\overline{y} \rangle$$

where $\langle \overline{y},\overline{z},\overline{\rho},\overline{x} \rangle \cong \langle \overline{z},\overline{xu},\overline{\rho_1},\overline{y} \rangle \cong S_4$.

Since $N_{\overline{H}}(\overline{X}) = \overline{N_H(X)}$ for all 2-subgroups X of H, the above gives a picture of the structures "mod cores" of the 2-local subgroups of $\overline{H}$.

Suppose that we have (III) $|\overline{N_G(S)}| = 6$. Then $P = O_3(\overline{N_G(S)}) \neq 1$ and hence $Z(S) = C_{Z(S)}(P) \times [Z(S),P]$ where $C_{Z(S)}(P)$ and $[Z(S),P]$ are $\overline{N_G(S)}$ invariant and $[Z(S),P] \cong E_4$. Thus $N_G(S)$ has 3 orbits on $Z(S)^{\#}$ which is false since

$$|t^{\overline{N_G(S)}}| = 6 .$$

Thus case (III) does not hold.

Next suppose that we have (II) $|\overline{N_G(S)}| = 2$. Then, up to choice of notation, there are two possibilities for $t^{\overline{N_G(S)}}$. One possibility leads fairly easily to S_8 and $SO(5,3)$ and the other possibility is

$$t^{\overline{N_G(S)}} = \{t, tz\} .$$

I shall now describe how we treated this possibility.

Choose $W \in Syl_2(N_G(S))$, thus S is of index 2 in W and $N_G(S) = O(N_G(S))W$. Our first step here is to demonstrate:

1) $A \underset{G}{\not\sim} B$ and $N_G(S) = N_G(A) \cap N_G(B)$.

PROOF: Assume that $A \underset{N_G(S)}{\sim} B$. Then $v \in W - S$ implies that $v : A \leftrightarrow B$. Hence

$$< I(C_S(v)) > = < C_{A \cap B}(v) > = < C_{< t,u,z >}(v) > \cong E_4$$

since $t^v = tz$. Thus $J_e(W) = S$ and hence $W \in Syl_2(G)$ which, since $|W| = 2^6 < 2^{10} < |G|_2$, is false. Thus A and B are both normal in $N_G(S)$. Hence $N_G(A)$ is transitive on $t^G \cap A$ and $N_G(B)$ is transitive on $t^G \cap B$. Since $A \underset{C_G(t)}{\not\sim} B$, it follows that $A \underset{G}{\not\sim} B$. Thus (1) holds.

Next we investigate the subgroup $N_G(A)$. Similar considerations will also clearly apply to the subgroup $N_G(B)$.

Set $N = N_G(A)$ and $C = C_G(A) = O(N) \times A$ and let $\overline{N} = N/C$. Since $\overline{N}$ is transitive on $t^G \cap A$, $|t^G \cap A| = 4$ and $C_{\overline{N}}(t) \cong S_3$, we have $|\overline{N}| = |t^G \cap A||C_N(t)| = 3 \cdot 2^3$. Clearly the 3-element $\rho \in N_H(A) - O(N)$ and satisfies $\rho^x = \rho^{-1}$, $\rho^3 \in O(N)$ and $C_A(\rho) = \langle t,u \rangle$. Since $C_A(\rho) = \langle t,u \rangle$ is not normal in N, we conclude that $O_3(\overline{N}) = 1$. Since $C_{\overline{N}}(t) \cong S_3$, we conclude that $\overline{N} \cong S_4$. Also $t^N = t^{\overline{N}} = t\langle y,z \rangle$ and hence $E_4 \cong F = \langle y,z \rangle \lhd N$.

Next let V denote the unique subgroup of N such that $C = O(N) \times A \leqslant V \leqslant N$ and $\overline{V} = O_2(\overline{N}) \cong E_4$ and let $\widetilde{N} = N/(O(N) \times F)$.

Clearly $|\widetilde{V}| = 2^4$, $\widetilde{t} \in I(\widetilde{V})$ and $S_3 \cong \langle \widetilde{\rho}, \widetilde{x} \rangle$ acts faithfully on $\widetilde{V}$. Since all involutions of tF are conjugate in N to t and t is not a square in G, it follows that $\widetilde{V}$ is not isomorphic to $Z_4 \times Z_4$. Hence

2) $\widetilde{V} \cong Z_2 \times Q_8$ or $\widetilde{V} \cong E_{16}$.

It is fairly easy to prove:

3) If $\widetilde{V} \cong Z_2 \times Q_8$, then N contains a Sylow 2-subgroup of G and hence $|N|_2 = |G|_2 = 2^7$. Thus $\widetilde{V} \cong E_{16}$, as $2^{10} < |G|_2$.

Since $\widetilde{t} \in C_{\widetilde{V}}(\widetilde{\rho})$, we conclude:

4) Either (α): $([V,\rho]O(N))/O(N) \cong E_{16}$ or

(β); $([V,\rho]O(N))/O(N) \cong Z_4 \times Z_4$.

Our study of both of these two subcases has been completed. To illustrate this work, I shall describe how we treated subcase (α).

5) If $([V,\rho]O(N))/O(N) \cong E_{16}$, then $|G|_2 \leqslant 2^9$.

PROOF: Assume that $([V,\rho]O(N))/O(N) \cong E_{16}$ and let $S \leqslant U \in Syl_2(N)$. We claim:

$\mathcal{E}_{32}(U)$ contains a unique element E. Also E is such that:

(i) $C_E(t) = \langle \tau, F \rangle$ for a unique $\tau \in \{u,tu\}$,

(ii) $U = E\langle x,t \rangle$, and

(iii) $|C_E(x)| = |C_E(xt)| = 2^3$.

To prove this, it is clear that we may assume that $O(N) = 1$. Thus $E_{16} \cong V_1 = [V,\rho] \lhd N = O_2(N)\langle \rho,x \rangle$, $V = O_2(N) = V_1\langle t,u \rangle$ and $F = C_{V_1}(t)$. Also $\langle t,u \rangle \times \langle \rho,x \rangle$ acts on V_1 with $C_{V_1}(\rho) = 1$ and $\langle \rho,x \rangle \cong S_3$. Hence

$$z \in C_{V_1}(x) \cong E_4 .$$

Also $\langle t,u \rangle$ centralizes $C_{V_1}(t) = F$ and acts on $C_{V_1}(x) \neq F$. Thus there exists a unique $\tau \in \{u,tu\}$ centralizing $C_{V_1}(x)$. Since $E_8 \cong \langle C_{V_1}(t), C_{V_1}(x) \rangle \leqslant C_{V_1}(\tau)$ and $C_{V_1}(\tau)$ is ρ-invariant, we conclude that $[V_1,\tau] = 1$. Let $E = V_1 \times \langle \tau \rangle$. Clearly (ii) holds. Since $C_E(x) = \langle \tau \rangle \times C_{V_1}(x)$, $\langle \rho,xt \rangle \cong S_3$ and $C_E(xt) = \langle \tau \rangle \times C_{V_1}(xt)$, (iii) also holds. Now (i) is clear and E is unique in U.

Since $I(tE) = tF \cup tuF$, $t^E = tF$ and $t \not\sim_G tu$, we conclude that $t^G \cap tE = tF = t^E$. Set $M = N_G(E)$ and choose $M \in Syl_2(M)$ such that $U \leqslant M$.

Next we claim that $E \in Syl_2(C_G(E))$ and $C_G(E) = O(M) \times E$. To see this, assume that $E < M \cap C_G(E) = E^* \lhd M$ where $E^* \in Syl_2(C_G(E))$. Hence there is an element $f \in E^* - E$ such that $[t,f] \in E$. Thus $t^f \in t^E$ and hence there is an element $e \in E$ such that $fe \in C_G(t) \cap C_G(E) \cap M = C_S(E) = F \leqslant E$. This implies that $f \in E$, which is false and our claim is proved.

Let $\overline{M} = M/C_G(E)$. Then $\overline{M}$ is isomorphic to a subgroup of $Aut(E) \cong GL(5,2)$. Also, the same argument as above implies that $C_{\overline{M}}(\overline{t}) = \langle \overline{t},\overline{x} \rangle$. Thus $\overline{M}$ is dihedral or semi-dihedral by [4, III, 14.23 and 11.9]. Since the exponent of a Sylow 2-subgroup of $GL(5,2)$ is 8, we conclude that $|\overline{M}| \leqslant 2^4$ and hence $2^7 \leqslant |M| \leqslant 2^9$.

Now suppose that $M \notin Syl_2(G)$. Then there exists a 2-element $\eta \in N_G(M) - M$ such that $\eta^2 \in M$. Then $E \neq E_1 = E^\eta < M$. Setting $\widetilde{M} = M/E$, we have $1 \neq \widetilde{E}_1 < \widetilde{M}$. Clearly $t \notin E_1$, $C_{\widetilde{M}}(\widetilde{t}) = \langle \widetilde{x},\widetilde{t} \rangle$ and $C_M(E) = E$. Also $C_E(\widetilde{t}) \neq C_E(\widetilde{x})$ and $|C_E(\widetilde{t})| = |C_E(\widetilde{x})| = |C_E(\widetilde{tx})| = 2^3$. It follows that $\widetilde{t} \notin \widetilde{E}$ and $\widetilde{x} \in \widetilde{E}_1$ or $\widetilde{tx} \in E_1$ and hence that $\widetilde{E}_1 \cong E_4$. Thus $\widetilde{M} \cong D_8$, $E_1 \cap C_E(t) = \langle \tau,z \rangle$ and $|E \cap E_1| = 2^3$. Let $\{\widetilde{x}_1\} = \{\widetilde{x},\widetilde{tx}\} \cap \widetilde{E}_1$. Then $E \cap E_1 = C_E(\widetilde{x}_1)$ and hence $x \in E_1$ or $tx \in E_1$. Letting $\{x_1\} = \{x,xt\} \cap E_1$ and

$$E_1 = \langle E \cap E_1, x_1, v_3 \rangle$$

for some involution v_3, we have v_3: $tE \leftrightarrow tx_1E$ and $\langle \widetilde{x}_1 \rangle = Z(\widetilde{M})$. But $I(M-EE_1) = I(tE) \cup I(tx_1E) = t^M \cup (tu)^M$ and $t \underset{G}{\not\sim} tu$. Letting $T_1 = \langle M, \eta \rangle$, it follows that

$$(T_1 - M) \cap C_G(t) \neq \phi .$$

Since $S \leqslant M$ and $S \in Syl_2(C_G(t))$, this is impossible. Thus case (4α) has been eliminated.

The remaining case (II) $|\overline{N_G(S)}| = 4$ seems to be the most difficult case. Then

$$t^{\overline{N_G(S)}} = t\langle u, z \rangle$$

and setting $N = N_G(A)$ and $\overline{N} = N/O(N)$, we have

$$|\overline{N}/\overline{A}| \in \{12, 24, 48\} .$$

It is fairly easy to eliminate the case $|\overline{N}/\overline{A}| \neq 48$. Then, assuming that $|\overline{N}/\overline{A}| = 48$, we conclude that $\overline{N}/\overline{A} \cong Z_2 \times S_4$. Setting $\overline{V} = O_2(\overline{N})$ and $\overline{V}_1 = [\overline{V}, \overline{\rho}]$, we have

$$\overline{V}_1 \triangleleft \overline{N} = \overline{V}\langle \overline{\rho}, \overline{x} \rangle$$

and $\overline{V}_1$ has the following 5 possible isomorphism types:

(i) $Q_8 * Q_8$,

(ii) E_{16},

(iii) $Z_4 \times Z_4$,

(iv) $Q_8 \times E_4$, and

(v) a maximal subgroup of a Sylow 2-subgroup of Sz(8).

Moreover, there always exists a subgroup $\overline{Y}$ of $C_{\overline{V}}(\overline{V}_1)$ of order 4 such that $\overline{Y} \lhd \overline{N}$ and $\overline{V} = \overline{V}_1\overline{Y}\langle \overline{t} \rangle$. Since there are two possibilities for the isomorphism type of $\overline{Y}$, there are 10 possibilities for the structure of $\overline{V}$. Many of these 10 cases have already been eliminated.

Finally, in surveying our methods and arguments, we have not specifically used the fact that the 2-component of $C_G(t)$ is $A_6 \cong PSL(2,9)$. Thus we seem to be classifying finite groups G satisfying conditions (1) and (2) of our general problem and such that:

$$|L|_2 = 2^3 \quad \text{and} \quad |N_G(L)/LC_G(L/O(L))|_2 = 2 .$$

Added in proof: We have now completely classified finite groups satisfying (*) .

REFERENCES

1. M. Aschbacher, *On finite groups of component type*, to appear.

2. D. Gorenstein and K. Harada, *Finite groups whose* 2-*sub-groups are generated by at most* 4 *elements*, Mem. Amer. Math. Soc. 147 (1974).

3. D. Gorenstein and J. H. Walter, *Balance and generation in finite groups*, J. Algebra 33 (1975), 224-287.

4. B. Huppert, *Endliche Gruppen I*, Springer-Verlag, Berlin (1967).

5. A. MacWilliams, *On* 2-*groups with no normal abelian sub-groups of rank* 3, *and their occurence as Sylow* 2-*sub-groups of finite simple groups*, Trans. Amer. Math. Soc. 150 (1970), 345-408.

6. J. G. Thompson, *Notes on the* B-*conjecture*, September, 1974.

UNIVERSITY OF MINNESOTA
MINNEAPOLIS, MINNESOTA

SOME CHARACTERIZATIONS BY CENTRALIZERS OF ELEMENTS OF ORDER THREE

BY

MICHAEL E. O'NAN

Great progress has been made recently in the problem of classifying those finite simple groups in which the centralizer of some involution is not 2-constrained. Part of the reason for this is a recent theorem of Aschbacher which gives a normal form for the centralizer of an involution in a large class of such groups. However, in a group in which the centralizers of all involutions are 2-constrained, no such canonical form is anticipated. Accordingly, in these circumstances it seems to be appropriate to study instead the centralizers of elements of order three. Here we study one of the simplest cases in which the element of order three has a non-3-constrained centralizer, and obtain the following theorem.

THEOREM: Let G be a finite simple group and a be an element of G of order 3. Suppose that the group $C_G(a)/\langle a\rangle$ is isomorphic to one of the groups $PSL(2,q)$, $PGL(2,q)$, or $P\Sigma L^*(2,q)$, where q is the power of an odd prime. Suppose in addition there is a subgroup P of G , where P is an elementary abelian group of order 3^2 and all non-identity elements of P are conjugate to the element a . Then, G is isomorphic to one of the following groups:

(1) PSU(3,5)

(2) PSL(3,7)

(3) M_{22}

(4) M_{23}

(5) M_{24}

(6) HS , the Higman-Sims group

(7) R , the Rudvalis group

Note that no assumption is made about the splitting of the extension of $C_G(a)$ over $\langle a \rangle$. In fact, in the groups M_{24} and R the extension in question does not split.

While the hypotheses of this theorem concern the 3-local structure of the group G , the proof of the theorem mostly involves 2-local analysis. By our hypothesis there is a subgroup P of G with P an elementary abelian group of order 3^2 and all elements of $P^{\#}$ fused to a . Without loss we may suppose that a is in P . By the structure of $C_G(a)$, there is a subgroup V of $C_G(a)$ where V is elementary abelian of order 2^2 and V is normalized by P . Moreover, $C_G(V) \cap C_G(a) = \langle a \rangle \times V$.

We begin by studying the group $C_G(V)$. It is immediate in the group $C_G(V)/V$ that the element $\bar{a}$ is self-centralizing. Therefore, by a theorem of Feit and Thompson on self-centralizing elements of order three, it follows that $C_G(V)$ has a nilpotent normal subgroup F such that $C_G(V)/F$ is isomorphic to one of the groups Z_3, S_3, A_5, or $L_3(2)$. Now P normalizes V, $C_G(V)$, and F. Moreover, all elements of $P^{\#}$ fuse to a. By the structure of $C_G(a)$, it follows that if b belongs to $P^{\#}$, then the centralizer of b on $O_2(F)$ is of order 1 or 2^2. Consequently the order of $O_2(F)$ is at most 2^8. Thus, for the group $C_G(V)/O(C_G(V))$ there are at most a finite number of possible structures. The remainder of this proof consists of analyzing the various possibilities.

Another fact which is basic to the subsequent analysis is that there are at most two conjugacy classes of fours subgroups of G which are normalized by the group P. Moreover, if there are exactly two conjugacy classes of fours groups normalized by P, then $C_G(a)$ is isomorphic to $Z_3 \times PSL(2,q)$. This follows without much difficulty by studying more carefully the structure of the group $N_G(\langle a \rangle)$.

Now the analysis of particular configuration begins. We begin by studying the case in which the order of $O_2(F)$ is 2^8. In this case the structure of $C_G(a)$ forces quickly that $F = O_2(F)$. Also $C_G(V)/F$ is isomorphic to Z_3 or

S_3. A more detailed analysis using the action of P on F shows that there are exactly four possible structures for F. Each case is studied individually.

First we treat the case in which F is elementary abelian. Then let a_1, a_2, a_3, a_4 represent the four subgroups of P of order 3, and set $V_i = C_F(a_i)$. Then $F = V_1 \times V_2 \times V_3 \times V_4$. Our first goal is to determine the structure of $N_G(F)$. Our knowledge of $N_G(V_1)$ gives a bound on the order of $N_G(F)/F$. Moreover, by earlier paragraphs certain of the V_i's are conjugate to others, and it is not hard to show that the conjugacy occurs in $N_G(F)$. These two statements are sufficient to force $N_G(F)/F$ to be a subgroup of $P\Gamma L(2,9)$. From this it follows easily that F is the unique subgroup of $N_G(F)$ of order 2^8. Therefore $N_G(F)$ contains a Sylow 2-subgroup of G. Next we eliminate the possibility that G is simple by fusion analysis. To do this one first determines the conjugacy classes on involutions in $N_G(F) - F$ and a certain amount of information on the centralizers of these involutions in $N_G(F)$. By Goldschmidt's theorem some involution j in $N_G(F) - F$ must fuse into an element k of F. It follows that for some $g \in G$ we have $C_G(j)^g \subseteq C_G(k)$. An explicit study of the structures of the Sylow 2-subgroups of these two two groups yields a contradiction.

The remaining cases in which F is of order 2^8 are

handled in a similar fashion. In these cases either a contradiction is obtained or one shows that the Sylow 2-subgroup of G is that of M_{24}. A theorem of Schoenwaelder then shows that G is isomorphic to M_{24}.

Next we treat those cases in which the order of $O_2(F)$ is 2^6. Three separate cases arise. In one, $O_2(F)$ is isomorphic to the Sylow 2-subgroup of $L_3(4)$. In the remaining two cases $O_2(F)$ is elementary abelian but the structure of $C_G(V)/F$ is different.

The most difficult case occurs when F is elementary abelian of order 2^6 and $C_G(F)/F$ is isomorphic to S_3. It is this case in which the Rudvalis group arises. One first shows, as in the earlier case in which F is elementary abelian, that $N_G(F)/F$ is isomorphic to the group $G_2(2)$. Then in a separate paper we show:

THEOREM: Let G be a finite simple group having an elementary abelian 2-subgroup F of order 2^6 such that F is a Sylow 2-subgroup of $C_G(F)$ and $N_G(F)/C_G(F)$ is isomorphic to $G_2(2)$ or $G_2(2)'$. Then G is isomorphic to the Rudvalis group.

The proof of this theorem is quite tedious.

Having completed these cases it is not hard to show that in all remaining cases the group G has a subgroup L where $A_4 \times A_4 \subseteq L \subseteq S_4 \times S_4$, such that if V is a normal fours group of L, then a Sylow 2-subgroup of $C_G(V)$ lies in

L. We treat this as a separate problem.

Set $E = O_2(L)$. First one proves without much difficulty that $N_G(E)/E$ is isomorphic to S_6, A_6, or a subgroup of $S_3 \text{ wr } Z_2$. In the last case it is not hard to show that a Sylow 2-subgroup of $N_G(E)$ lies in the group $D_8 \text{ wr } Z_2$. Then one proves the following theorem about 2-groups:

THEOREM: Let E be an elementary abelian subgroup of order 16 in a 2-group P. Suppose that $N_P(E)$ is isomorphic to some subgroup of $D_8 \text{ wr } Z_2$. Then, P is of sectional 2-rank at most four.

Then, using the theorem of Gorenstein and Harada, G is known.

If $N_G(E)/E$ is isomorphic to A_6, G is known by a theorem of Harada. When $N_G(E)/E$ is isomorphic to S_6, a rather tedious "pushing up" argument shows that the Sylow 2-subgroup of G has order at most 2^{12} and a very restricted structure. Fusion analysis eliminates many of the remaining cases. In the others it turns out that G is of type HS or $L_5(q)$ for a suitable congruence on q. Then, theorems of Gorenstein and Harris are applicable to identify G in the first case, and Mason to identify G in the second case.

RUTGERS UNIVERSITY
NEW BRUNSWICK, NEW JERSEY

A CHARACTERIZATION OF $PSp_4(3^m)$ BY THE CENTRALIZER OF AN ELEMENT OF ORDER THREE

BY

JOHN L. HAYDEN

This paper is a continuation of [7] and [8] and deals with the case $q = 3^{2n}$. The main theorem of the paper characterizes $PSp_4(3^m)$ by the centralizer of an element of order three. The proof uses a characterization of $L_2(3^{2n})$ as a $C\theta\theta$-group which is stated below as a conjecture. Indeed, much work has been done in this area and recent results [9] and [2] suggest that the following conjecture is true.

CONJECTURE. Let G be a finite group which satisfies the following:

i) G contains an elementary abelian 3-group D of order $q = 3^{2n}$ with $C_G(d) = D$ for all $d \in D^{\#}$.

ii) G has at least 2 classes of 3-elements.

iii) D is not normal in G.

Then $[N_G(D):D] = \frac{q-1}{2}$ and $G \cong L_2(q)$.

[9] shows that a group satisfying the hypothesis of the conjecture is simple with $[N_G(D):D] = \frac{q-1}{2}$ and [2] shows $G \cong L_2(q)$ provided there is a nontrivial 2-group normalized by some 3-element. In [10], the conjecture has been proven for $q = 9$.

As an application of this conjecture the following main theorem is proved.

THEOREM 1. *Let* C *be the centralizer in* $PSp_4(3^m)$, $m > 1$, *of an element of order* 3 *in the center of some Sylow* 3-*subgroup. Let* G *be a finite group satisfying:*

(a) G *contains an element* α *of order* 3 *with* $C_G(\alpha) \cong C$.

(b) *For all* $z \in Z(C_G(\alpha))$, $C_G(z) = C_G(\alpha)$.

(c) G *has at least* 2 *classes of central* 3-*elements.*

Then either $C_G(\alpha) \trianglelefteq G$ *or* $G \cong PSp_4(3^m)$.

From the remarks of the first paragraph we will assume $q = 3^{2n}$ for all further arguments. It will be apparent that the conjecture will be used to establish only the nonsimple case of the above theorem.

1. THE NONSIMPLE CASE.

It is assumed that the reader is familiar with [8] so

that the structure of $C_G(\alpha)$ and its subgroups listed in section 1 of [8] may be used. We will assume $C_G(\alpha) = UL$, $U \cap L = 1$, $|U| = q^3$, $L \cong SL(2,q)$ and P is a Sylow 3-subgroup of $C_G(\alpha)$ with maximal abelian subgroup M. As in [8], the simplicity of G depends upon whether $N_G(M)$ is 3-closed or not. Consequently, we make the hypothesis that $N_G(M)$ is 3-closed throughout this section. This implies $N_G(M) = N_G(P)$ and because M is the unique abelian subgroup of P of order q^3, no element of $Z = Z(P)$ is conjugate in G to an element of $M - Z$.

Let Q be a Sylow 2-subgroup of L. Because Q is quaternion with central involution t, property (b) of Theorem 1 implies $C_G(z,Q) = Z\langle t \rangle$ for all $z \in Z^{\#}$. Let $X = C_G(Q)/\langle t \rangle$. The image $\overline{Z}$ of Z in X is a $C\theta\theta$-group and (c) implies $|N_X(\overline{Z}):\overline{Z}| \neq q - 1$. If $\overline{Z} \lhd X$, $Z \lhd C_G(Q)$ and $C_G(Q)$ is 3-closed. If $C_G(Q)$ is not 3-closed, $X \cong L_2(Q)$ by the conjecture. We have proven the following.

(1.1) *If* $C_G(Q)$ *is not* 3-*closed,* $C_G(Q)/\langle t \rangle \cong L_2(q)$.

(1.2) *If* $C_G(Q)$ *is not* 3-*closed,* $C_G(Q)$ *has a quaternion Sylow* 2-*subgroup.*

PROOF. (1.1) implies $C_G(Q) \cap N_G(Z) = ZK_2$, $|K_2| = q-1$ with $K_2/\langle t \rangle$ cyclic. Let Q, $P \cap L$, K_1 be chosen as in section 3 of [8] so that $(P \cap L)K_1$ is a Sylow 3-normalizer of L and Q contains the quaternion group generated by

$$\begin{pmatrix} 1 & 1 \\ 1 & -1 \end{pmatrix} \quad \text{and} \quad \begin{pmatrix} 0 & 1 \\ -1 & 0 \end{pmatrix} .$$

Because K_2 normalizes L and centralizes Q, it follows that K_2 normalizes $P = U(P \cap L)$ and K_1. Consequently, $K_1K_2 = K$ is a subgroup of $N_G(P)$ of order $\frac{(q-1)^2}{2}$ and (c) of theorem 1 implies $N_G(P) = PK$.

K_2 leaves invariant the subgroup V of P' containing all elements inverted by t. For $k \in K_2$, $v \in V$, k centralizes v implies that k centralizes $[v, v^y]$ where $y \in Q$ inverts K_1. From the structure of L, $[v, v^y] = z \neq 1$ whenever $v \neq 1$. However $C_{K_2}(z) = \langle t \rangle$ so that K_2 acts regularly on V. Because $K_2/\langle t \rangle$ contains no four-group, [4, 4.11, pg. 200] implies that K_2 has a cyclic Sylow 2-group. Hence K_2 is cyclic of order $q - 1$.

Let S be a Sylow 2-group of $C_G(Q)$ containing $S \cap K_2$. Because $q \equiv 1 \pmod 8$, $q - 1 = 2^r e$, $(2,e) = 1$, $r \geqslant 3$ and $|K_2 \cap S| = 2^r$. Since $S/\langle t \rangle$ is isomorphic to a Sylow 2-group of $L_2(q)$, $|S/\langle t \rangle| = 2^r$ and $|S| = 2^{r+1}$. Therefore S contains a cyclic subgroup $\langle s \rangle$ of index 2. [4, 4.4, pg. 193] now implies S is quaternion, semidihedral or dihedral. If S is dihedral, [5], together with the fact that $C_G(Q)$ has no subgroup of odd order, implies that $C_G(Q)$ contains a subgroup isomorphic to $L_2(q)$. Then $C_G(Q) = F \times \langle t \rangle$ with $F \cong L_2(q)$. But $|\langle s \rangle \cap F| \geqslant 2^{r-1}$, so $t \in F$, which is impossible. If S is semidihedral, Proposition 1 of [1] implies that $C_G(Q)$ has a normal subgroup of index 2. From the structure of $C_G(Q)$, $C_G(Q) = F \times \langle t \rangle$, $F \cong L_2(q)$, which we have seen is impossible. Therefore, $C_G(Q)$ has a quaternion Sylow 2-group S, $|S| = |Q| = 2^{r+1}$ and $S \cong Q$.

(1.3) *Assume* $C_G(Q)$ *is not* 3-*closed and let* S *be a Sylow* 2-*subgroup of* $C_G(Q)$. *Then* QS *is a Sylow* 2-*subgroup of* G.

PROOF. (1.2) implies that S is quaternion, $S \cong Q$, $S \cap Q = \langle t \rangle$, $[S,Q] = 1$. We have chosen Q in such a way that Q contains the quaternion group

$$Y = \left\langle \begin{pmatrix} 1 & 1 \\ 1 & -1 \end{pmatrix}, \begin{pmatrix} 0 & 1 \\ -1 & 0 \end{pmatrix} \right\rangle,$$

$K_2 \cap S$ is cyclic of order 2^r and $P \cap L$ corresponds with the matrix group

$$\left\langle \begin{pmatrix} 1 & 0 \\ \lambda & 1 \end{pmatrix} \middle| \lambda \in F_q \right\rangle .$$

Then

$$x = \begin{pmatrix} 1 & 0 \\ 1 & 1 \end{pmatrix} \in P \cap L$$

normalizes Y. Clearly $C_G(Q) \subseteq C_G(Y)$ and, as

$$C_G(Y) \cap C_G(z) = Z \langle t \rangle$$

for all $z \in Z^{\#}$, $\overline{Z}$ is a $C\theta\theta$-subgroup of $F = C_G(Y)/\langle t \rangle$. Because F is not 3-closed and contains a nontrivial 2-group normalized by an element of order 3, [2] implies $F \cong L_2(q)$ and we conclude $C_G(Y) = C_G(Q)$. Hence $C_G(Q)$ admits $x \in P \cap L$ as an automorphism, K_2 normalizes L and centralizes Q so that the structure of aut L forces K_2 to centralize x. Consequently x induces an automorphism of $C_G(Q)$ which centralizes the Sylow 3-normalizer ZK_2. The structure of $C_G(Q)$ implies that x centralizes $C_G(Q)$.

Let $R = QS$. We first show that no element of Q of order 4 is conjugate in G to an element of S. Indeed, suppose that $y \in Q$ has order 4 and $g^{-1}yg \in S$ for some $g \in G$. Then $C_G(g^{-1}yg)$ contains $g^{-1}Zg$ and the preceding paragraph shows $x \in C_G(g^{-1}yg)$, $x \in P \cap L$.

A Sylow 3-subgroup of $C_G(y)$ which contains Z is abelian and is a subgroup of $C_G(y,z) = ZC_L(y)$. Hence Z is a Sylow 3-subgroup of $C_G(y)$. It follows that x is conjugate to an element $z \in Z$. However, $x,z \in M$ with $x \in M - Z$. The remarks preceding (1.1) imply x and z are not conjugate in G and it follows that no element of Q of order 4 is conjugate in G to an element of S.

The elements of QS of order 4 belong to Q or S or have the form ab, $a \in Q$, $b \in S$. In the latter case, a^2b^2 is an involution of QS different from t. Now

$$Z(R) = Z(Q)Z(S) = \langle t \rangle$$

so t is characteristic in R. It follows that the elements of $N_G(R)$ permute the elements of order 4 with square t. Because no such element of Q is conjugate to an element of S, $N_G(R)$ permutes the elements of Q of order 4. This implies $N_G(R) \subseteq N_G(Q)$.

For $g \in N_G(Q)$, $g^{-1}Zg = a^{-1}Za$ for some $a \in C_G(Q)$. Hence

$$ga^{-1} \in N_G(Z) \cap N_G(Q)$$

and

$$N_G(Q) = (N_G(Z) \cap N_G(Q))C_G(Q) .$$

As $N_G(P) = PK$,

$$|K| = |K_1K_2| = \frac{(q-1)^2}{2}, \; N_G(Z) = C_G(Z)K_2$$

and

$$N_G(Q) = ZQK_2C_G(Q) = QC_G(Q) .$$

Since S is a Sylow 2-subgroup of $C_G(Q)$, $R = QS$ is a Sylow 2-subgroup of $N_G(Q)$. We conclude that R is a Sylow 2-subgroup of $N_G(R)$ and consequently a Sylow 2-subgroup of G.

(1.4) $C_G(Q) = ZC_K(Q)$.

PROOF. Let us suppose that $C = C_G(Q)$ is not 3-closed so that by (1.3) $R = QS$ is a Sylow 2-subgroup of G.

Let V be any quaternion subgroup of Q of order 8 and let $\beta \in L$ be an element of order 3 which normalizes V.

The argument (1.3) shows $C_G(V) = C_G(Q) = C$ and it follows that β normalizes C. This implies that β leaves the Sylow 3-normalizer ZK_2 of C invariant and centralizes Z. For $z \in Z$, $\beta^{-1}K_2\beta = z^{-1}K_2 z$ so βz^{-1} normalizes K_2. The structure of C forces βz^{-1} to be the trivial automorphism of C and we conclude that $< Q, \beta z^{-1} > \subseteq C_G(C)$. Because V was chosen arbitrarily, all elements of Q of order 4 are conjugate in $C_G(C)$. Indeed, βz^{-1} has the same action as β on V, and if we set

$$Q = < x_1, y_1 \mid x_1^{2^{r-1}} = t = y_1^2,\ y_1^{-1}x_1y_1 = x_1^{-1} >,$$

every element of Q of order 4 is conjugate in $C_G(C)$ to the element $x_1^{2^{r-2}}$. Let

$$S = < x_2, y_2 \mid x_2^{2^{r-1}} = y_2^2 = t >$$

where y_2 inverts x_2. Then every involution of $QS - < t >$ is conjugate in $C_G(C)C$ to $v = (x_1x_2)^{2^{r-2}}$.

Let us assume v and t are conjugate in G. Then there exists a Sylow 2-subgroup $F \subseteq C_G(v)$ containing

$$W = < x_1, x_2, y_1y_2 > .$$

Comparing orders, W is a normal subgroup of F of index 2. Since $W' = < x_1^2, x_2^2 >$ and the elements of W' of order 2^{r-1} have t as a power while the elements with v as a power are

$(x_1^2x_2^2)^i$ and have order at most 2^{r-2}, t is characteristic in W'. Consequently $\langle t \rangle$ is characteristic in W and t lies in the center of F. But $\langle t,v \rangle \subseteq Z(F)$ contrary to the fact that $Z(QS) = \langle t \rangle$ with $F \cong QS$. We conclude that t and v are not conjugate.

[3] implies that $G = C_G(t)O(G)$. Because $C_G(t)$ has a Sylow 3-subgroup T of order q^2 ([8], pg. 632), $P \cap O(G)$ is a nontrivial normal subgroup of P and thus $Z \cap O(G) \neq 1$. This implies $C_G(Q)$ has a nontrivial normal subgroup of odd order contrary to the structure of $C_G(Q)$. We conclude that $C_G(Q)$ is 3-closed and $Z \lhd C_G(Q)$.

Let $y \in C_G(Q)$. Because $Z \lhd C_G(Q)$, y induces an automorphism of L centralizing Q. The structure of aut L forces y to be a field automorphism so that y leaves $P \cap L$ and K_1 invariant. Thus $y \in N(P) \cap C(t) = C_P(t)K$. Because $C_P(t)K \cap N(K_1) = ZK$, $C_G(Q) \subseteq ZK$ and finally $C_G(Q) = ZC_K(Q)$.

(1.5) $N_G(Q) = ZQN_K(Q)$.

PROOF. Let $g \in N_G(Q)$. (1.4) implies that g normalizes Z, hence $g \in N_G(Z) \cap C(t) = ZLK$. Because $q - 1 = 2^re$, $(2,e) = 1$, $2^r \geqslant 8$, we may choose Q to be the Sylow 2-subgroup of L generated by

$$x = \begin{pmatrix} \alpha & 0 \\ 0 & \alpha^{-1} \end{pmatrix}, \quad \alpha \in F_q, \quad |\alpha| = 2^r,$$

and by

$$y = \begin{pmatrix} 0 & 1 \\ -1 & 0 \end{pmatrix} .$$

We may assume $g = \ell k$, $\ell \in L$, $k \in K$. Because $K_1 = C_K(Z) \lhd K$ and $\langle x \rangle$ is the unique cyclic subgroup of K_1 of order 2^r, k leaves $\langle x \rangle$ fixed. Moreover, $\langle x \rangle$ is characteristic in Q so that g leaves $\langle x \rangle$ invariant and we conclude that $\ell \in N_L\langle x \rangle$. This implies that $g \in Z(\langle y \rangle K_1)K = ZQK$ and that $N_G(Q) = ZQN_K(Q)$

Let V be the subgroup of P' generated by all elements of P' inverted by t. For $v \in V^{\#}$, $k \in K$, $k^{-1}vk = k_1^{-i}vk_1^i$ for $k_1^i \in K_1$. Hence $kk_1^{-i} \in C_K(v)$ and $K = K_1C_K(v)$. If R is a Sylow 2-subgroup of $N_K(Q)$, R contains $K \cap Q = \langle x \rangle$ and $R \subseteq K_1C_K(v)$. A Sylow 2-group of $K_1C_K(v)$ containing R has the form $\langle x \rangle W$ where $W \subseteq C_K(v)$. Hence $R = \langle x \rangle(R \cap W)$. Now $R \cap W \subseteq C_K(v)$ so that

$$\langle x \rangle \cap (R \cap W) = 1 .$$

In fact, $R \cap W$ acts regularly on Z so that $R \cap W$ is cyclic or quaternion.

We conclude that a Sylow 2-subgroup of $N_K(Q)$ has the form $\langle x \rangle A$, $\langle x \rangle \cap A = 1$ where A is cyclic or quaternion and acts regularly on Z. (1.5) implies that $Q\langle x \rangle A = QA$ is a Sylow 2-subgroup of $N_G(Q)$, $Q \cap A = 1$. We now show that QA is a Sylow 2-subgroup of G.

(1.6) *Let* $S = QA$ *be a Sylow* 2-*subgroup of* $N_G(Q)$, $Q \cap A = 1$. *Then* S *is a Sylow* 2-*subgroup of* G.

PROOF. The remarks preceding (1.6) show that $A \subseteq K$ and that A is cyclic or quaternion. Let us assume $A \neq 1$ and let τ be a central involution of A. If τ centralizes an element $\ell \in P \cap L$, consider the four-group $< t,\tau >$. Then $U = C_U(t)C_U(t\tau)C_U(\tau)$ so that τ or $t\tau$ centralizes $w \in U - P'$. Because $< t,\tau >$ centralizes ℓ, τ or $t\tau$ centralizes $[w,\ell] = vz$, $z \in Z^\#$, v an element of P' inverted by t. Then $t\tau$ or τ centralizes z as $< t,\tau >$ leaves Z invariant. We conclude that τ inverts $P \cap L$ and normalizes K_1 so that the structure of aut L forces τ to have the same action on L as does $\begin{pmatrix}1 & 0\\ 0 & -1\end{pmatrix}$. Hence τ centralizes $< x >$ and inverts y.

As $Q \lhd S$, t is central in S. Let us suppose that $Z(S)$ contains another involution $\mu = ba$, $b \in Q$, $a \in A$. The fact that τ centralizes μ forces $b^\tau = b$ so that

$$b \in < x > = C_Q < \tau > .$$

Hence $\mu = x^i a$ and μ normalizes P. The automorphism of L induced by μ centralizes Q and leaves $P \cap L$ and K_1 invariant, so must be a field automorphism. Consequently, μ centralizes an element $\ell \in P \cap L$. Applying the argument of the preceding paragraph to the four-group $\langle t,\mu\rangle$, we see that μ cannot exist. Hence $Z(S)$ has a unique involution t.

Let $b \in Q$, $a \in A$, and suppose that $ba \in S$ has order 4 with square t. Clearly $b \neq 1$ and $(Qba)^2 = Qa^2 = Q$ so $a^2 \in Q \cap A = 1$. Hence $a = 1$ or $a = \tau$. If $a = 1$, $ba \in Q$, and otherwise $ba = x^i\tau$ or $x^iy\tau$. In the latter case $(x^iy\tau)^2 = (x^iy\tau)(x^iy\tau) = x^iyx^i\tau t\tau = x^iyx^iy^{-1} = x^ix^{-i} = 1$ as τ inverts y. Because τ centralizes x, we conclude that $ba = x^{2^{r-2}}\tau$ is the only element of $S - Q$ of order 4 with square t.

Let $b \in Q$, $|b| = 4$. $C_G(b,Z) = ZC_L(b)$ so that Z is a Sylow 3-subgroup of $C_G(b)$. Because $P \cap L$ is centralized by $x^{2^{r-2}}\tau$, and Z, $P \cap L$ are subgroups of M not conjugate in G, $x^{2^{r-2}}\tau$, b are not conjugate in G. We conclude that $N_G(S)$ permutes the elements of Q of order 4 among themselves and thus leaves Q invariant. Hence $N_G(S) \subseteq N_G(Q)$ and (1.5) implies that $S = QA$ is a Sylow 2-subgroup of G.

If $A = 1$, Q is a Sylow 2-subgroup of $N_G(Q)$ and hence a Sylow 2-subgroup of G.

(1.7) $G = C_G(Z)K$ *and* $C_G(Z) \trianglelefteq G$.

PROOF. (1.6) implies that $S = QA$ is a Sylow 2-subgroup of G. Let μ be an involution of S different from t and suppose t and μ are conjugate in G. Then $u = ba$,

$b \in Q$, $a \in A$ and $\mu^2 = 1$ forces $a^2 \in Q \cap A = 1$. Hence $a = \tau$ and $\mu = \tau$, $t\tau$ or $x^i y\tau$ for some integer i . Since τ inverts y , $y^{-1}\tau y = y^{-1}y^{-1}\tau = t\tau$ so that if $\mu = \tau$ or $t\tau$, t and τ would be conjugate in G . But τ centralizes $v \in P' - Z$ and $C_G(t)$ has $T = C_M(t)$ as a Sylow 3-subgroup so that if t and τ were conjugate, v would be conjugate to an element $m \in T$ in $N_G(M) = PK$. This is impossible because the elements of $P' - Z$ are left invariant under the action of PK and t centralizes no element of $P' - Z$.

We may assume that $\mu = x^i y\tau$ is conjugate to t . As $\tau\mu\tau = t\mu$, $U = C_U(t)C_U(t\mu)C_U(\mu)$, we conclude that μ or $t\mu$ centralizes an element of $U - Z$. Every element of $U - Z$ is conjugate in $C_G(Z)$ to an element $w \in P' - Z$ so that since t and μ are conjugate, w is conjugate to an element of $C_M(t) = T$. The preceding paragraph shows that w can be conjugate to no element of T so we conclude t and μ cannot be conjugate.

Finally, t is conjugate to no involution of $S - \langle t \rangle$ so $G = C_G(t)O(G)$. Then $O(G) \cap P \neq 1$ so $O(G) \neq 1$. Let B be a minimal characteristic subgroup of $O(G)$. If $(|B|,3) = 1$, $B = \Pi C_B(z)$, $z \in Z^{\#}$ so that $B \subseteq C_G(Z)$. This is impossible as $C_G(Z)$ contains no such subgroup. Hence B is a 3-group with $B \lhd P$ and $Z \cap B \neq 1$. For $g \in G$, $z \in B \cap P$,

$$g^{-1}zg \in B \subseteq P$$

and because $g^{-1}zg$ has order 3, it is an element of U or M. However z is conjugate to no element of $U - Z$ or $M - Z$ and we conclude $g^{-1}zg \in B \cap Z$. Then there exists $k \in K$ such that $g^{-1}zg = k^{-1}zk$ or $gk^{-1} \in C_G(z)$. Then $g \in C_G(Z)K$ and $G = C_G(Z)K$.

2. The Simple Case.

Let G be a finite group satisfying (a), (b), (c) of Theorem 1. If $N_G(M)$ is 3-closed, the results of section 1 imply that $G = C_G(Z)K$ and $C_G(Z) \trianglelefteq G$. Hence we will assume $N_G(M)$ is not 3-closed and the structure of $N_G(M)$ is given by [8, pg. 630]. Let τ be the involution of $C_G(t)$ such that Z^τ is a Sylow 3-subgroup of L. $C_G(Z,t) = ZL$, $L \cong SL(2,q)$ so that L contains an element c of order 4 inverting K_1. Because L is the union of the double cosets $Z^\tau K_1$ and $Z^\tau K_1 c Z^\tau K_1 = Z^\tau c z^\tau K_1$, [8, (4.4)] shows that $[L,L^\tau] = 1$. Hence $LL^\tau\langle \tau \rangle$ is a subgroup of $C_G(t)$ with Sylow 2-subgroup $S = QQ^\tau\langle \tau \rangle$, and $L \cap L^\tau = \langle t \rangle$.

(2.1) $S = QQ^\tau\langle \tau \rangle$ *is a Sylow 2-subgroup of* G, *where* $[Q,Q^\tau] = 1$ *and* $Q \cap Q^\tau = \langle t \rangle$.

PROOF. The structure of S implies $Z(S) = \langle t \rangle$. Suppose $s \in S$, $s^2 = t$. It is easy to see that if $s \in QQ^\tau$,

$s \in Q$ or Q^τ. If $s = ab\tau$, $a \in Q$, $b \in Q^\tau$, then

$$s^2 = (ab^\tau)(ba^\tau) = t$$

so that

$$(ab^\tau) \in Q \cap Q^\tau = \langle t \rangle .$$

Should $ab^\tau = 1$, then $a = \tau b^{-1}\tau$ and $s = \tau b^{-1}\tau b\tau$ so that $s^2 = 1$. If $ab^\tau = t$, then $b\tau = \tau a^{-1}t$, and $s^2 = 1$. We conclude that all elements of S of order 4 with square t belong to Q or Q^τ.

Let $X = C_G(Q)/\langle t \rangle$. Because X is a $C\theta\theta$-group and the image of L^τ in X is isomorphic to $L_2(q)$, [2] implies that $X \cong L_2(q)$. Consequently $C_G(Q) = L^\tau$. Let $y \in N_G(S)$ and let a, b, be elements of Q of order 4 such that $Q = \langle a,b \rangle$. Then a^y, b^y are noncommuting elements of S with square t so that $Q^y = Q$ or Q^τ.

Thus y or $y\tau$ normalizes Q. Assuming $y \in N_G(Q)$, Z^y is a Sylow 3-subgroup of $C_G(Q) = L^\tau$, so that $Z^{yw} = Z$ for some $w \in L^\tau$. This implies $yw \in N(Z) \cap C(t) = ZLK$ so $y \in LL^\tau \cap N(Q) = QL^\tau$. Because y normalizes S, $y \in QQ^\tau$. If $y\tau$ normalizes Q, $y \in QQ^\tau\langle \tau \rangle = S$, so that $y \in S$ in either case. Consequently $N_G(S) = S$ and S is a Sylow 2-subgroup of G.

(2.2) $G \cong PSp_4(q)$.

PROOF. (2.1) implies that G has a Sylow 2-subgroup of type $PSp_4(q)$, $q \equiv 1 \pmod 8$. If G has a normal subgroup X of index 2, $H = C_G(Z) \subseteq X$ as H contains no subgroup of index 2. Hence X satisfies the hypothesis of Theorem 1 and has a Sylow 2-group of type $PSp_4(q)$. This is impossible and we conclude that G has no subgroup of index 2. Suppose that $O(G)$ is nontrivial. If $O(G)$ is a 3'-group, Z leaves $O(G)$ invariant and this forces $O(G) \subseteq C_G(Z)$ which is impossible. This implies $P \cap O(G) \neq 1$ so that

$$O(G) \cap Z \neq 1 .$$

Because L^τ has Z as a Sylow 3-subgroup, this is impossible. We conclude that G is a fusion simple group so that [6, Theorem B] $C_G(t)$ contains a normal subgroup which is the central product of two subgroups L_1, L_2 isomorphic to $SL(2,q)$. It follows that $LL^\tau \trianglelefteq C_G(t)$ and that $C_G(t) = LL^\tau\langle \tau \rangle$ (see [8,(4.12)]). Finally, an application of [11] implies that $G \cong PSp_4(q)$.

The results of sections 1 and 2 may now be combined to complete the proof of Theorem 1.

REFERENCES

1. J. L. Alperin, R. Brauer and D. Gorenstein, *Finite groups with quasi-dihedral and wreathed Sylow 2-subgroups*, Trans. Amer. Math. Soc., vol. 151(1970), pp. 1-260.

2. L. R. Fletcher, *A transfer theorem for* $C\theta\theta$-*groups*, Quart. J. Math. Oxford (2), vol. 22(1971), pp. 505-533.

3. G. Glauberman, *Central elements in core-free groups*, J. Algebra, vol. 4(1966), pp. 403-420.

4. D. Gorenstein, *Finite Groups*, Harper and Row, New York, 1968.

5. __________ and J. H. Walter, *On finite groups with dihedral Sylow 2-subgroups*, Illinois J. Math., vol. 6 (1962), pp. 553-593.

6. D. Gorenstein and Koichiro Harada, *Finite groups with Sylow 2-subgroups of type* PSp(4,q), q *odd*, Tokyo Univ. Faculty of Science J., vol. 20(1973), pp. 341-372.

7. J. Hayden, *A characterization of the finite simple group* $PSp_4(3)$, Canadian J. Math., vol. 25 (1973), pp. 539-553.

8. J. Hayden, *A characterization of the finite simple groups* $PSp_4(3^m)$, m *odd,* Illinois J. Math., vol. 18 (1974), pp. 622-648.

9. M. Herzog, *On finite groups which contain a Frobenius subgroup,* J. Algebra, vol. 6(1967), pp. 192-221.

10. G. Higman, *Odd characterizations of finite simple groups,* (lecture notes, University of Michigan, 1968).

11. W. J. Wong, *A characterization of the finite projective symplectic groups* $PSp_4(q)$, Trans. Amer. Math. Soc., vol. 139(1969), pp. 1-35.

BOWLING GREEN STATE UNIVERSITY
BOWLING GREEN, OHIO

CHARACTERIZATION OF ${}^3D_4(q^3)$, $q = 2^n$ BY ITS SYLOW 2-SUBGROUP

BY

ANTHONY HUGHES

Let X be a finite group with Sylow 2-subgroup U. We say that a finite group G is of type X if a Sylow 2-subgroup of G is isomorphic to U. Our main result is the following:

THEOREM A.

Let G be a finite simple group of type ${}^3D_4(q^3)$, $q = 2^n$. Then G is isomorphic to ${}^3D_4(q^3)$.

Now ${}^3D_4(q^3)$, $q = 2^n$ has precisely four parabolic subgroups; that is, subgroups containing $B = N(U)$ for U a Sylow 2-subgroup of ${}^3D_4(q^3)$. These are B, ${}^3D_4(q^3)$ itself and two others, say P_1 and P_2. For $i = 1,2$, P_i is a 2-local subgroup with $O(P_i) = 1$. Let t be an involution of $Z(U)$. Then only one of the P_i, P_2 say, contains $C(t)$. Finally, if we set $F_i = O_2(P_i)$, then $O^{2'}(P_1)$ is a split extension of F_1 by $SL(2,q)$ and $O^{2'}(P_2)$ is a split extension of F_2 by $SL(2,q^3)$.

Now let G be an arbitrary finite group of type ${}^3D_4(q^3)$, $q = 2^n$, with Sylow 2-subgroup S. Then $S \cong U$. Let M and D be subgroups of S which correspond to F_1 and F_2 respectively. We establish Theorem A by studying the subgroups $N_G(M)$ and $N_G(D)$. Indeed, Theorem A is an immediate consequence of the following result:

THEOREM B.

Let G be a finite group of type ${}^3D_4(q^3)$, $q = 2^n$, with Sylow 2-subgroup S. Then at least one of the following holds:

(a) $O^{2'}(N_G(M))$ is solvable of 2-length one and $G = O(G)N_G(D)$.

(b) $O^{2'}(N_G(D))$ is solvable of 2-length one and $G = O(G)N_G(Y)$, where $Y = Z(M)$.

(c) $O^{2'}(G/O(G)) = {}^3D_4(q^3)$.

OUTLINE OF PROOF OF THEOREM B.

The structure of S is analyzed and the automorphism group of S is found. We then determine G in the case when G is 2-constrained and show that $O_2(G) = M$, D or S in this case. We now show that either $N_G(M)$ is solvable of 2-length one or $O^{2'}(N_G(M))/O(N_G(M)) \cong SL(2,q)$. We obtain a similar result for $N_G(D)$. If $N_G(M)$ is solvable of 2-length one, then we show that $Z(S)$ is a strongly closed abelian subgroup of S. A recent result of Goldschmidt now gives (a). A similar argument yields (b) if $N_G(D)$ is solvable of 2-length one. Finally, we show that (c) holds if neither $N_G(D)$ nor $N_G(M)$ is solvable of 2-length one.

UNIVERSITY OF ILLINOIS AT CHICAGO CIRCLE
CHICAGO, ILLINOIS

SIGNALIZER FUNCTORS

BY

GEORGE GLAUBERMAN

Suppose r is a prime, A is an Abelian r-subgroup of a finite group G, and θ is a solvable A-signalizer functor on G. Assume that $m(A) \geqslant 3$. Then θ is complete.

This result extends Goldschmidt's results for the cases

(a) $m(A) \geqslant 4$, and

(b) $m(A) \geqslant 3$ and $r = 2$.

(J. Algebra 21 (1972), 137-148, 321-340).

An expository account of this work will be published in the Proceedings of the International Symposium on Finite Groups in Sapporo, Japan, September, 1974. The work itself will appear in the Proceedings of the London Mathematical Society.

At the Conference, the author also mentioned some progress in simplifying the Odd Order Paper of Feit and Thompson, specifically Chapter IV, in which the maximal subgroups of a minimal counterexample are investigated (without character theory). In his talk, Professor David Sibley also mentioned progress on Chapter V, in which the investigation is continued by means of character theory.

University of Chicago
Chicago, Illinois

STRONGLY CLOSED 2-SUBGROUPS OF FINITE GROUPS

BY

DAVID GOLDSCHMIDT

DEFINITION: S is a strongly closed subgroup of G if $N_{S^g}(S) \subseteq S$ for all $g \in G$.

If S is a p-group, the condition is equivalent to: $S^g \cap P \subseteq S$ for all $g \in G$, where $S \subseteq P \in Syl_p(G)$. We obtain the following results:

THEOREM A: Suppose the finite group G contains a direct product of two strongly closed 2-subgroups $S_1 \times S_2$. Then $[< S_1^G >, < S_2^G >] \subseteq O(G)$.

THEOREM B: Suppose S is a strongly closed p-subgroup of G for some prime p, and $S \subseteq P \in Syl_p(G)$. Then:

a) If U_1 and U_2 are G-conjugate subsets of P, then SU_1 and SU_2 are $N_G(S)$-conjugate.

b) $(G' \cap P)S = (N_G(S)' \cap P)\, S$.

c) $C_G(S)' \cap P$ is strongly closed.

There are various corollaries to the above, including:

COROLLARY A1: (Product fusion) Suppose a Sylow 2-subgroup of G is the direct product of two strongly closed subgroups $S_1 \times S_2$. Then $< S_1^G > \cap < S_2^G > \subseteq O(G)$.

COROLLARY B1: Suppose S is a strongly closed 2-subgroup of G. Then $C_G(S)^{(\infty)}\, O(G) \trianglelefteq G$.

COROLLARY B3: Suppose S is a strongly closed 2-subgroup of G and $N_G(S)/C_G(S)$ is a 2-subgroup. Then

$$S \in Syl_2(< S^G >) \ .$$

UNIVERSITY OF CALIFORNIA
BERKELEY, CALIFORNIA

PART II

THE KNOWN SIMPLE GROUPS

THE STRUCTURE OF THE "MONSTER" SIMPLE GROUP

BY

ROBERT L. GRIESS JR.

We present some evidence for the existence of a simple group F, called the "monster." It was discovered independently by Fischer and Thompson, and by the author. Our approach was to work from the following hypotheses [1]:

(A) F is a simple group containing nonconjugate involutions t,z.

(B) $C = C_F(z)$ is a 2-constrained group with structure $O_2(C) \cong 2_+^{1+24}$, $C/O_2(C) \cong \cdot 1$ (Conway's simple group).

(C) $H = C_F(t)$ has structure $H = H'$, $Z(H) = \langle t \rangle$ and $H/\langle t \rangle \cong F_2$, Fischer's $\{3,4\}$-transposition group (the "baby monster").

In our notation, 2_ε^{1+2n} denotes an extra special 2-group of order 2^{1+2n} and type $\varepsilon = +, -$. In (C), F_2 is a simple group of order $2^{41}\cdot 3^{13}\cdot 5^6\cdot 7^2\cdot 11\cdot 13\cdot 17\cdot 19\cdot 23\cdot 29\cdot 31\cdot 47$. We assume the properties of F_2 claimed by Fischer.

Fischer had suggested in a lecture at Bielefeld that the following configuration in a simple group might be interesting: ρ is an element of order 3 conjugate to ρ^{-1}, $O_3(C(\rho))$ is extra special of order 3^{13} and exponent 3, $C(\rho)$ is 3-constrained and $C(\rho) = O_3(C(\rho))\cdot S$, where S is a full covering group of Suz and $Z(S) = \langle \rho \rangle \times \langle z \rangle$ where $|z| = 2$ and z inverts $O_3(C(\rho))/\langle \rho \rangle$. Thompson then determined that $C(z)$ must have shape as in (B) above, and proceeded to derive other local information as consequences. In particular, there is actually a quadruple of "new" simple groups which arise in this setting. Namely, F contains elements ω_i of order $i = 1, 2, 3, 5$ such that $F_i = C_F(\omega_i)/\langle \omega_i \rangle$ is a new simple group. The exact sequence $1 \to \langle \omega_i \rangle \to C_F(\omega_i) \to F_i \to 1$ splits for $i \neq 2$ and is nonsplit for $i = 2$. The group F_3 has been studied in detail by Thompson (denoted E in his notation [3]) and F_5 has been studied in detail by Harada (denoted F in his notation [2]). We remark that, using computer techniques, existence has been proven for F_3 and F_5 by P. Smith and S. Norton, respectively. The existence and uniqueness questions for F_2 and $F = F_1$ remain to be settled.

Our results may be summarized as follows:

THEOREM 1. F has exactly two classes of involutions.

THEOREM 2. $|F| = 2^{46}\cdot 3^{20}\cdot 5^{9}\cdot 7^{6}\cdot 11^{2}\cdot 13^{3}\cdot 17\cdot 19\cdot 23\cdot 29\cdot 31\cdot 41\cdot 47\cdot 59\cdot 71$.

THEOREM 3. F has trivial multiplier and trivial outer automorphism group. Also, F_2 has multiplier of order at most 2 (and is equal to 2 if F exists) and trivial outer automorphism group.

THEOREM 4. If $\chi = 1$ is an irreducible complex character of F, then $\chi(1) \geqslant 196883 = 47\cdot 59\cdot 71$.

CONJECTURE. F exists and there is an irreducible character of degree 196883.

Theorem 3 implies that the isomorphism type of H in (C) is unique. However, as we shall see, there are precisely two isomorphism types of abstract groups as in (B). Only one can live as the centralizer of an involution in a simple group. The proof of these statements about C is outlined in the following two results.

LEMMA. Let V be an irreducible 24 dimensional $F_2(\cdot 0)$-module which is irreducible and supports a nondegenerate quadratic form preserved by $\cdot 0$. Then $V \cong \Lambda/2\Lambda$, Λ the Leech lattice.

PROPOSITION. Let C be as in (B). Set $W = O_2(C)$.

(i) As a module for $C/W \cong \cdot 1$, $W/W' \cong \Lambda/2\Lambda$. Also $H^1(\cdot 1, \Lambda/2\Lambda) = 0$.

(ii) The extension $1 \to W/W' \to C/W' \to C/W \to 1$ is nonsplit (so that $H^2(\cdot 1, \Lambda/2\Lambda) \neq 0$) and the isomorphism type of C/W' is uniquely determined by (B).

(iii) $H^2(C/W', F_2) \cong F_2 \oplus F_2$.

(iv) The isomorphism type for $C = C_F(z)$ is uniquely determined by the cocycle class in $H^2(C/W', F_2)$ which does not restrict to 0 on W/W' and on $C_{C/W'}(\rho)$, where $|\rho| = 3$ and $C_{W/W'}(\rho) = 1$.

Next, we sketch some local subgroup information about F. We use the imprecise notation $X \cdot Y$ to denote a group with normal subgroup isomorphic to X with factor group isomorphic to Y. By $\widehat{X \cdot Y}$ we mean a nonsplit extension and by p^n we mean an elementary abelian p-group of rank n.

SOME 2-LOCALS	SOME 3-LOCALS
$\widehat{2 \cdot F_2}$	$3^{1+12}\ \widehat{2 \cdot Suz}\ 2$
$(2_+^{1+24}) \cdot (\cdot 1)$	$\widehat{3 \cdot M(24)}$
$[2^2 \cdot 2^{11} \cdot (2^{11} \times 2^{11})](\Sigma_3 \times M_{24})$	$\Sigma_3 \times F_3$
$2^{10} \cdot 2^{16} \cdot D_5(2)$	$\widehat{3^8 \cdot \Omega^-(8,3)} \cdot 2$
$\widehat{2^2 \cdot {}^2E_6(2)} \cdot \Sigma_3$	$[3^2 \cdot 3^5 \cdot (3^5 \times 3^5)](GL(2,3) \times M_{11})$

Notice that certain 3-locals seem to be analogues of 2-locals and vice versa. Normalizers of subgroups of prime order which lie in C or H include, for example, the following groups:

$$5 \times F_5 , \quad 5^{1+6}\,\widehat{2HJ}\cdot 4 , \quad (7 \times \text{Held})6 , \quad 7^{1+4}\cdot\widehat{2A_7}\cdot 6 .$$

Also, we mention that a Sylow 11-subgroup of F is elementary of order 11^2 and its normalizer contains a normal subgroup of index 5 which is a Frobenius group with complement isomorphic to SL(2,5).

Finally, we give a few results about representations of F. Take $\chi \neq 1$ a character of F. The proof of the following lemma relies on the earlier Proposition about C.

LEMMA. A character of C with kernel $1, \langle z \rangle, W$ has degree $\geq 24\cdot 2^{12} = 98304$, $24(2^{12}-1) = 98280$, 276 respectively. Also a character of C/W of degree > 276 has degree ≥ 299.

Now, $\chi|_C$ must have a constituent with kernel 1. Since z fuses in F to elements of $W - \langle z \rangle$, there must also be constituents with kernel $\langle z \rangle$. Since $\chi(1)$ divides $|F|$, the smallest possible candidate for $\chi(1)$ is

$$196883 = 98304 + 98280 + 299 = 47\cdot 59\cdot 71$$

where the summands have the significance they do in the Lemma. A few interesting consequences of the existence of χ are the following. Take $\theta \in F$, $|\theta| = 3$, $K = C_F(\theta)$ a covering group of $M(24)'$. Then θ has eigenvalues $1, \omega, \omega^{-1}$ with multiplicities 66149, 65367, 65367, $\chi(\theta) = 782$ and $\chi|_K$ has a faithful constituent of degree 783. This is the smallest possible degree of a faithful ordinary or projective representation of $M(24)'$. Now take $t \in F$ with $C_F(t) = H$. Then $\chi(t) = 4371$ and $\chi|_H$ breaks up into constituents of degrees 1, 4371, 96255 and 96256. The only constituent faithful on $\langle t \rangle$ is the last one, and 4371 is the smallest degree of an ordinary or faithful representation of F_2 .

REFERENCES

1. R. Griess, *On the subgroup structure of the group of order* $2^{46}\cdot 3^{20}\cdot 5^{9}\cdot 7^{6}\cdot 11^{2}\cdot 13^{3}\cdot 17\cdot 19\cdot 23\cdot 29\cdot 31\cdot 41\cdot 47\cdot 59\cdot 71$, to appear.

2. K. Harada, *On the simple group* F *of order* $2^{14}\cdot 3^{6}\cdot 5^{6}\cdot 7\cdot 11\cdot 19$, these proceedings.

3. J. Thompson, Sapporo Lectures, 1974.

RUTGERS UNIVERSITY
NEW BRUNSWICK, N.J.

UNIVERSITY OF MICHIGAN
ANN ARBOR, MICHIGAN

ON THE SIMPLE GROUP F OF ORDER $2^{14} \cdot 3^6 \cdot 5^6 \cdot 7 \cdot 11 \cdot 19$

BY

KOICHIRO HARADA

In the fall of 1973, B. Fischer raised the following question: Is there any simple group M containing an element x of order 3 such that $C_M(x)$ is an extension of an extra-special group of order 3^{13} by the double cover of the Suzuki group Suz?

Thompson took up the problem and concluded that, under some "reasonable" assumption, $C_M(j)$ is an extension of an extra-special group of order 2^{25} by the Conway group .1 where j is the involution in the center of the double cover of Suz.

Using the 24-dimensional representation of .1 over GF(2) which had been worked out by Conway, Thompson was able to determine the centralizers of p-elements of M hence the p-share of the prime p where $p = 2, 3, 5, 7, 11, 13, 17, 19, 23, 29, 31$ and 47: i.e.,

$$|M| = 2^{46} \cdot 3^{20} \cdot 5^{9} \cdot 7^{6} \cdot 11^{2} \cdot 13^{3} \cdot 17 \cdot 19 \cdot 23 \cdot 29 \cdot 31 \cdot 47 \cdot g'$$

where

$$(g' , 2 \cdot 3 \cdot 5 \cdot 7 \cdot 11 \cdot 13 \cdot 17 \cdot 19 \cdot 23 \cdot 29 \cdot 31 \cdot 47) = 1 .$$

Using the Sylow theorems, Conway showed that $g' = 41 \cdot 59 \cdot 71$ is the minimal possible number. The normalizers of Sylow p-subgroups of all the prime divisors of $|M|$ have been constructed (assuming $g' = 41 \cdot 59 \cdot 71$). It appears that $g' = 41 \cdot 59 \cdot 71$, but no proof is known.

Thompson found two more possible new simple groups E and F. Namely, E is the centralizer of an element of order 3 in M and F is that of an element of order 5. The existence of E and F has been established, respectively, by J. G. Thompson-P. Smith and by S. Norton-P. Smith.

Griess has independently been working on a finite group G satisfying the following condition:

(i) G contains involutions z and t,

(ii) $C_G(z)$ is an extension of an extra-special group of order 2^{25} by the Conway group $.1$, and

(iii) $C_G(t)$ is the nonsplit extension of Z_2 by Fischer's "baby monster" B.

Griess has reached the same group M above.(*)

This paper deals with the structure of F .

THEOREM. *Let* G *be a finite simple group containing an involution* t *such that* $C_G(t) \cong \hat{A}ut(HS)$ *or* $\hat{HS}$ *where* HS *is the Higman-Sims group and* ^ *denotes the double cover. Then* G *is a simple group of order* $2^{14} \cdot 3^6 \cdot 5^6 \cdot 7 \cdot 11 \cdot 19$.

1. THE STRUCTURES OF HS, AUT (HS), $\hat{HS}$ AND $\hat{A}$UT (HS)

(A) Let $\bar{T}$ be a Sylow 2-subgroup of HS , then $\bar{T}$ is generated by elements

$$\bar{z},\ \bar{\ell},\ \bar{\alpha}_1,\ \bar{\alpha}_2,\ \bar{\alpha}_3,\ \bar{\alpha}_4,\ \bar{x},\ \bar{y} \text{ and } \bar{a}$$

subject to the relations

$$\bar{\ell}_1^{\,2} = \bar{\alpha}_1^{\,2} = \bar{\alpha}_2^{\,2} = \bar{\alpha}_3^{\,2} = \bar{\alpha}_4^{\,2} = \bar{z}\ ,$$

$$\bar{z}^2 = \bar{x}^2 = \bar{y}^2 = \bar{a}^2 = 1\ ,$$

$$[\bar{\alpha}_1, \bar{\alpha}_3] = [\bar{\alpha}_2, \bar{\alpha}_4] = [\bar{x}, \bar{\alpha}_1] = [\bar{x}, \bar{\alpha}_2] =$$

$$[\bar{y}, \bar{\alpha}_2] = [\bar{a}, \bar{\ell}] = \bar{z}\ ,$$

$$[\bar{x}, \bar{\alpha}_3] = \bar{\alpha}_1,\ [\bar{x}, \bar{\alpha}_4] = \bar{\alpha}_2,\ [\bar{y}, \bar{\alpha}_3] = \bar{\alpha}_1\bar{\alpha}_2\bar{\ell}\ ,$$

$$[\bar{y}, \bar{\alpha}_4] = \bar{\alpha}_1\bar{\ell}\,\bar{z},\ [\bar{a}, \bar{\alpha}_1] = [\bar{a}, \bar{\alpha}_2] = \bar{\alpha}_1\bar{\alpha}_2\bar{z}\ ,$$

---

*Griess uses the names F_1, F_2, F_3, F_5 for M, B, E, F above.

$$[\bar{a}, \bar{\alpha}_3] = [\bar{a}, \bar{\alpha}_4] = \bar{\alpha}_3\bar{\alpha}_4\bar{z},\ [\bar{a}, \bar{y}] = \bar{x}\ \bar{\alpha}_1\bar{\alpha}_2\bar{\ell}\ \bar{z}\ ,$$

with all remaining commutators of pairs of generators being trivial [9].

(B) $|\bar{T}| = 2^9$, $Z(\bar{T}) = \langle \bar{z} \rangle$, $Z_2(\bar{T}) = \langle \bar{\ell}, \bar{\alpha}_1\bar{\alpha}_2 \rangle$,

$Z_3(\bar{T}) = \langle \bar{\ell}, \bar{\alpha}_1, \bar{\alpha}_2 \rangle$, $\bar{T}' = \Phi(\bar{T}) = \langle \bar{\ell}, \bar{\alpha}_1, \bar{\alpha}_2, \bar{\alpha}_3\bar{\alpha}_4, \bar{x} \rangle$.

(C) Set $\bar{E} = \langle \bar{\ell}, \bar{\alpha}_1, \bar{\alpha}_2, \bar{\alpha}_3, \bar{\alpha}_4 \rangle$

and

$$\bar{F} = \langle \bar{\ell}, \bar{\alpha}_1, \bar{\alpha}_2, \bar{x}, \bar{y} \rangle .$$

Then

$$\bar{E} \cong \bar{F} \cong Z_4 * Q_8 * Q_8 \cong Z_4 * D_8 * Q_8 .$$

All square roots of $\bar{z}$ are contained in $\bar{E} \cup \bar{F}$.

(D) $\bar{T}$ has nine conjugacy classes of involutions. The number of involutions in each class and a representative is given below.

$\bar{z}$,	$\overline{\alpha_1\alpha_2}$,	$\overline{\alpha_1\ell}$,	$\overline{\alpha_3\alpha_4}$,	$\overline{\alpha_1\alpha_4}$,	$\bar{a}$,	$\bar{x}$,	$\bar{y}$,	$\overline{a\ell}$
1,	2,	4,	8,	16,	16,	8,	16,	16

(E) Let $\bar{B} = \langle \bar{z}, \overline{\alpha_1\alpha_2}, \overline{\alpha_1\ell} \rangle$. Then $\bar{B} \triangleleft \bar{T}$, $C_{\bar{T}}(\bar{B}) \cong Z_4 \times Z_4 \times Z_4$ and $N_{HS}(\bar{B})/C_{\bar{T}}(\bar{B}) \cong GL(3,2)$.

Let $\bar{A}_1 = \langle \bar{z}, \overline{\alpha_1\alpha_2}, \overline{\alpha_3\alpha_4}, \bar{a} \rangle$. Then $N_{HS}(\bar{A}_1)$ is a non-split extension of $\bar{A}_1$ by S_6.

(F) The involution fusion pattern of HS is

$$\overline{z} \sim \overline{\alpha_1\alpha_2} \sim \overline{\alpha_1\ell} \sim \overline{\alpha_3\alpha_4} \sim \overline{\alpha_1\alpha_4} \sim \overline{a} \mid \overline{x} \sim \overline{y} \sim \overline{a\ell}$$

(G) The structure of the nonsplit extension H of Z_2 by Aut(HS) is not unique. In this paper we are interested only in the case where $H - H'$ contains an involution, in which case the structure of H is uniquely determined. In this section H always denotes such an extension:

$H \cong \hat{Aut}$ (HS) *and* $H - H'$ *contains an involution.*

Let S be a Sylow 2-subgroup of H and $T = S \cap H'$. We set $\langle t \rangle = Z(H)$ and $\overline{H} = H/\langle t \rangle$. Hence $\overline{T}$ is a Sylow 2-subgroup of $\overline{H}' \cong HS$. [z, ℓ, etc. are inverse images of $\overline{z}$, $\overline{\ell}$, etc. in T.] The conjugacy classes and the character tables of Aut(HS) and $H\hat{S}$ have been determined, respectively, by Frame [5] and by Rudvalis [15].

$\overline{H} \cong$ Aut(HS) has 39 conjugacy classes. 21 of them are represented in $\overline{H}'$. Of the 21 classes, 12 classes split into pairs of classes of H. Hence H has $12 + 12 + 9 = 33$ conjugacy classes represented in H'. Of the $39 - 21 = 18$ classes of $\overline{H}$ not represented in $\overline{H}'$, 6 classes split into pairs of classes of H. Hence H has $6 + 6 + 12 + 33 = 57$ conjugacy classes.

(H) Conjugacy classes of H.

CLASSES REPRESENTED IN H'

x	$C_H(x)$	Order in H/Z(H)	Powers
1	$2^{11}\cdot 3^2\cdot 5^3\cdot 7\cdot 11$	1	
$2_1 = t$	$2^{11}\cdot 3^2\cdot 5^3\cdot 7\cdot 11$	1	
$2_2 = z$	$2^{11}\cdot 3\cdot 5$	2	
$2_3 = tz$	$2^{11}\cdot 3\cdot 5$	2	
3	$2^5\cdot 3^2\cdot 5$	3	
4_1	2^9	4	$4_1^2 = 2_2$
4_2	2^7	4	$4_2^2 = 2_3$
$4_3 = \ell$	$2^9\cdot 3\cdot 5$	4	$4_3^2 = 2_2$
$4_4 = x$	$2^7\cdot 3^2\cdot 5$	2	$4_4^2 = 2_1$
5_1	$2^4\cdot 3\cdot 5^2$	5	
5_2	$2^2\cdot 5^2$	5	
5_3	$2^4\cdot 5^3$	5	
6_1	$2^5\cdot 3^2\cdot 5$	3	$6_1 = 2_1\cdot 3$
6_2	$2^5\cdot 3$	6	$6_2 = 2_2\cdot 3$
6_3	$2^5\cdot 3$	6	$6_3 = 2_3\cdot 3$
7	$2^2\cdot 7$	7	
8_1	2^5	8	$8_1^2 = 4_1$
8_2	2^4	8	$8_2^2 = 4_2$
10_1	$2^4\cdot 3\cdot 5^2$	5	$10_1 = 2_1\cdot 5_1$
10_2	$2^2\cdot 5^2$	5	$10_2 = 2_1\cdot 5_2$

CLASSES REPRESENTED IN H' (Continued)

x	$C_H(x)$	Order in H/Z(H)	Powers
10_3	$2^4 \cdot 5^3$	5	$10_3 = 2_1 \cdot 5_3$
10_4	$2^4 \cdot 5$	10	$10_4 = 2_2 \cdot 5_3$
10_5	$2^4 \cdot 5$	10	$10_5 = 2_3 \cdot 5_3$
11	$2 \cdot 11$	11	
12_1	$2^3 \cdot 3^2$	6	$12_1 = 3 \cdot 4_4$
12_2	$2^3 \cdot 3$	12	$12_2 = 3 \cdot 4_3$
14_1	$2^2 \cdot 7$	7	$14_1 = 2_1 \cdot 7$
15	$2^2 \cdot 3 \cdot 5$	15	
20_1	$2^3 \cdot 5$	10	$20_1 = 4_4 \cdot 5_1$
20_2	$2^3 \cdot 5$	20	$20_2 = 4_3 \cdot 5_3$
20_3	$2^3 \cdot 5$	20	$20_3 = 4_3 \cdot 5_3$
22	$2 \cdot 11$	11	
30_1	$2^2 \cdot 3 \cdot 5$	15	$30_1 = 2_1 \cdot 15$

CLASSES NOT REPRESENTED IN H'

x	$C_H(x)$	Order in H/Z(H)	Powers
$2_4 = v_1$	$2^8 \cdot 3^2 \cdot 5 \cdot 7$	2	
$2_5 = v_2$	$2^8 \cdot 3 \cdot 5$	2	
4_5	$2^6 \cdot 3$	4	$4_5^2 = 2_3$
4_6	$2^7 \cdot 5$	4	$4_6^2 = 2_2$

CLASSES NOT REPRESENTED IN H' (Continued)

x	$C_H(x)$	Order in H/Z(H)	Powers
6_4	$2^3 \cdot 3^2$	6	$6_4 = 2_4 \cdot 3$
6_5	$2^4 \cdot 3$	6	$6_5 = 2_5 \cdot 3$
6_6	$2^4 \cdot 3^2 \cdot 5$	6	$6_6 = 2_4 \cdot 3$
8_3	2^6	8	$8_3^2 = 4_1$
8_4	2^6	8	$8_4^2 = 4_1$
8_5	$2^4 \cdot 5$	4	$8_5^2 = 4_4$
10_6	$2^2 \cdot 3 \cdot 5$	10	$10_6 = 2_4 \cdot 5_1$
10_7	$2^2 \cdot 5$	10	$10_7 = 2_5 \cdot 5_2$
10_8	$2^2 \cdot 5$	10	$10_8 = 2_5 \cdot 5_2$
12_3	$2^3 \cdot 3$	12	$12_3 = 3 \cdot 4_5$
14_2	$2^2 \cdot 7$	14	$14_2 = 2_4 \cdot 7$
14_3	$2^2 \cdot 7$	14	$14_3 = 2_4 \cdot 7$
20_4	$2^3 \cdot 5$	20	$20_4 = 4_6 \cdot 5_3$
20_5	$2^3 \cdot 5$	20	$20_5 = 4_6 \cdot 5_3$
20_6	$2^3 \cdot 5$	20	$20_6 = 4_6 \cdot 5_3$
20_7	$2^3 \cdot 5$	20	$20_7 = 4_6 \cdot 5_3$
30_2	$2^2 \cdot 3 \cdot 5$	30	$30_2 = 2_4 \cdot 15$
30_3	$2^2 \cdot 3 \cdot 5$	30	$30_3 = 2_4 \cdot 15$
40_1	$2^3 \cdot 5$	20	$40_1 = 5_1 \cdot 8_5$
40_2	$2^3 \cdot 5$	20	$40_2 = 5_1 \cdot 8_5$

The following are the properties needed later. Some of the proofs are omitted, as they are straightforward.

(I) $S' = T'$. This holds also when $H - H'$ contains no involutions.

PROOF: It suffices to prove that $\bar{S}/\bar{T}'$ is abelian. Let v be an element of $S - T$. By (B), $\bar{T}/\bar{T}'$ is elementary of order 8. Since $< \bar{T}', \bar{a} >$ is the unique subgroup containing 55 involutions, we have $[\bar{v}, \bar{a}] \subseteq \bar{T}'$. By (C), $< \bar{E}, \bar{F} >^{\bar{v}} = < \bar{E}, \bar{F} > = < \bar{T}', \bar{y}, \bar{\alpha}_3 >$. On the other hand, by (F) we see that

$$|< \bar{T}', \bar{y} > \cap \bar{z}^{\bar{H}'}| = 15, \quad |< \bar{T}', \bar{y} > \cap \bar{x}^{\bar{H}'}| = 24,$$

$$|< \bar{T}', \bar{\alpha}_3 > \cap \bar{z}^{\bar{H}'}| = 31, \quad |< \bar{T}', \bar{\alpha}_3 > \cap \bar{x}^{\bar{H}'}| = 8 ,$$

$$|< \bar{T}', \overline{y\alpha_3} > \cap \bar{z}^{\bar{H}'}| = 15, \quad |< \bar{T}', \overline{y\alpha_3} > \cap \bar{x}^{\bar{H}'}| = 8 .$$

Hence $[\bar{v}, < \bar{y}, \bar{\alpha}_3 >] \subseteq \bar{T}'$. This proves $\bar{S}' = \bar{T}'$ as required.

(J) z is an involution and x is an element of order 4.

(K) $E = < t, \ell, \alpha_1, \alpha_2, \alpha_3, \alpha_4 > \cong Z_2 \times Z_4 * Q_8 * Q_8$.

PROOF: It suffices to prove that $t \notin \Phi (E)$. Let σ be an element of order 5 in $N_H(E)$. Then $C_E(\sigma) = < t, \ell > \cong Z_2 \times Z_4$. As $C_E(C_E(\sigma))$ is σ-invariant, we conclude that

$C_E(\sigma) = Z(E)$. Since $\overline{[E, \sigma]} \cong D_8 * Q_8$, $|[E, \sigma]| = 32$ or 64. If $|[E, \sigma]| = 32$, then $[E, \sigma] \cong D_8 * Q_8$. Suppose that $Z([E, \alpha]) \neq \langle \ell^2 \rangle$. Then the number of involutions $i(E)$ in E is $23 = 4 \cdot 12/2 - 1$. On the other hand, as $\overline{E}$ contains 31 conjugates of $\overline{z}$ in $\overline{H}$, $i(E) = 63$. This proves that $t \notin \Phi(E)$ when $|[E, \sigma]| = 32$. Suppose next that $|[E, \sigma]| = 64$. Then by [16, Lemma 5.26], $[E, \sigma]$ is of type $U_3(4)$ and $E = \langle \ell \rangle * [E, \sigma]$. In this case, $i([E, \sigma]) = 3$, whereas $i(\overline{[E, \sigma]}) = 11$. This proves (K).

(L) Set $z = \ell^2$. Then by (K) z is the uniquely determined inverse image of $\overline{z}$. Also $Z(S) = \langle t, z \rangle$.

(M) $F = \langle t, \ell, \alpha_1, \alpha_2, x, y \rangle$ contains 15 involutions all of which are contained in $\langle t, z, \ell\alpha_1, \alpha_1\alpha_2 \rangle \cong E_{16}$. F contains 16 square roots of z, 48 of t, and 48 of tz.

PROOF: Let ρ be an element of order 3 in $N_H(F)$. Then by the structure of H, ρ centralizes $\langle t, \ell \rangle$ and acts fixed-point-free on $F/\langle t, \ell \rangle$. Clearly then $\langle t, \ell \rangle = C_F(\rho) = Z(F)$. Hence F is a central product of $C_F(\rho)$ with $[F, \rho]$. As $\overline{[F, \rho]} \cong Q_8 * Q_8$, $|[F, \rho]| = 32$ or 64. Suppose that $|[F, \rho]| = 32$. Then $[F, \rho] \cong Q_8 * Q_8$. Hence $F \cong Z_4 \times Q_8 * Q_8$ or $Z_2 \times Z_4 * Q_8 * Q_8$ and so $i(F) \geqslant 19 = i(Q_8 * Q_8)$. On the other hand, $\overline{F}$ contains exactly 7 involutions conjugate to $\overline{z}$ in $\overline{H}$ and so $i(F) = 15$. This proves that $|[F, \rho]| = 64$. This forces $\Phi(F) = \Phi([F, \rho]) = \langle t, z \rangle$. Hence $F = \langle \ell \rangle * [F, \rho]$. Since $\langle t, z, \ell\alpha_1, \alpha_1\alpha_2 \rangle \cong E_{16}$, all involutions of F are contained in $\langle t, z, \ell\alpha, \alpha_1\alpha_2 \rangle = \langle t, z \rangle \cdot [\langle t, \ell, \alpha_1, \alpha_2 \rangle, \rho] \subseteq [F, \rho]$. Moreover, $[F, \rho]/\langle t \rangle \cong Q_8 * Q_8$ and ρ acts fixed-point-free on $[F, \rho]/\langle t, z \rangle$. The structure of such a 2-group $[F, \rho]$ is studied in [8, Part III, Section 5]. In fact, $[F, \rho]$ is isomorphic to the inverse image in $\hat{M}_{12}$ of a subgroup of M_{12} isomorphic to $Q_8 * Q_8$. By [8, page 152], we see that t has 24 square roots in $[F, \rho]$. If z had a square root in $[F, \rho]$, then $i(F) > 15$. Hence z has no square roots. Hence again by [8, page 152], tz has 24 square roots in $[F, \rho]$. The assertion follows now immediately.

(N) $T' = S' = \langle t, \alpha_1, \alpha_2, \ell, \alpha_3\alpha_4, x \rangle$ contains 31 involutions, 16 square roots of t, 32 square roots of z and 16 square roots of tz.

PROOF: If $w \neq t$ is an involution of S', then $\bar{w}$ is an involution of $\bar{H}'$ conjugate to $\bar{z}$. There are 15 such involutions in $\bar{S}'$. Hence $i(S') = 31$. If w is a square root of t, then $\bar{w} \sim \bar{x}$ in $\bar{H}'$. Since $|\bar{x}^{\bar{H}'} \cap \bar{S}'| = 8$, t has 16 square roots. If w is a square root of z, then $w \in E \cup F$ by (C). As all 16 square roots of z in F are contained in $\langle t, \ell, \alpha_1, \alpha_2, \rangle \cong Z_4 \times Z_2 \times Z_2 \times Z_2$, we conclude that $w \in E \cap S' = \langle t, \alpha_1, \alpha_2, \ell, \alpha_3\alpha_4 \rangle$. By (K), we immediately conclude that z has 32 square roots in S'. Again by (C), all square roots of tz are contained in $E \cup F$, hence in F - E. F - E contains 48 square roots of tz and 16 of them are contained in $S' \cap (F - E)$. Hence (N) holds.

(N') $T' \cdot \Omega_1(T)$ contains 48 square roots of t and 16 of tz.

(O) T contains 80, 32 or 48 elements of order 4 whose squares are t, z or tz respectively.

(P) S contains exactly two elementary abelian subgroups of order 64. They are $A = \langle t, z, \alpha_1\alpha_2, \alpha_3\alpha_4, a, v_1 \rangle$ and A^y, where $v_1 \in H - H'$. $N_H(A)/A \cong S_6$. The orbit lengths of the involutions of A under the action of $N_H(A)$ are 1, 15, 15, 12, and 20. 12 involutions are conjugate to 2_5 of (H) and 20 involutions to 2_4 of (H).

(Q) $C_H(2_4) \cong Z_2 \times Z_2 \times A_8$, $C_H(2_5) \cong Z_2 \times Z_2 \times E_{16} \cdot A_5^{(1)}$ where $E_{16} \cdot A_5^{(1)}$ is a split extension of E_{16} by A_5 in which A_5 acts nontrivially but intransitively on the involutions of E_{16}. We set $2_4 = v_1$, $2_5 = v_2$.

(R) $C_H(3)/\langle 3 \rangle$ contains a normal subgroup of index 2 isomorphic to $Z_2 \times Z_2 \times A_5$. The Sylow 2-subgroups of $C_H(3)$ are $(Z_2 \times Z_2) \wr Z_2$.

(S) $C_{H'}(<2_1, 2_2>)$ is an extension of $E = <t, \ell, \alpha_1, \alpha_2, \alpha_3, \alpha_4> \cong Z_2 \times Z_4 * D_8 * D_8$ by S_5 (for a suitable choice of 2_2). $C_H(<2_1, 2_2>)'E - E$ contains no involutions.

(T) $H'/Z(H) \cong HS$ contains only one conjugacy class of subgroups isomorphic to A_8. If $\overline{K} \cong A_8$, then all involutions of $\overline{K}$ are conjugate to $\overline{z}$.

PROOF: See [13].

(U) $C_H(5_1) \cong Z_5 \times SU^{\pm}(2, 5)$,

$C_H(5_2) \cong Z_5 \times Z_2 \times D_{10}$ with $<2_1, 2_5>$ being a Sylow 2-subgroup,

$C_H(5_3)$ is an extension of an extra-special group of order 5^3 by $Z_2 \times Q_8$.

(V) $C_H(7) = <7> \times <2_1, 2_4>$ for a suitable choice of the element 7 in (H).

(W) For a suitable choice of the element 11 in (H), $<2_1, 2_5> \cong Z_2 \times Z_2$ is a Sylow 2-subgroup of $N_H(<11>)$.

LEMMA 1.1. *Let* X *be a nontrivial extension of* E_{32} *by* A_5. *Suppose that* $Z(X) \cong Z_2$ *and* $X/O_2(X)$ *acts intransitively on* $(O_2(X)/Z(X))^{\#}$. *Then the* A_5 *acts completely reducibly on* $O_2(X)$.

PROOF: We may assume that X is embedded in a split

extension of E_{32} by $GL(5, 2)$. As $Z(X) \cong Z_2$, X is embedded in $Y = E_{32} \cdot E_{16} \cdot A_8 \subset E_{32} \cdot GL(5, 2)$. Let σ be an element of order 3 in X. Then by the intransitivity $C_{O_2(X)}(\sigma) \cong E_8$. We observe directly that $O_2(Y)$ is an extra-special group $Q_8 * Q_8 * Q_8 * Q_8$. Hence $C_{O_2(Y)}(\sigma) \cong Q_8 * Q_8$. Thus a subgroup A of X isomorphic to A_5 acts intransitively on $(O_2(Y)/O_2(X))^{\#}$. Since all complements of $A \cdot O_2(Y)/O_2(X)$ to $O_2(Y)/O_2(X)$ are conjugate to $A \cdot O_2(X)/O_2(X)$ in this case, we conclude that there is only one possible action of A on $O_2(X)$. This proves the complete reducibility.

2. Fusion-simple groups having $\hat{Aut}(HS)$ or $\hat{HS}$ as the centralizer of an involution.

Throughout the rest of the paper G always denotes a finite group containing an involution t such that $C_G(t) \cong \hat{Aut}(HS)$ or $\hat{HS}$ and $G \neq O(G)C_G(t)$. We shall show at the end of section 4 that G is a simple group of order $2^{14} \cdot 3^6 \cdot 5^6 \cdot 7 \cdot 11 \cdot 19$.

Set $H = C_G(t)$ and let S be a Sylow 2-subgroup of H. Then $T = S \cap H'$ is of order 2^{10} and its structure is uniquely determined, while that of S is not, in general, uniquely determined. We shall use the same notation for the generators of T as in the previous section; i.e., $T = \langle t, z, \ell, \alpha_1$,

$\alpha_2, \alpha_3, \alpha_4, x, y, a >$.

We first prove:

PROPOSITION 2.1. *S is not a Sylow 2-subgroup of G.*

PROOF: Suppose false. By (1), (I) and (N), $Z(S) = < t, z >$ and the numbers of square roots of z and t in $S' = T'$ are different. Hence $N(S) \subseteq H$. In particular, $t \not\sim tz \not\sim z \not\sim t$ in G . As any involution of T is conjugate to t, tz or z, t must be conjugate in G to an involution v of $S - T$ by Glauberman's theorem. Hence $H \cong \hat{\mathrm{Aut}}(HS)$ and $H - H'$ contains an involution. The result of the previous section is thus applicable. By (P), we may assume that $v \in A = < t, z, \alpha_1\alpha_2, \alpha_3\alpha_4, a, v > \cong E_{2^6}$. By (P) again, S contains exactly two elementary abelian subgroups of order 64 and they are conjugate in S . Hence $v \sim t$ in $N_G(A)$. We shall contradict this fact by showing that $N_G(A) \subseteq H$.

By (P), $N_H(A)/A \cong S_6$ and S_6 acts on A with orbit lengths 1, 15, 15, 12 and 20. As $|N_G(A)|_2 = |N_H(A)|_2 = 2^{10}$, $k = |N_G(A): N_H(A)|$ is odd. As $t \not\sim z$ and $t \not\sim zt$, only $k = 13$, 21 or 33 are possible. Since $|GL(6, 2)| = 2^{14} \cdot 3^4 \cdot 5 \cdot 7^2 \cdot 31$, $k = 21$ must hold. As the Sylow 2-subgroups of $\overline{N} = N_G(A)/A$ are isomorphic to $Z_2 \times D_8$, $0^2(\overline{N}) \subset \overline{N}$ and so $0^2(\overline{N})$ has dihedral Sylow 2-subgroup of order 8. One concludes easily that $0^2(\overline{N}) \supseteq Z_7 \times A_6$ or $Z_3 \times A_6$. In the first case, a

7-element of $\overline{N}$ must centralize the subgroup of A of order at most 4 centralized by A_6. Hence the 7-element also centralizes the subgroup of A, whereas $k = 21$. Hence $O^2(\overline{N}) \not\supset Z_7 \times A_6$ and so $\overline{N} \supseteq \overline{N}_1 \cong S_7$. As t is isolated in $A \cap H'$, $A \cap H' \not\leq N_1$, the inverse image of $\overline{N}_1$ in $N_G(A)$. Hence z or tz has $15 + 12 = 27$ conjugates in N_1. But then correspondingly tz or z has exactly 15 conjugates in N_1. Again by $A \cap H' \ntrianglelefteq N_1$, N_1 must act on an elementary abelian group of order 16 in $A \cap H'$. As $\overline{N}_1 \cong S_7 \not\subset GL(4, 2)$, we obtain a contradiction. This completes the proof of Proposition 2.1.

By the previous proposition, G contains a 2-group S_1 such that $S_1 \supset S$ and $|S_1 : S| = 2$. Let $u \in S_1 - S$. Then $t^u = tz$ or z.

LEMMA 2.2. *The following conditions hold:*

(i) $t^u = tz$, $z^u = z$, $t \not\sim z$ *in* G

(ii) $H \cong \hat{Aut}(HS)$, $H - H'$ *contains an involution, and the structure of* H *is uniquely determined.*

PROOF: By (0), T contains 80 square roots of t, 32 of z and 48 of tz. Hence $T^u \neq T$ and so $[S: T] = 2$. Thus $H \cong \hat{Aut}(HS)$. By (N), $S' = T'$ contains 16 square roots of t and of tz, and 32 of z. Hence (i) holds. By (N') we conclude similarly that $\Omega_1(T)^u \not\subseteq T$. Hence $S - T$ contains an involution. (ii) follows immediately. This proves the lemma.

If we choose involutions v_1 and v_2 as in (Q), then

$$C_H(v_1) \cong Z_2 \times Z_2 \times A_8$$

and

$$C_H(v_2) \cong Z_2 \times Z_2 \times E_{16} \cdot A_5^{(1)} .$$

Since $\Omega_1(T)^u \not\subseteq T$, there is an involution i in T such that $i^u \in S - T$. Then $(it)^u = i^u \cdot tz \in S - T$. Since every involution of T is conjugate in H to i, it or t, we conclude that one of the following two cases holds:

(a) $v_1 \sim t \not\sim v_2 \sim z$ in G or

(b) $v_2 \sim t \not\sim v_1 \sim z$ in G.

Taking $t \sim tz$ into account, we see that t has 36 conjugates in A if (a) holds and 28 if (b) holds.

We shall rule out the case (b) in two lemmas. As in Prop. 2.1, we set $A = \langle t, z, \alpha_1\alpha_2, \alpha_3\alpha_4, a, v_1 \rangle \cong E_{2^6}$. We see as before that $v_1 \sim tv_1 \sim tz \sim t$ or $v_2 \sim tv_2 \sim tz \sim t$ in $N_G(A)$.

LEMMA 2.3. *The following condition holds:*

(i) $N_G(A)/A \cong PSp(4, 3)$ *if the fusion pattern* (a) *holds*

(ii) $N_G(A)/A \cong A_8$ *if* (b) *holds.*

PROOF: Suppose first that (a) holds. Then $|N_G(A): N_H(A)| = 36$. As $N_H(A)/A \cong S_6$, $|\overline{N}| = 2^6 \cdot 3^4 \cdot 5 =$

$|PSp(4, 3)|$, where $\overline{N} = N_G(A)/A$. By a theorem of Brauer [2] all simple groups of order $2^a \cdot 3^b \cdot 5$ are known. Hence if $\overline{N} \not\cong PSp(4, 3)$, then A_6 is the unique nonsolvable composition factor of $\overline{N}$. Clearly then $\overline{N}^{(\infty)} = \overline{N_H(A)'} \cong A_6$. As $\overline{N}$ acts on the fixed elements of A under the A_6 , t cannot have 36 conjugates. Hence (i) holds.

Suppose next that (b) holds. Then as above $|\overline{N}| = 2^6 \cdot 3^2 \cdot 5 \cdot 7 = |A_8|$. Again $\overline{N}$ has a unique nonsolvable composition factor. The composition factor is isomorphic to A_8 , A_7 or A_6 by [10]. Here we have used the fact that $L_3(4)$ does not contain S_6 . Hence $\overline{N}^{(\infty)} = \overline{L} \cong A_8$, A_7 or A_6. One can rule out the last case as above. Suppose that $\overline{L} \cong A_7$ or A_8 . Let $\overline{P}$ be a Sylow 3-subgroup of $\overline{N \cap H}$. Then $\overline{P} \subseteq \overline{L}$. $\overline{P}$ centralizes a four subgroup of A consisting of t and two other involutions conjugate to v_1 . As $t \not\sim v_1$, $N_{\overline{N}}(\overline{P}) \subseteq \overline{H \cap N}$. As $\overline{N} = N_{\overline{N}}(\overline{P})\overline{L}$ and $(\overline{H \cap N})' \subseteq \overline{L}$, $[\overline{N}: \overline{L}] \leqslant 2$. Hence $\overline{L} \cong A_8$ must hold.

LEMMA 2.4. The case (ii) of Lemma 2.3 does not occur.

PROOF: Suppose that (ii) held. Let a be an element of order 3 of $N = N_G(A)$ which corresponds to a 3-cycle of A_8. We shall prove that $C_G(a) \cong Z_3 \times PSp(4, 5)$. We then investigate the fusion of elements of order 3 of $C_G(a)$ and prove that it is not compatible with the fusion of elements of order

3 of $H \cong \hat{Aut}(HS)$.

If we choose a suitably then $C_A(a) \supseteq \langle t, v_1 \rangle$. Suppose $C_A(a) = \langle t, v_1 \rangle$. Then $C_N(t)$ would involve $Z_3 \times A_5$. This is impossible as $\overline{N \cap H} \cong S_6$. Hence $C_A(a) \cong E_{16}$ and so $C_N(a) = \langle a \rangle \times F$ where F is an extension of E_{16} by A_5. The A_5 also contains a 3-cycle of A_8 and it centralizes a four subgroup of $O_2(F) \cong E_{16}$. Thus $F \cong E_{16} \cdot A_5^{(1)}$. By (R), we know the structure of $C_G(a) \cap H/\langle a \rangle$, which is an extension of $Z_2 \times Z_2 \times A_5$ by Z_2 with Sylow 2-subgroups isomorphic to $(Z_2 \times Z_2) \wr Z_2$.

F acts on $O_2(F)^{\#} = C_A(a)^{\#}$ with orbit lengths 5 and 10. As $|F|_2 = 64$ while $|C_F(t)|_2 = 32$, t has 10 conjugates and v_1 (hence z) has 5 in $O_2(F)$. F has Sylow 2-subgroup of type A_8 , as such an extension is uniquely determined [8,Part II, Lemma 2.6]. Hence $O_2(F)$ is weakly closed in a Sylow 2-subgroup of F . If $|C_G(a)|_2 > 64$, then $|C_H(a)|_2 \geq 64$, contradicting the fact that $|C_H(a)|_2 = 32$. Hence $C_G(a)$ has Sylow 2-subgroup of type A_8 . Since $C_G(a)$ contains F and $C_H(a)$ properly, we conclude by [7] that $C_G(a)/\langle a \rangle \cong PSp(4,5)$ and so $C_G(a) = \langle a \rangle \times K$ where $K \cong PSp(4, 5)$. Let P_1 be a Sylow 3-subgroup of K . We set $P = \langle a \rangle \times P_1$. Then $P \cong Z_3 \times Z_3 \times Z_3$. Since $N_K(P_1)/C_K(P_1) \cong D_8$ and there is an involution which inverts $\langle a \rangle$, $|N_G(P)/C_G(P)|_2 \geq 16$. Set $N_1 = N_G(P)/C_G(P)$. Then N_1 is isomorphic to a subgroup of $GL(3,3)$, which is of order $2^5 \cdot 3^3 \cdot 13$. If $13 \mid |N_1|$, then $N_1 \supseteq SL(3, 3)$ and N_1 is transitive on $P^{\#}$. But then $|C_G(a)|_3 \geqslant 3^4$, which is not the case. Hence N_1 is a 2, 3-group.

We need to count the number of conjugate elements of a in P. There are two cyclic subgroups of order 3 in P_1 which is centralized by "noncentral" involutions of K. As a "noncentral" involution of K is conjugate to t and all elements of order 3 of H are conjugate in H, $2 + 6 + 6 = 14$ elements of P are visibly conjugate to a. They are conjugate in $N_G(P)$ as P is a Sylow 3-subgroup of the centralizer of such an element of order 3. As $|N_G(P) \cap C_G(a)/C_G(P)| = 2^3$, $|N_1| \geqslant 14 \cdot 2^3$. As N_1 is a 2, 3-group, only $|N_1| = 18 \cdot 2^3$ is possible. On the other hand, a Sylow 3-subgroup of H admits a quasi-dihedral 2-subgroup of order 16 in H. Hence N_1 has Sylow 2-subgroups isomorphic to QD_{16}. Clearly then all elements of order 3 of K must be conjugate in $N_G(\langle a \rangle)$. But this is false, as K has two conjugacy classes of elements of order 3 and if σ_1 and σ_2 are the representatives, then $C_K(\sigma_1)$ or $C_K(\sigma_2)$ has, respectively, quaternion or dihedral Sylow 2-subgroups. This completes the proof.

Thus $N_G(A)/A \cong PSp(4, 3)$. Set $M = C_G(z)$ and $R = O_2(N_M(A)')$.

Lemma 2.5. *The following holds:*

(i) $z \sim v_2$ *in* $N_G(A)$;

(ii) $N_M(A)/A \cong E_{16} \cdot A_5^{(1)}$;

(iii) $|N_M(A): N_M(A)'| = 2$,

(iv) $R = O_2(N_M(A)') \cong D_8 * D_8 * D_8 * D_8$, *the extra-special group of order* 2^9 *of the plus type; and*

(v) $N_M(A)/R \cong Z_2 \times A_5$ *and the extension splits over* R .

Proof: Clearly $N_G(A)$ cannot act on $< z^{N_H(A)} > \cong E_{16}$ or E_{32} . Hence z is conjugate to v_2 by an element of $N_G(A)$. This is (i). z has thus 27 conjugates in A and so $|N_M(A)/A| = 2^6 \cdot 3 \cdot 5$. Since $PSp(4, 3)$ does not contain an element of order 15, $N_M(A)/A$ is nonsolvable and hence involves A_5 . If $N_M(A)/A$ was non-2-constrained, $N_M(A)/A$ would be contained in the centralizer of an involution of $N_M(A)/A$. One cannot, however, find such an involution in $PSp(4, 3)$. Hence $N_M(A)/A$ is an extension of E_{16} by A_5 . As $PSp(4, 3)$ has Sylow 2-subgroups of type A_8 , (ii) holds.

Let B be the inverse image of $O_2(N_M(A)/A)$ in $N_M(A)$. Then $|B| = 2^{10}$ and $B/A \cong E_{16}$. As $|N_H(A) \cap M/A| = |S_6|/15 = 48$ and $|N_H(A) \cap C_G(v_2)/A| = |S_6|/12 = 60$, all 36 conjugates of t in A are divided by the action of $N_M(A)$ into two orbits of length $16 \cdot |A_5|/48 = 20$ and $16 \cdot |A_5|/ 60 = 16$. Here we have used $v_2 \sim z$ in $N_G(A)$.

Let $C = Z(B \bmod \langle z \rangle)$. The previous argument shows that $C \subset A$. Indeed, the orbit of length 16 cannot lie in C. As the A_5 acts on C and A/C, $|C| = 4$ or 32. Suppose that $C \cong Z_2 \times Z_2$. Then $C \subseteq Z(N_M(A))$. In particular, a 5-element σ of $N_M(A)$ centralizes C. On the other hand, $|N_G(A)|$ is divisible by 5 to the first power and as $N_H(A)/A \cong S_6$, $C = C_A(\sigma)$ must contain a conjugate of t. This conflicts with the orbit length of the conjugates of t under $N_M(A)$ given above. Thus $C \cong E_{32}$.

Let v be an element of A - C such that $v \sim t$ in $N_G(A)$ and $|v^{N_M(A)}| = 16$. Clearly then $A - C = v^{N_M(A)} \cup (vz)^{N_M(A)}$. Furthermore $C_G(v) \cap N_M(A)/A \cong A_5$. This implies that $\Phi(B) \subseteq C$.

We next consider $W = N_M(A)/C$. W is an extension of $B/C \cong E_{32}$ by A_5. As the A_5 centralizes $A/C \cong Z_2$ and acts intransitively on B/A, the A_5 is completely reducible on B/C by Lemma 1.1. Let R/C be a complement to A/C in B/C. Then $|R| = 2^9$ and is A_5-admissible. As $\langle t, v_2 \rangle \sim \langle v, z \rangle$ in $N_G(A)$ and $N_G(A) \cap H \cap C_G(v_2)$ is a split extension of E_{64} by A_5 by (Q), we conclude that $N_M(A)$ is a split extension of R by $\langle v \rangle \times A_5$ and $R = O_2(N_M(A)')$.

We now prove that R is extra-special. We have shown that t has 20 conjugates in A under $N_M(A)$. Those 20 conjugates must lie in C. As the A_5 in $N_M(A)$ can give rise to only 10 conjugates of t, R acts nontrivially on C. This forces that $Z(R) = \langle z \rangle$.

$PSp(4, 3)$ has exactly two conjugacy classes of involutions and we see easily that there are no four subgroups in $PSp(4, 3)$ which consist of three "central" involutions. As $N_M(A)$ contains a subgroup isomorphic to A_5, "noncentral" involutions of $PSp(4, 3) \cong N_G(A)/A$ split off over A. Let σ be an element of order 3 of $N_G(A)$ acting fixed-point-free on A. $C_{PSp(4,3)}(\sigma)$ contains a "central" involution of $PSp(4,3)$. Hence such an involution also splits off over A. Hence each coset of RA over A contains an involution.

As R acts regularly on $v^{N_M(A)}$ and $(vz)^{N_M(A)}$, no element of $R - A$ centralizes an element of $A - C$. This implies that every involution of $R - A$ lies in $R - C$. Hence each coset of R over C contains an involution. By the irreducible action of the A_5 on R/C and $C/\langle z \rangle$, we conclude that $R/\langle z \rangle$ is elementary. Together with $Z(R) = \langle z \rangle$, we see that R is extra-special. Since the 2-rank of R is at least 5, (iv) holds. (v) is trivial by the preceding paragraph.

LEMMA 2.6. *Let* $V = \langle t, z \rangle \cong Z_2 \times Z_2$. *Then* $O_2(N_G(V)) = O_2(N_M(A')) = R \cong D_8 * D_8 * D_8 * D_8$.

PROOF: By (S), $O_2(N_{H'}(V)) \cong Z_2 \times Z_4 * D_8 * D_8$. Since

$N_{H'}(V)/O_2(N_{H'}(V)) \cong S_5$, $O_2(N_H(V))$ has order 2^8. As $u \in S_1 - S$ normalizes V and $[u, V] \neq 1$, $Q = O_2(N_G(V))$ is of order 2^9 and $N_G(V)/Q \cong S_5$. Let σ be an element of order 5 in $N_G(V) \cap H'$ then $Q = C_Q(\alpha)[<\alpha>, Q]$ and

$$O_2(N_{H'}(V)) \supset [<\alpha>, Q] \cong D_8 * Q_8 .$$

By Thompson's $A \times B$ lemma, $C_Q(\alpha)$ centralizes $[<\alpha>, Q]$. Hence $Q = C_Q(\alpha) * [<\alpha>, Q]$ and also $C_Q(\alpha) \trianglerighteq <t, \ell> \cong Z_2 \times Z_4$.

By Lemma 2.5, $R = O_2(N_M(A)') \cong D_8 * D_8 * D_8 * D_8$. Clearly $R \subseteq N_G(V)$. Hence $\overline{S_1} = S_1/<z>$ must contain an elementary abelian group $\overline{R}$ of order 2^8. We shall argue that $\overline{R} = \overline{Q}$ is the unique possibility.

As $S_1/Q \cong D_8$, $|\overline{R} \cap \overline{Q}| \geqslant 2^6$. Set

$$\overline{W} = <\overline{t}> \times \overline{[<\alpha>, Q]} .$$

Then $\overline{W} \subseteq Z(\overline{Q})$ and $\overline{W} \cong E_{32}$. Suppose $m(\overline{Q}) \leqslant 7$. Then $\overline{R} - \overline{Q}$ contains an involution $\overline{r}$. We have $[\overline{r}, \overline{R} \cap \overline{Q}] = 1$ and $|\overline{R} \cap \overline{Q}| \geqslant 2^6$. Hence $|\overline{R} \cap \overline{Q} \cap \overline{W}| \geqslant 2^4$ and so $|C_{\overline{W}}(\overline{r})| \geqslant 2^4$. But by (A) (applied mod $<t>$), this is impossible. Hence $\overline{Q} \cong E_{2^8}$.

Set $\overline{S_1} = <\overline{Q}, \overline{x}, \overline{y}, \overline{a}>$ where x, y and a are the same elements as in (G). As any involution of $\overline{S_1} - \overline{Q}$ centralizes at most 4-dimensional space on $<\overline{t}, \overline{\ell}, \overline{[<\alpha>, Q]}> = \overline{W}_1 \cong E_{2^6}$, if $\overline{R} \neq \overline{Q}$, then $|\overline{R} \cap \overline{Q}| = 2^6$. Hence $\overline{x} \in \overline{RQ}$ and so $\overline{x}$ centralizes a 6-dimensional space of $\overline{Q}$. On the other hand we know by (H) that $|C_G(x)|_2 = 2^7$. Hence $C_Q(x) < x >$ is a Sylow 2-subgroup of $C_G(x)$. By (A), $[x, < a, y >] \subseteq < t >$ and as $< a, x, y >$ covers $S_1/Q \cong D_8$, $C_{< a, y >}(x)$ cannot lie in $C_Q(x) < x >$. This contradiction shows that $\overline{R} = \overline{Q}$ is the unique elementary abelian subgroup of order 2^8 in $S_1/< z >$. This proves our lemma.

LEMMA 2.7. $N_G(R)/R \cong A_5 \wr Z_2$.

PROOF: By the structure of $N_M(A) \triangleright R$ and $N_G(V) \triangleright R$, we obtain the following information on the structure of $N_G(R)/R = \overline{N}$.

(i) $\overline{N}$ *contains at least two conjugacy classes of subgroups isomorphic to* A_5 *such that the actions of 5-elements on* R *are different.*

(ii) $|N_G(R): N_G(R) \cap H| \leqslant 270 - 30 - 1$.

(i) and (ii) follow from the fact that a 5-element of $N_M(A)$ and that of $N_G(V)$ have distinct actions on R. The condition (ii) is obtained by computing the maximal possible number of the conjugates of t in R under $N_G(R)$. We know that E (see (C), (F)) contains 30 involutions conjugate to z in G. If E were normalized by $N_G(R)$, there would exist an element of order 5 which centralizes E hence R. Thus R contains more than 30 conjugates of z in G. Since there are exactly 270 involutions in R we obtain (ii).

(iii) $\overline{N}$ *contains an involution* $\overline{v}$ *with* $C_{\overline{N}}(\overline{v}) \cong Z_2 \times A_5$.

If v is the same element as in Lemma 2.5, then $C_{\overline{N}}(\overline{v}) < v > = A$. Hence the inverse image of $C_{\overline{N}}(\overline{v})$ normalizes A . Hence $C_{\overline{N}}(\overline{v}) = C_{\overline{N_M(A)}}(\overline{v}) \cong Z_2 \times A_5$.

The condition (ii) implies that $|\overline{N}| < |S_5| \cdot 120$. Let $\overline{L} = \overline{N}^{(\infty)}$. Suppose that the maximal solvable normal subgroup $S(\overline{L})$ of $\overline{L}$ is not in the center of $\overline{L}$. Then $|S(\overline{L})| \geq 16$ and so $|\overline{N}/S(\overline{L})| < |S_5| \frac{15}{2} = 900$. Only possible nonsolvable composition factors of $\overline{N}/S(\overline{L})$ are A_5 or A_6 so (i) cannot hold. Hence $S(\overline{L}) \subseteq Z(\overline{L})$. Suppose that $S(\overline{L}) > 1$. By the structure of $D_4(2) \cong Out(R)'$, $S(\overline{L}) \cong Z_2, Z_3$ or Z_5 . By (i) $\overline{N}/S(\overline{L})$ must be divisible by 5^2, but there exists no such element in $Out(R)$. Hence $S(\overline{L}) = 1$. Thus $\overline{L}$ is semi-simple. If $\overline{L}$ has two components, then $\overline{N} \cong A_5 \int Z_2$ holds. Suppose that $\overline{L}$ is simple. Since $|\overline{L}| < 120^2$ and $5^2 \mid |\overline{L}|$, $\overline{L} \cong L_2(25)$ is the unique possibility by [10]. This is a contradiction as $13 \mid |\overline{L}|$ and $13 \nmid |D_4(2)|$. This completes the proof of Lemma 2.7.

We shall next investigate the action of $A_5 \int Z_2$ on R. We use the list of conjugacy classes of $D_4(2)$ given in [4]. It is easy to see that $A_5 \int Z_2$ has two conjugacy classes $< \sigma_1 >$ and $< \sigma_2 >$ of cyclic groups of order 3 and three classes $<\rho_1>$, $<\rho_2>$, and $<\rho_3>$ of order 5. We may choose $<\sigma_1>$ and $<\rho_1>$ so that each is contained in a direct component of $A_5 \times A_5$. We call $<\sigma_2>$, $<\rho_2>$, and $<\rho_3>$ "diagonal" subgroups. By [4], one sees that Out(R) has two conjugacy classes of $A_5 \times A_5$:

Case [I]. $C_R(\sigma_1) \cong D_8 * Q_8 * D_8$, $C_R(\sigma_2) = D_8 * D_8$,
$C_R(\rho_1) \cong D_8 * Q_8$, $C_R(\rho_2) = C_R(\rho_3) = <z>$.

Case [II]. $C_R(\sigma_1) \cong <z>$, $C_R(\sigma_2) \cong Q_8 * Q_8$,
$C_R(\rho_1) = <z>$, $C_R(\rho_2) \cong D_8 * Q_8$,
$C_R(\rho_3) = <z>$.

We can visualize [I] easily by considering:

$$\{(D_8 * Q_8)A_5\} * \{(D_8 * Q_8)A_5\} .$$

[II] can be obtained by twisting [I] by the triality automorphism of $D_4(2)$. The triality automorphism maps the class IV of [4] to LVI and XII to LX.

By the structure of $N_G(V)/R \cong S_5$, we see that a 5-element of $N_G(V)$ centralizes t . Since any direct component of $A_5 \times A_5$ cannot be embedded in S_5 in $A_5 \wr Z_2$, we must have the case [II].

We have thus proved:

LEMMA 2.8. *If* σ_1, σ_2, ρ_1, ρ_2 *and* ρ_3 *are the elements defined as above, then*

$$C_R(\sigma_1) = C_R(\rho_1) = C_R(\rho_3) = \langle z \rangle, \; C_R(\sigma_2) \cong Q_8 * Q_8$$

and

$$C_R(\rho_2) \cong D_8 * Q_8 .$$

LEMMA 2.9. *The following condition holds.*

(i) $R^{\#}$ *has four conjugacy classes of elements under the action of* $N_G(R)$: *the central involution* z, 150 *conjugates of* w *with* $w \sim z$ *in* G, 120 *conjugates of* t, *and* 240 *conjugates of elements of order* 4 .

(ii) $N_G(R) - N_G(R)'$ *contains exactly two conjugacy classes in* $N_G(R)$ *of involutions represented by* v *and* vz *where* $v \sim t$, $vz \sim z$ *in* G .

PROOF: As $|N_G(R) : N_H(R)| = 120$, t has 120 conjugates in R. t is centralized by an element of order 5 of $N_G(R)$ which without loss we may assume is ρ_2. As ρ_1 acts on $C_R(\rho_2) \cong D_8 * Q_8$ nontrivially, any involution of R centralized by an element of order 5 in $N_G(R)$ is conjugate to t. Let w be an involution of $R - \langle z \rangle$ conjugate to z in G. By the argument above and by Lemma 2.8, $|w^{N_G(R)}|$ is divisible by $2 \cdot 3 \cdot 5^2 = 150$. Hence $|w^{N_G(R)}| = 150$.

As $\langle \rho_2 \rangle$ is contained in S_5 of $N_G(R)/R$,

$$N_G(R) \cap N_G(\langle \rho_2 \rangle)/C_R(\rho_2)\langle \rho_2 \rangle$$

is a Frobenius group of order 20. It is now easy to see that all elements of R of order 4 centralized by an element of order 5 are conjugate in $N_G(R)$. Suppose that there is an element of R of order 4 not centralized by any 5-element of $N_G(R)$. Then there are at least $2\cdot 3\cdot 5^2 = 150$ such elements. Hence there are at most 90 remaining elements. On the other hand, we know that $C_G(\ell) \cap N_G(R)$ involves A_5. The A_5 must be "diagonal" in $A_5 \times A_5 \subseteq N_G(R)/R$ and hence maximal in $A_5 \times A_5$. Hence ℓ has at least $120 = 2 \times 60$ conjugates in R. This contradiction establishes (i) of Lemma 2.9.

We recall the involution v in Lemma 2.5. By the structure of $N_G(R)/R \cong A_5 \wr Z_2$, the involution $v \sim t$ is in $N_G(R) - N_G(R)'$. Clearly all involutions of $N_G(R)/R - N_G(R)'/R$ are conjugate to v mod R. As $C_R(v) \cong E_{32}$ and $vz \sim z$, $C_{\overline{R}}(\overline{v}) \cong E_{16}$ where $\overline{R} = R/\langle z \rangle$. Hence all involutions of $\langle v,R \rangle - \langle z \rangle$ are conjugate to v mod $\langle z \rangle$. This establishes (ii).

LEMMA 2.10. *The following holds:*

(i) *Let* $\overline{i}$ *be an involution of* $\overline{N} - \overline{R}$, *then* $[\overline{R},\overline{i}] \cong E_{16}$ *where* $\overline{N} = N_G(R)/\langle z \rangle$.

(ii) $M = N_G(R) = C_G(z)$.

PROOF: If $\overline{i} \in \overline{N} - \overline{N}'$, (i) was shown in Lemma 2.9. $\overline{N}'/\overline{R}$ has two conjugacy classes of involutions in $\overline{N}/\overline{R}$. One is contained in a direct factor A_5 and the other is a "diagonal". Both involutions invert cyclic groups of order 5 which act fixed-point-free on $\overline{R}$. Hence $C_{\overline{R}}(\overline{i}) = [\overline{R}, \overline{i}] \cong E_{16}$.

We apply a theorem of Goldschmidt [6] to prove (2). Corollary 4 of [6] states: Assume

(1) T is a Sylow 2-subgroup of a finite group X

(2) W is a weakly closed subgroup of T (with respect to X)

(3) A is an abelian normal subgroup of $N_X(W)$ and $A \subseteq C_T(W)$

(4) $\mathcal{S} = \{B \subseteq T, \mid B$ is conjugate in X to a subgroup of A, $B \not\subseteq A\}$

(5) $r = \max\{m(B/C_B(W)) \mid B \in \mathcal{S}\}$.

Then either X is of "known" type or the following hold:

(I) There exists $B \in \mathcal{S}$ such that $m(B) + r \geqslant m(A)$

(II) Let t be an involution in T conjugate in X to an involution of A. Then $m([A, t]) \leqslant 2r$, and if $B/C_B(W)$ is elementary for all $B \in \mathcal{S}$ which satisfy (1), then $m([A, t]) \leqslant r$.

Now let $X = C_G(z)/\langle z \rangle$, $A = W = \overline{R}$. We first prove that $\overline{R}$ is the unique elementary abelian group of order 2^8 in a Sylow 2-subgroup $\overline{T}$ of $\overline{N} \cong N_G(R)/\langle z \rangle$. This will imply (1) and (2). By Lemma 2.10 (i) if $\overline{R}_1 \cong E_{2^8}$ and $\overline{R}_1 \neq \overline{R}$, then $\overline{R}_1 \cap \overline{R} = E_{16}$. Hence $\overline{N}'$ splits over $\overline{R}$. Hence N contains a subgroup N_1 isomorphic to $A_5 \times A_5$ or to

$$SL(2, 5) * SL(2, 5) .$$

As $C_N(t)$ involves A_5 , $C_{N_1}(t)' \cong A_5$. Hence $C_G(< t, z >)'E$ splits over $E = < i, \ell, \alpha_1, \alpha_2, \alpha_3, \alpha_4 >$. On the other hand (S) implies that there is no involution in $C_G(< t, z >)'E$. This is a contradiction. Hence (1), (2) hold. (3) is trivial. Suppose that X is of "unknown" type. Then there exists $\overline{B}$ which satisfies [I] and [II]. By Lemma 2.10(i) and [II], $r \geqslant 4$. As $T/R \cong (Z_2 \times Z_2) \wr Z_2$,

$$r = \max\{m(\overline{B}/C_{\overline{B}}(\overline{R})) = m(\overline{B}/\overline{B} \cap \overline{R})\} \leqslant 4 .$$

Hence $r = 4$. Thus $\overline{N}'$ again splits over $\overline{R}$ and we can obtain a contradiction as above. Hence X is of "known" type.

If $O(M) > 1$, then every conjugate of t in R must invert $O(M)$. Hence $[t_1 t_2, O(M)] = 1$ where $t_2 \neq t_1 \sim t_2 \sim t$ in $N_G(R)$. As $N_G(R)$ is irreducible on $\overline{R}$, $[R, O(M)] = 1$. Hence $O(M) = 1$.

By the theorem of Goldschmidt [6], (A) $\overline{R}^X$ is a central product of an abelian 2-group and quasi-simple groups whose central factor groups are isomorphic to $L_2(2^n)$, $n \geqslant 3$, $Sz(2^{2n+1})$, $n \geqslant 1$, $U_3(2^n)$, $n \geqslant 2$, $L_2(q)$, $q \equiv 3, 5 \pmod 8$, or the simple groups of JR-type; and (B) $\overline{R} = O_2(\overline{R}^X)\Omega_1(\overline{T}_1)$ for some Sylow 2-subgroup $\overline{T}_1$ of $\overline{R}^X$ containing $\overline{R}$.

By the irreducible action of $N_G(R) \subseteq M$ on $\overline{R}$ and the structure of $N_G(R)/R$, we conclude immediately that $\overline{R} = \overline{R}^x$. Hence $M = N_G(R)$.

We recall that σ_2 is an element of order 3 in $N_G(R) = M$ not contained in a direct factor of $M'/R \cong A_5 \times A_5$. Also v_1 is an involution of $H - H'$ $(H = C_G(t) \cong \hat{\mathrm{Aut}}\ (HS))$ such that $C_{H'}(v_1) \cong Z_2 \times A_8$. We know that $v_1 \sim t$ in G.

LEMMA 2.11. *The following condition holds:*

(1) $N_G(< v_1, t >) = (< v_1, t, \sigma > \times K)< u >$ *where* $< v_1, t, \sigma > \cong A_4$, $\sigma^3 = 1$, $K \cong A_8$, $< v_1, t, \sigma, u > \cong S_4$, $u^2 = 1$, $< \sigma, u > \cong S_3$, $K< u > \cong S_8$.

(2) *"Central" involutions of* K *are conjugate in* G *to* z, *i.e.,* $(12)(34)(56)(78) \sim z$ *in* G. *"Non-central" involutions of* K *are conjugate in* G *to* t: *i.e.,* $(12)(34) \sim t$ *in* G.

(3) *The element* σ *of* (1) *is conjugate to an element of order* 3 *in* H. *Hence* $\sigma \sim \sigma_2$ *in* G.

(4) $C_G(\sigma) = < \sigma > \times K_1$ *where* $K_1 \cong A_9$. *Further,* $N_G(< \sigma >) = (< \sigma > \times K_1)< u >$ *where* u *is the element in* (1), *and* $K_1< u > \cong S_9$.

PROOF: As $C_G(< v_1, t >) \cong Z_2 \times Z_2 \times A_8$ and the centralizer of v_1 in $H/< t >$ is isomorphic to $Z_2 \times S_8$,

$$N_H(< v_1, t >) = (< v_1, t > \times K)< u >$$

where $u^2 = 1$, $< v_1, t, u > \cong D_8$, $K \cong A_8$ and $K< u > \cong S_8$.

Consider $C_G(v_1) \sim H$. As $[t, K] = 1$, $t \notin C_G(v_1)'$. As $2_4 \cdot 2_1 \sim 2_4$ in H by (H), there is a 2-element in $C_G(v_1)$ acting on $< v_1, t >$ nontrivially. (1) follows immediately from this.

Let $L = < v_1, t, \alpha > \times K \cong A_4 \times A_8$. We shall show that a "non-central" involution of K is conjugate in G to t . L contains an elementary abelian B of order 64. By (P), $A \sim B$ in H where $A = < t, z, \alpha_1\alpha_2, \alpha_3\alpha_4, a, v_1 >$. Let σ' be a 3-element of K which corresponds to a 3-cycle of K with $B^\sigma = B$. Then $< \sigma, \sigma' > \cong Z_3 \times Z_3$ and centralizes a four subgroup U of $B \cap K$. As $|N_G(A) \cap C_G(z)| = 2^{12} \cdot 3 \cdot 5$ by Lemma 2.5 (ii), U cannot contain any conjugate of z . Hence U consists of three conjugates of t , as required. In particular, σ is conjugate to a 3-element of H. Thus the latter half of (2) and (3) hold.

We next argue that the "central" involutions of K are conjugate in G to z . By (T), all involutions of $\overline{K}$ are conjugate in $\overline{H}'$ where $\overline{H} = H/< t >$. Without loss we can assume that $\overline{z}$ is a "central" involution of $\overline{K}$. Then

$$C_{\overline{K}}(\overline{z})'' \cong Q_8 * Q_8 .$$

By (F), $C_{\overline{K}}(\overline{z})'' \subseteq \overline{E} = < \overline{\ell}, \overline{\alpha}_1, \overline{\alpha}_2, \overline{\alpha}_3, \overline{\alpha}_4 >$. As z is the unique involution which has a square root in E by (K), we see that the first half of (2) must hold.

Now let σ'' be a 3-element of H. Then by (R),

$$C_G(\sigma'') \cap H/\langle \sigma'' \rangle = (\langle v_1, t \rangle \times F)\langle s \rangle$$

where $\langle v_1, t, s \rangle \cong D_8$, $s^2 = 1$, $F \cong A_5$, $F\langle s \rangle \cong S_5$. Set $\overline{D} = C_G(\sigma'')/O(C_G(\sigma''))$. We already know by Lemma 2.11 (1)(3) that $\overline{D} \supseteq \overline{D}_1 \cong A_8$. A Sylow 2-subgroup of $\overline{D}_1$ is of type A_8 and so contains a unique elementary abelian group $\overline{C}$ of order 16. $N_{\overline{D}_1}(\overline{C})$ divides the involutions of $\overline{C}$ into two orbits of size 9 and 6. By (2) above, the 9 involutions are conjugate to $\overline{z}$ and 6 to $\overline{t}$. As $|C_G(\sigma'') \cap H|_2 = 2^5$, we conclude immediately that $|\overline{D}|_2 = |\overline{D}_1|_2 = 2^6$. We now see by [7] that $\overline{D} \cong A_9$ as $\overline{C_G(\sigma'') \cap H} \cong C_{A_9}((12)(34))$ and $\overline{D} \supset \overline{D}_1 \cong A_8$. Clearly then $O(C_G(\sigma'')) = 1$ and so (4) holds. This completes the proof of Lemma 2.11.

LEMMA 2.12. *Let* L *be a subgroup isomorphic to* A_5 *in* $C_G(\langle v_1, t \rangle) \cong Z_2 \times Z_2 \times A_8$ *such that* L *contains a* 3-*cycle of* A_8. *Then* $C_G(L) \cong A_7$.

PROOF: We first show that a Sylow 2-subgroup of $C_G(L)$ is dihedral of order 8. The conjugacy classes of elements of order 5 in L are uniquely determined in H. Let ρ be a 5-element of L. Then by (U), $C_H(\rho) \cong Z_5 \times SU^{\pm}(2, 5)$. If σ is an element of order 3 in H, then by (R), $C_H(\sigma)$ has Sylow

2-subgroup isomorphic to $(Z_2 \times Z_2) \wr Z_2$. By the structure of $N_H(< v_1, t >)$ described in the first paragraph of the previous lemma, $C_H(L)$ contains $(< v_1, t > \times \sigma')< u >$ where

$$< v_1, t, u > \cong D_8 ,$$

$|\sigma'| = 3$ and $< \sigma', u > \cong S_3$. Clearly then $C_H(L)$ has to have Sylow 2-subgroups isomorphic to D_8 . As $C_H(\rho)$ does not contain A_5 , $C_H(L) = (< v_1, t > \times \sigma')< u > = C_G(L) \cap H$. This forces

$$|C_G(L)|_2 = |C_H(L)|_2 = 8 .$$

By the structure of $N_G(< v_1, t >)$, $C_G(L)$ contains a subgroup isomorphic to $(A_4 \times Z_3)Z_2$ where $A_4 \cdot Z_2 \cong S_4$ and $Z_3 \cdot Z_2 \cong S_3$. This implies first that $O(C_G(L)) = 1$ and that $C_G(L)' \supset A_4 \times Z_3$. Hence $C_G(L) \cong A_7$, as desired.

PROPOSITION 2.13. G *contains a subgroup* G_0 *isomorphic to* A_{12} .

PROOF: Let $U = < v_1, t >$. Then by Lemma 2.11 (1) , $N_G(U) = (D \times F)< u >$ where $U \subset D \cong A_4$, $F \cong A_8$, $u^2 = 1$, $D< u > \cong S_4$, $F< u > \cong S_8$. Let $< \sigma >$ be a cyclic subgroup of order 3 of D normalized by u . Then by Lemma 2.11 (4) , $C_G(\sigma) \cong Z_3 \times A_9$ and $N_G(< \sigma >)/< \sigma > \cong S_9$. Without loss we can assume that u acts on F as a transposition.

Let $u = t_1, t_2, \ldots, t_7$ be a set of canonical generators of $F\langle t_1\rangle \cong S_8$. As

$$C_G(\sigma)'\langle t_1\rangle \cong S_9$$

and

$$C_G(\sigma)'\langle t_1\rangle \supset F\langle t_1\rangle,$$

the set of involutions can be extended to a set of canonical generators of S_9 ; $t_0, t_1, \ldots, t_7$, where $t_i^2 = 1, (t_i t_{i+1})^3 = 1$, $(t_i t_j)^2 = 1$, $j > i + 1$. t_1 centralizes a unique involution t' of $U = \langle v_1, t\rangle$. Then $[\langle t'\rangle, F\langle t_1\rangle] = 1$. We shall argue that

$$\sigma, t', t_0, t_1, \ldots, t_7$$

is a set of canonical generators of A_{12}. Namely,

(1) $\sigma^3 = 1$, $t'^2 = 1$, $t_i^2 = 1$, $0 \leq i \leq 7$;

(2) $(\sigma t')^3 = 1$, $(\sigma t_i)^2 = 1$, $0 \leq i \leq 7$;

(3) $(t't_0)^3 = 1$;

(4) $(t't_i)^2 = 1$, $1 \leq i \leq 7$;

(5) $(t_i t_{i+1})^3 = 1$, $0 \leq i \leq 6$;

(6) $(t_i t_j)^2 = 1$, $0 \leq i < j-1 \leq 6$.

By our choice of the elements all relations except (3) obviously hold. Let $L = \langle t_4t_5, t_4t_6, t_4t_7\rangle$. Then $L \cong A_5$. By Lemma 2.12, $C_G(L) \cong A_7$. We have $C_G(L) \supseteq \langle \sigma, t', t_0, t_1, t_2\rangle$.

Since $\langle \sigma, t'\rangle \times \langle t_1t_2\rangle \cong A_4 \times Z_3$ and such subgroups form a unique conjugacy class in A_7, we can assume that $t' = (12)(34)$, $t_1t_2 = (567)$. Applying a suitable permutation on $\{1, 2, 3, 4\}$ we can also assume that $\sigma = (123)\sigma'$ where σ' is a power of (567).

As t_1 inverts σ and centralizes t',

$$t_1 = (12)(3)(4)\ldots .$$

Without loss $t_1 = (12)(56)$. Then $t_2 = (12)(67)$. As t_0 inverts σ, and centralizes t_2, $t_0 = (12)(67)$ or $(12)(45)$. Since $< t_0, t_1, t_2 > \cong S_4$, $t_0 = (12)(45)$. Clearly then $(t't_0)^3 = 1$, as required. This completes the proof.

LEMMA 2.14. *If we use the standard notation of the elements of* $A_{12} \cong G_0$, *then*

(i) $z \sim (12)(34)(56)(78)$

(ii) $t \sim (12)(34) \sim (12)(34)(56)(78)(9,10)(11,12)$

(iii) $\sigma_1 \sim (123)(456)(789)$

(iv) $\sigma_2 \sim (123) \sim (123)(456) \sim (123)(456)(789)(10,11,12)$.

PROOF: (i) and $t \sim (12)(34)$ are proved in Lemma 2.11. Let $A_1 = < (12)(34), (13)(24), (56)(78), (57)(68), (9,10)(11,12), (9,11)(10,12) > \cong E_{64}$. Then

$$A_1 \sim A = < t, z, \alpha_1\alpha_2, \alpha_3\alpha_4, a, v_1 >$$

in G. One can see easily that A_1 contains precisely 27 elements which have cyclic decompositions like (12)(34)(56)(78). As all the remaining elements are conjugate in $N_G(A_1)$, (ii) must hold.

$<(123),(456)>$ is centralized by $(78)(9,10) \sim t$; and $(123)(456)(789)(10,11,12)$ is centralized by

$$(14)(25)(36)(7,10)(8,11)(9,12) \sim t .$$

Hence (iv) holds. As $(123)(456)(789)$ is centralized by $(14)(25)(36)(10,11) \sim z$, (iii) must hold. This completes the proof.

LEMMA 2.15. $C_G(\sigma_1) \cap M \cong SL(2,5)$.

PROOF: Suppose false. Then by the structure of $M = N_G(R)$, we see that $C_G(\sigma_1) \cap M = < z > \times F$ where $F \cong A_5$. Let $C = C_G(\sigma_1)$.

By [11], C is of sectional 2-rank 4 and so we can apply the main theorem of [8]. Set $\overline{C} = C/O(C)$ and $\overline{L} = \overline{C}^{(\infty)}$. If $O_2(\overline{C}) > 1$, then $O_2(\overline{C}) = < \overline{z} >$ and so $\overline{C} = < \overline{z} > \times \overline{F}$. Suppose that $O_2(\overline{C}) = 1$. Then $\overline{L}$ is the direct product of simple groups. If $\overline{z} \in \overline{L}$, then $\overline{z}$ must be in a component of $\overline{L}$ and hence $\overline{L} \cong J_1$. Thus $\overline{C} \cong J_1$. Next assume that $\overline{z} \notin \overline{L}$. If $\overline{L}$ is not simple, then clearly $< \overline{z}, \overline{L} > \cong A_5 \wr Z_2$. Hence $\overline{C} \cong A_5 \wr Z_2$. Suppose finally that $\overline{L}$ is simple. Then by [8, Main Theorem] we can check readily that only $L_2(16)$, $L_3(4)$ and $U_3(4)$ admit an involutive automorphism whose stabilizer is A_5. Since all involutions of $< \overline{z}, \overline{L} > - \overline{L}$ are conjugate in our present case, we have $\overline{C} = < \overline{z} > \overline{L}$ and $\overline{L} \cong L_2(16)$, $L_3(4)$ or $U_3(4)$.

Thus we have four cases to consider:

[I] $\overline{C} = <\overline{z}> \times \overline{F}$,

[II] $\overline{C} \cong J_1$,

[III] $\overline{C} \cong A_5 \int Z_2$, and

[IV] $\overline{C} = <z>\overline{L}$ where $\overline{L} \cong L_2(16)$, $L_3(4)$ or $U_3(4)$.

We first eliminate [III] and [IV]. Suppose [III] or [IV] holds. As z inverts $O(C)/<\sigma_1>$, the product of two involutions conjugate to z centralizes $O(C)/<\sigma_1>$. Hence $\overline{L}$ centralizes $O(C)$. Thus $L = C^{(\infty)} \cong A_5 \times A_5$, $L_2(16)$, $L_3(4)$, a 3-fold covering of $L_3(4)$ or $U_3(4)$, and $O(C)L = O(C) * L$. Since $<\sigma_1, \sigma_2> \sim <(123), (456)(789)>$ of G_0 ,

$$|C_G(<\sigma_1, \sigma_2>)| = 2 \cdot 3^4.$$

Since $<\sigma_1, \sigma_2> \subset <\sigma_1> \cdot L$ and $|L|_3 \leqslant 3^2$, we have

$$|C_G(\sigma_1)|_3 = 3^4 .$$

This is impossible as $\sigma_1 \sim (123)(456)(789)$ of G_0 .

Next suppose that $\overline{C} \cong J_1$. Then again $L = C^{(\infty)} \cong J_1$ and $C = O(C) \times J_1$. Hence $|O(C)| = 3$. This again contradicts $|C_G(\sigma_1)| \geqslant 3^5$.

Finally suppose $\overline{C} \cong Z_2 \times A_5$. Without loss we may assume that $\sigma_1 = (123)(456)(789)$ and $\sigma_2 = (123)$. Hence

$$<(123), (456), (789), (10,11,12)>$$

is a Sylow 2-subgroup of $C_G(<\sigma_1, \sigma_2>)$. We first conclude that $O(C)$ must contain a conjugate σ_2' of σ_2 in G, as $(123) \sim (123)(456)$ in G. Since $O(C)/<\sigma_1>$ is abelian and $|C_G(\sigma_2')|_3 = 3^5$, $O(C)$ is a 3-group of order at most 3^5.

Let $i \neq z$ be an involution of $<z> \times F$. Since σ_1 is not conjugate to σ_2 in G, i is not conjugate to t. If $i \sim z$ in G, then $C_G(i) \cap C_G(\sigma_1) \cong Z_2 \times A_5$, which is impossible by the structure of $C = C_G(\sigma_1)$. Hence i is neither conjugate to z nor to t. This forces C to be 3-constrained. Hence $|O(C)| = 3^5$. Clearly then $O(C)$ is extra-special, as otherwise $O(C)$ would be abelian and a Sylow 3-subgroup of $C_G(\sigma_2')$ would be fully contained in $O(C)$. Thus $<z> \times F \cong Z_2 \times A_5$ is isomorphic to a subgroup of $Sp(4, 3)$. But $Sp(4, 3)$ does not contain even A_4! This can easily be seen by the fusion pattern of $PSp(4, 3)$ or by appealing directly to Brauer-Wielandt formula. This contradiction establishes the lemma.

LEMMA 2.16. *The following condition holds:*

(i) $O(C_G(\sigma_1))$ *is an extra-special group of order* 3^5 *and* $C_G(\sigma_1)/O(C_G(\sigma_1)) \cong SL(2, 5)$.

(ii) $|G|_3 = 3^6$,

(iii) *all involutions of* $M' - R$ *are conjugate in* M *and are conjugate to* z *in* G.

PROOF: By the previous lemma, a Sylow 2-subgroup of

$C_G(\sigma_1)$ is a quaternion group of order 8. Hence

$$C_G(\sigma_1) = O(C_G(\sigma_1))(M \cap C_G(\sigma_1)) .$$

As z inverts $O(C_G(\sigma_1))/\langle \sigma_1 \rangle$, the latter portion of the proof of the previous lemma clearly applies to conclude that $O(C_G(\sigma_1))$ is an extra-special group of order 3^5. This establishes (i). (ii) is an immediate consequence of (i).

$M'/R \cong A_5 \times A_5$ has two conjugacy classes of involutions in M/R. Together with Lemma 2.10(i), (i) above implies that any involution of $M' - R$ lies "diagonally" in $M'/R \cong A_5 \times A_5$. Again by Lemma 2.10(i), we see that every involution of $M' - R$ is conjugate to i or iz where i is a fixed involution of $M' - R$.

To complete the proof, it suffices to show that $i \sim iz \sim z$ in G. This can be seen by inspecting the structure of $C_{A_{12}}((12)(34)(56)(78)) = X$. We see that $X = R_1P_1Q_1$ where

$R_1 = \langle (12)(34)(56)(78), (12)(34), (34)(56), (13)(24)(57)(68),$
$(15)(26)(37)(48), (9,10)(11,12), (9,11)(10,12) \rangle ,$

$P_1 = \langle (135)(246), (9,10,11) \rangle ,$

$Q_1 = \langle (13)(24), (78)(9,10) \rangle .$

As R_1 is invariant under P_1 and is generated by the centralizers of $(135)(246)$ and $(9,10,11)$, $R_1 \subseteq O_2(M)$ under the identification $z = (12)(34)(56)(78)$. Since Q_1 is a four subgroup, $Q_1 \not\subseteq M'$. If $j \in Q_1 - M'$, we know that $jz \not\sim j$ in

G . Hence $(13)(24)(78)(9,10) \in M' - O_2(M)$. As

$$(13)(24)(78)(9,10) \sim (13)(24)(78)(9,10)(12)(34)(56)(78)$$
$$= (14)(23)(56)(9,10)$$

in $C_{A_{12}}(z)$, the lemma is proved.

LEMMA 2.17 *The following condition holds:*

(1) $|G|_5 = 5^6$,

(2) *Let* $5_1, 5_2, 5_3$ *be the representatives of the conjugacy classes of elements of order* 5 *of* H. *(Chosen as in* (H)*). Then* $C_G(5_1) \cong Z_5 \times U_3(5)$, $C_G(5_2)$ *is an extension of a group of order* 5^4 *by* $Z_2 \times Z_2$, *and* $C_G(5_3)$ *is an extension of an extra-special group of order* 5^5 *by* F *with*

$$O_2(F) \cong D_8 * Q_8 \ , \ |F| = 2^5 \cdot 5 \ .$$

(3) $M = C_G(z)$ *has three conjugacy classes of subgroups of order* 5 . *One is conjugate in* G *to* $< 5_2 >$, *another to* $< 5_3 >$. *Let* $< 5_4 >$ *be the remaining one. Then* $C_G(5_4)$ *is an extension of an elementary abelian group of order* 5^3 *by* SL(2, 5) .

PROOF: We may choose $5_1, 5_2, 5_3$ in such a way that $|< 5_1, 5_2, 5_3 >| = 5^3$. Then $< 5_3 > = Z(< 5_1, 5_2, 5_3 >)$. As

$C_H(5_1) \cong Z_5 \times SU^{\pm}(2, 5)$ and 5_1 is the only class of elements of order 5 centralized by a 3-element, $5_1 \sim (12345)$ of $G_0 \cong A_{12}$. Clearly then $C_G(5_1)$ contains A_7. It is now easy to see that $O(C_G(5_1)) = 1$, and G has no subgroup of index 2 and that $C_G(5_1)/< 5_1 >$ is simple. By [1], $C_G(5_1) \cong Z_5 \times U_3(5)$. As $C_G(5_1)$ has only one conjugacy class of involutions, z does not centralize any conjugate to 5_1.

By (U), $|C_H(5_2)| = 5^2 \cdot 2^2$ and a Sylow 2-subgroup of $C_H(5_2)$ is the four group $< v_2, t >$. As $v_2 t \sim v_2$, $< v_2, t >$ is a Sylow 2-subgroup of $C_G(5_2)$ and

$$C_G(5_2) = O(C_G(5_2))< v_2, t > .$$

Since $v_2 \sim z$, 5_2 must be conjugate to an element of M. Clearly $< 5_2 > \sim < \rho_3 >$ of Lemma 2.8. We conclude by Brauer-Wielandt's formula that $O(C_G(5_2))$ is a 5-group of order $\leqslant 5^4$.

Next consider $C_G(5_3)$. By (U), $C_H(5_3)$ is of order $5^3 2^4$, and $< t, z >$ centralizes 5_3 for a suitable choice of 5_3. Clearly $< 5_3 > \sim < \rho_2 >$ of Lemma 2.8. Hence

$$C_G(5_3) \cap M \cong Z_5 \times F$$

where F is an extension of $D_8 * Q_8$ by Z_5. As z is isolated in a Sylow 2-subgroup of $C_G(5_3)$, we have

$$C_G(5_3) = O(C_G(5_3)) \cdot F .$$

Since $t \sim tz$ in F ,

$$|C_G(t) \cap O(C_G(5_3))| = |C_G(tz) \cap O(C_G(5_3))| = 5^3 .$$

Together with $|C_G(z) \cap O(C_G(5_3))| = 5$, the information above implies that $|O(C_G(5_3))| = 5^5$. Since

$$5_2 \in H \cap C_G(5_3) \subseteq O(C_G(5_3))$$

and $O(C_G(5_3))/< 5_3 >$ is abelian, $|O(C_G(5_2))| = 5^4$ must hold. Also $O(C_G(5_3))$ is a nonabelian group of order 5^5. As F acts irreducibly on the Frattini factor group, $O(C_G(5_3))$ is extra-special. This forces $|G|_5 = 5^6$. Also $O(C_G(5_2))$ is the direct product of Z_5 and an extra-special group of order 5^3.

Finally consider $C_G(5_4)$. Clearly $< 5_4 > \sim < \rho_1 >$ of Lemma 2.8. Hence $C_M(5_4) \cong SL(2, 5)$ and so

$$C_G(5_4) = O(C_G(5_4)) \cdot C_M(5_4) .$$

As $|G|_5 = 5^6$ and $< 5_4 > \not\sim < 5_3 >$, $|O(C_G(5_4))|_5 \leqslant 5^4$. Since $O(C_G(5_2)) \subset O(C_G(5_3))$ and $5_4 \in O(C_G(5_2))$, $5_4 \in O(C_G(5_3))$. This in turn implies that $5_3 \in O(C_G(5_4))$. Hence

$$5^3 \leqslant |O(C_G(5_4))| \leqslant 5^4 .$$

As z inverts $O(C_G(5_4))/< 5_4 >$, $|O(C_G(5_4))| = 5^3$. Clearly then $O(C_G(5_4))$ must be elementary. This completes the proof.

LEMMA 2.18. $|G|_7 = 7$. G *contains exactly one conjugacy class of elements of order* 7. *If* 7_1 *is a representative, then* $C_G(7_1) \cong Z_7 \times A_5$, *and* $[N_G(< 7_1 >): C_G(< 7_1 >)] = 6$.

PROOF: Let P be a Sylow 7-subgroup of H. By (V), $C_H(P) = \langle t, v_1 \rangle \times P$ for a suitable choice of P. Since

$$N_G(\langle t, v_1 \rangle)' \cong A_4 \times A_8 ,$$

all three involutions of $\langle t, v_1 \rangle$ are conjugate in $C_G(P)$. Clearly then $\langle t, v_1 \rangle$ is a Sylow 2-subgroup of $C_G(P)$. By Proposition 2.13, $C_G(P)$ contains a subgroup isomorphic to A_5. Hence $C_G(P) \cong Z_7 \times A_5$, as required.

LEMMA 2.19. $|G|_{11} = 11$. G *has exactly one conjugacy class of elements of order* 11. *If* 11_1 *is a representative, then* $C_G(11_1) \cong Z_2 \times Z_{11}$ *and*

$$[N_G(\langle 11_1 \rangle): C_G(\langle 11_1 \rangle)] = 10 .$$

PROOF: Let P_0 be a Sylow 11-subgroup H. By (H), we have that $C_H(P_0) = \langle t \rangle \times P_0$, and $|N_H(P_0): C_H(P_0)| = 10$. By (W) we may assume that v_2 inverts P_0 and so $\langle t, v_2 \rangle$ is a Sylow 2-subgroup of $N_G(P_0)$. We also have that

$$C_G(P_0) = O(C_G(P_0)) \rtimes \langle t \rangle .$$

Suppose that $|O(C_G(P_0))| > 11$. Then $O(C_G(P_0)) \cap C_G(v_2)$ or $O(C_G(P_0)) \cap C_G(v_2 t)$ is nontrivial. As $v_2 \sim v_2 t \sim z$, M must contain an element of odd order > 1 whose centralizer has an element of order 11. This is impossible by Lemmas 2.11(4), 2.15 and 2.16.

LEMMA 2.20. *$H = C_G(t)$ contains 7976 involutions conjugate in G to t and 51975 of z. $M = C_G(z)$ has 1080 conjugates of t and 8311 of z.*

PROOF: As $t \sim tz \sim v_1$ and $z \sim v_2$ in G, the first part of the lemma is clear:

$$7976 = 1 + |H|/|C_H(tz)| + |H|/|C_H(v_1)|$$

$$51975 = |H|/|C_H(z)| + |H|/|C_H(v_2)| .$$

By Lemma 2.9, $O_2(M) = R$ contains 151 conjugates in G of z and 120 of t. Let v be an involution of $M - M'$. Then $|C_M(v)| = |C_H(v_2)| = 2^8 \cdot 3 \cdot 5$. As $v \not\sim vz$ in G, z and t each has $2^6 \cdot 3 \cdot 5 = 960$ conjugates in $M - M'$. One sees that $|C_{\overline{M}}(\overline{i})| = 2^9$ where $\overline{M} = M/\langle z \rangle$ and i is an involution of $M' - O_2(M)$. As $i \sim iz$ in M by Lemma 2.16(iii), $|C_M(i)| = 2^9$. Hence $M' - O_2(M)$ contains 7200 involutions and all are conjugate to z in G. Our assertions follow from this information immediately.

3. THE ORDER OF G.

We have shown that $|G| = g = 2^{14} \cdot 3^6 \cdot 5^6 \cdot 7 \cdot 11 \cdot g'$ where $(g', 2 \cdot 3 \cdot 5 \cdot 7 \cdot 11) = 1$.

One can see easily by Lemmas 2.16 and 2.17 that the Sylow 3-normalizer is of order $3^6 \cdot 2^3$ and the Sylow 5-normalizer $5^6 \cdot 2^3$. Together with the information about the order of Sylow

7- and 11-normalizers of G, one obtains

$$\begin{aligned} g' &\equiv 1 \pmod{2} \\ &\equiv 1 \pmod{3} \\ &\equiv 4 \pmod{5} \\ &\equiv 5 \pmod{7} \\ &\equiv 8 \pmod{11} \end{aligned}$$

Hence $g' \equiv 19 \ (\mathrm{mod}\ 2\cdot 3\cdot 5\cdot 7\cdot 11 = 2310)$.

As G has two conjugacy classes of involutions, we can apply Thompson's order formula to get the precise order of G. The formula states:

$$g = |C_G(z)|a(t) + |C_G(t)|a(z)$$

where $a(t)$ denotes the number of ordered pairs (α, β) with $\alpha \sim t$, $\beta \sim z$ in G and some power of $\alpha\beta$ being t. $a(z)$ denotes the number of such pairs with respect to z.

Thus

$$g = 2^{14}\cdot 3^6\cdot 5^6\cdot 7\cdot 11\cdot g' = 2^{14}\cdot 3^2\cdot 5^2\cdot a(t) + 2^{11}\cdot 3^2\cdot 5^3\cdot 7\cdot 11\cdot a(z) .$$

On the other hand, we know by Lemma 2.20 that H contains 7976 conjugates of t in G and 51975 of z, while M has 1080 conjugates of t and 8311 of z. Hence

$$2^{14}\cdot 3^6\cdot 5^6\cdot 7\cdot 11\cdot g' \leqslant 2^{14}\cdot 3^2\cdot 5^2\cdot 7976\cdot 51975 + 2^{11}\cdot 3^2\cdot 5^3\cdot 7\cdot 11\cdot 1080\cdot 8311 .$$

Hence $g' \leqslant 106.347 + 110.814 = < 218$ and so $g' = 19$.

Thus we have shown:

PROPOSITION 3.1. $|G| = 2^{14} \cdot 3^6 \cdot 5^6 \cdot 7 \cdot 11 \cdot 19$.

LEMMA 3.2. The Sylow 19-normalizers are Frobenius groups of order $19 \cdot 9$.

PROOF: Trivial by Sylow's theorem.

4. THE CONJUGACY CLASSES AND THE CHARACTER TABLES OF SOME LOCAL SUBGROUPS

By Lemma 2.16 (1) and (2), we know that $|G|_3 = 3^6$ and the center of a Sylow 3-subgroup is conjugate to $< \sigma_1 > \cong Z_3$. We set $\sigma_1 = 3_B$.

By Lemma 2.17 (1) and (2), we know that $|G|_5 = 5^6$ and the center of a Sylow 5-subgroup is conjugate to $< 5_3 > \cong Z_5$. We set $5_3 = 5_C$. We also set $z = 2_B$.

In this section we determine the conjugacy classes and the character tables of $C_G(3_B)$, $N_G(< 5_C >)$ and $C_G(2_B)$. This information will be very useful to determine the conjugacy classes and the character table of G .

[1] THE CONJUGACY CLASSES AND THE CHARACTER TABLE OF $C_G(3_B)$ (SEE TABLE II).

Set $K = C_G(3_B)$. We know that $O(K)$ is an extra-special group of order 3^5 and K contains a subgroup $K_1 \cong SL(2, 5)$ such that $K = O_3(K) \cdot K_1$.

The degrees of the irreducible characters of $SL(2, 5)$ are known to be 1, 3, 3, 4, 5, 2, 2, 4, and 6, where the last four are the degrees of the faithful characters.

Set $\overline{K} = K/Z(K)$. If $\overline{a}$ is a nontrivial element of $O(\overline{K})$, then $|C_{\overline{K}}(\overline{a})| = 3^5$. From this, one sees that $\overline{K}$ acts on the set of hyperplanes of $O(\overline{K})$ with orbit lengths 40 and 40. Let λ be a linear character of $O(\overline{K})$. Then the stabilizer group of λ in $\overline{K}$ is of order 3^5. Hence by a theorem of Clifford (see [12, V.17.11]), we obtain three irreducible characters of $\overline{K}$ of degree $40 = |SL(2, 5)|/3$. Using a linear character belonging to the second orbit of length 40, we again obtain three new irreducible characters of degree 40.

As for faithful irreducible characters, we see, by [12, V.17.11], that faithful irreducible characters of $O(K)$ can be extended to those of K. We thus obtain two degrees 9 and 9. Again using [12, V.17.11], we easily obtain all the degrees of the irreducible characters of $\overline{K}$. Those are: 1, 3, 3, 4, 5, 2, 2, 4, 6, 40, 40, 40, 40, 40, 40, 9, 9, 27, 27, 27, 27, 36, 36, 45, 45, 18, 18, 18, 18, 36, 36, 54, and 54.

It is easy to find all 33 conjugacy classes of K. The values of the irreducible characters can be filled in without too much effort. We omit the details. We, however, remark that any faithful irreducible character is the tensor product of a character of degree 9 and a character of

$$K/O(K) \cong SL(2, 5) .$$

One more remark is that the author was unable to determine the values of the characters of degree 9 on the classes 3_4, 3_5 and 3_6 (see Table II) by any direct means. It is easy to show that the values must be

$$\begin{matrix} \{3, 3\omega, 3\omega^2\} \\ \{3, 3\omega^2, 3\omega\} \end{matrix} \quad \text{or} \quad \begin{matrix} \{-3, -3\omega \ \ -3\omega^2\} \\ \{-3, -3\omega^2, -3\omega\} \end{matrix} \quad \text{where} \quad \omega^2 + \omega + 1 = 0 .$$

The orthogonality relation with any row or any column will not determine the sign. The author calculated the group algebra constant:

$$c = \frac{|K|}{|C_K(3_4)|^2} \sum \frac{\chi(3_4)^3}{\chi(1)}$$

where χ ranges over all irreducible characters. One obtains that $c < 0$ if the latter set of values holds. Hence the former must hold.

[II] The conjugacy classes and the character table of $N_G(<5_C>)$ (see Table III).

Set $K = N_G(< 5_C >)$. Then $O(K)$ is an extra-special group of order 5^5 . K contains a subgroup K_1 of order $2^5 \cdot 5 \cdot 4$ such that $K = O(K) \cdot K_1$. $O_2(K_1) \cong D_8 * Q_8$ and $K_1/O_2(K_1)$ is a Frobenius group of order 20. We also know that the unique involution of $Z(O_2(K_2)) \cong Z_2$ inverts $O(K)/Z(K)$. It is straightforward to determine all degrees of irreducible characters of $K/O(K)$. Those are 1, 1, 1, 1, 4, 5, 5, 5, 5, 10, 10, 4, 4, 4, 4, and 16.

We next investigate the action of K on the linear characters of $O(K)/Z(O(K)) = O(\overline{K})$ where $\overline{K} = K/Z(O(K))$. One finds that K acts on those linear characters with orbit lengths 1, 80, 160, 320, 32, and 32. Correspondingly the stabilizer groups in $\overline{K}/O(\overline{K})$ are isomorphic to $\overline{K}/O(\overline{K})$, Z_8, $Z_2 \times Z_2$, Z_2, D_{10}, and D_{10}. One can then determine the degrees of all faithful irreducible characters of $\overline{K}$ by a theorem of Clifford. Those are eight 80's, four 160's, two 320's, eight 32's and two 128's.

Finally, we determine the degrees of faithful irreducible characters of K. The commutator subgroup K' of K is of order $5^5 \cdot 2^5 \cdot 5$ and Satz. V.17.11 of [12] is applicable for K'. Namely, any faithful irreducible character of $O(K)$ can be lifted to an irreducible character of K''. Hence K'' has $4 \cdot 16 = 64$ irreducible characters of degree 25. Hence one of them must be fixed by a 5-element in $K' - K''$. Hence K' has a faithful irreducible character of degree 25. One can now find all degrees of the faithful irreducible characters of K' by taking the tensor product of the faithful irreducible character of degree 25 of K' and each irreducible character of

$K'/O(K)$. There are twenty 25's, twelve 125's and twenty 100's.

Since the element 5_c is rational in $N_G(< 5_c >) = K$, four faithful irreducible characters of K' of the same degree must be put together to form a character of K. We thus obtain the degrees of all faithful irreducible characters of K. Those are five 100's, three 500's, and five 400's.

One can obtain all 53 conjugacy classes and the values of the characters by a standard method. The details are again omitted.

[III] THE CONJUGACY CLASSES AND THE CHARACTER TABLE OF $C_G(2_B)$ (SEE TABLE IV).

The character table of $C_G(2_B)$ is considerably harder to construct than those of $C_G(3_B)$ or $N_G(< 5_C >)$.

We again set $M = C_G(2_B)$. $O_2(M) \cong D_8 * D_8 * D_8 * D_8$ and $M/O_2(M) \cong A_5 \wr Z_2$. As in the preceding cases, it is not hard to find the degrees of all irreducible characters of $\overline{M} = M/< 2_B >$. There are 42 such characters.

One cannot determine the degrees of faithful irreducible characters of K or even the total number of such characters by the method employed in the cases $K = C_G(3_B)$ or $N_G(< 5_C >)$.

The author first found that M has either 56 or 57 conjugacy classes (the behaviour of the element of order 8 of of the last column of Table IV was ambiguous). Therefore M has either 14 or 15 faithful irreducible characters.

We shall first argue that the degree of every faithful irreducible character of $M' \cong 2 \cdot 2^8(A_5 \times A_5)$ is divisible by 64.

Let M_1 be a *normal* subgroup of M' such that

$$M_1 \supset O_2(M) = R$$

and $M_1/R \cong A_5$. By Lemma 2.15, M_1 contains $M_2 \cong SL(2, 5)$. Let a be an element of order 4 in M_2. Then $C_M(a)$ involves A_5. By the action of 3, 5-elements in $C_M(a)$ on R we see that $C_R(a)$ is an elementary abelian subgroup of order 32. As $C_{\overline{R}}(\overline{a}) \cong E_{16}$ where $\overline{M} = M/\langle z \rangle$, $a \not\sim at = a^{-1}$ in $\langle a, R \rangle$ but $a \sim az$ in M_1.

Let ϕ be the unique faithful irreducible character of R of degree 16. By [12, V.17.11], ϕ can be lifted to an irreducible character ϕ_1 of $\langle a, R \rangle$. Hence $\langle a, R \rangle$ has exactly two faithful irreducible characters ϕ_1, ϕ_2, both of which are of degree 16. We also have that $\phi_1(a) = \pm 4i$ and $\phi_2(a) = \mp 4i$. As $a \sim at = a^{-1}$ in a Sylow 2-subgroup $R_1 \supset \langle a, R \rangle$ of M_1, R_1 has a unique faithful irreducible character whose restriction on $\langle a, R \rangle$ is $\phi_1 + \phi_2$. Thus the degree of every faithful irreducible character of M_1 is divisible by 32. Let σ_1 be an element of order 3 in M_1. Then $|C_{M_1}(\sigma)| = 6$ while $|C_{\overline{M}_1}(\overline{\sigma})| = 3$. Hence there exist exactly three faithful irreducible characters of M_1 whose degrees are prime to 3. This shows that there are exactly four faithful irreducible characters of M_1 and 32, 32, 64, and 96 are the degrees.

Let $W = < w_1, w_2 > \cong Q_8$ be a subgroup of M' which maps into another *normal* A_5 of M'/R. Then

$$|M_1W| = 2^{13} \cdot 3 \cdot 5 .$$

By [12, V.17.11] again, 32, 32, 32, 32, 64, 64, 96, and 96 are the degrees of faithful irreducible characters of $M_1 < w_1 >$. It is easy to obtain the values (up to signs) of the characters above on w_1. These are $\pm 8i$, $\pm 8i$, $\pm 8i$, $\pm 8i$, $\pm 16i$, $\pm 16i$, $\pm 24i$, or $\pm 24i$, respectively. In particular, none of the characters vanish on w_1. As $w_1 \sim w_1 t = w_1^{-1}$ in M_1W, we conclude that 64, 64, 128, and 192 are the degrees of faithful irreducible characters of M_1W, as desired.

Let $64f_i$, $1 \leqslant i \leqslant 15$ be the degrees of faithful irreducible characters of M, where $f_i \geqslant 1$, $1 \leqslant i \leqslant 14$ and $f_{15} \geqslant 0$. Then

$$\sum 2^{12} f_i^2 = 2^{13} \cdot 3^2 \cdot 5^2 .$$

Hence $\sum f_i^2 = 450$. Clearly then $25 \nmid f_i$. If ρ is an element of order 5, then it is easy to see that $C_M(\rho)/< \rho >$ has no faithful irreducible character whose degree is divisible by 5. Hence $5 \nmid f_i$.

Since M' contains an element σ_1 of order 3 with $C_{M'}(\sigma_1) \cong Z_3 \times SL(2, 5)$ and $SL(2, 5)$ has an irreducible character of degree 6, M' has exactly two blocks of defect 1 for the prime 3 consisting of faithful characters. As those two blocks are conjugate in M, we can assume that $f_i = 2\cdot 3\cdot f_i'$, $i = 1, 2$ and 3. Moreover, f_i' are powers of 2 and $f_1' + f_2' = f_3'$. As $\sum f_i^2 = 450$, we must have $f_1' = f_2' = 1$ and $f_3' = 2$. Hence $f_1 = f_2 = 6$ and $f_3 = 12$ and so

$$\sum_{i=4}^{15} f_i^2 = 234 .$$

Suppose that $9 | f_i$ for some i. Then $f_i = 9$. Hence M' must have a character of degree $64\cdot 9$ and so M has at least two characters of degree $64\cdot 9$. Clearly then M has exactly two such characters. Hence

$$\sum_{f_i = 6}^{15} f_i^2 = 72$$

with all f_i's, $6 \leqslant i \leqslant 15$ being powers of 2. One can conclude that $f_{15} = 0$ and $\{64f_i, 6 \leqslant i \leqslant 14\}$ is {four 64's, one 128, and four 256's} or {six 128's and three 256's} .

Next suppose that no f_i is divisible by 9. Then f_i's, $4 \leqslant i \leqslant 15$, are powers of 2 . Thus we have

$$a + 4b + 16c + 64d = 234 ,$$

$$a + b + c + d = 11 \text{ or } 12 ,$$

$$a, b, c, d \geqslant 0 ,$$

where

$\{64f_i,\ 4 \leqslant i \leqslant 15\} = \{$a 64's, b 128's, c 256's, and d 512's$\}$.

Therefore, $3b + 15c + 63d = 223$ or 222 . Hence $f_{15} \neq 0$ and $b + 5c + 21d = 74$. We can obtain that $\{64f_i,\ 4 \leqslant i \leqslant 15\}$ is {two 64's, two 128's, six 256's and two 512's}, {two 64's, six 128's, one 256, and three 512's} or {six 64's, one 128, two 256's, and three 512's} .

Thus there are five possibilities for the set of numbers $\{64f_i\}$. To rule out the last four possibilities, we argue that if χ_i is a faithful irreducible character, then

$$\chi_i(5_C) \equiv \chi_i(1) \pmod{20} \quad \text{(see Table IV).}$$

$C_M(5_C) \supset \langle 5_C \rangle \times Q = W$ where $Q \cong D_8 * Q_8$. Let λ_1 be a nonprincipal linear character of $\langle 5_C \rangle$ and λ_2 be the unique irreducible character of Q of degree 4 . Then $\lambda_1\lambda_2$ is a character of W. We evaluate the inner product $\langle \chi_i, \lambda_1\lambda_2 \rangle_W$.

$$\begin{aligned} \langle \chi_i, \lambda_1\lambda_2 \rangle_W &= \frac{1}{|W|}\,(\chi_i(1) \cdot 4 + \chi_i(2_B)(-4) \\ &\quad + \chi_i(5_C)(-1)4 + \chi_i(5_C \cdot 2_B)(-1)(-4)) \\ &= \frac{1}{|W|}(8\chi_i(1) - 8\chi_i(5_C)) \ . \end{aligned}$$

Since $|W| = 5 \cdot 2^5$, we conclude that

$$\chi_i(1) \equiv \chi_i(5_C) \pmod{20} \ .$$

Since χ_1, χ_2, and χ_3 belong to a 3-block of defect 1, it is easy to complete the characters. In particular,

$$\chi_1(5_C) = \chi_2(5_C) = 4 \quad \text{and} \quad \chi_3(5_C) = 8 .$$

As $|C_M(5_C)| = 800$ and $|C_{\overline{M}}(\overline{5}_C)| = 400$, we have

$$\sum_{i=4}^{15} \chi_i(5_C)^2 = 400 - 16 - 16 - 64 = 304 .$$

Suppose that $\{64f_i, 6 \leqslant i \leqslant 14\}$ = {six 128's and three 256's} holds. Then

$$\sum_{i=6}^{14} \chi_i(5_C)^2 \geqslant 6 \cdot 64 + 3 \cdot 16 = 432 .$$

This is a contradiction. Similarly, we can rule out the last three possibilities. Hence 384, 384, 768, 576, 576, 64, 64, 64, 64, 128, 256, 256, 256, and 256 are the degrees of all faithful irreducible characters of M . Having found all degrees and classes, we can fill in the values of the character table by a standard method.

REMARK. Burgoyne has recently suggested that a full use of the paper of Clifford [3] would immediately give the degrees of the faithful characters of $M = C_G(2_B)$.

We know that the (unique) faithful irreducible character of $O_2(M)$ does not lift to M . We see easily that the Schur multiplier of $A_5 \int Z_2$ is of order 2, and that the degrees of the projective representations of $A_5 \int Z_2$ are four 4's, an 8, four 16's, two 24's, two 36's, and a 48. Hence, by

[3], four 64's, a 128, four 256's, two 384's, two 576's, and a 768 are the degrees of the faithful irreducible characters of M , consequently M has 56 conjugacy classes.

5. THE CONJUGACY CLASSES OF G .

In Prop. 2.13 , we have shown the existence of a subgroup G_0 isomorphic to A_{12} . We first determine the fusion of elements of G_0 in G .

G_0 has 43 conjugacy classes. These are as follows, where $a^\alpha \cdot b^\beta$ .... denotes an element of $G_0 \cong A_{12}$ with cycle decomposition having α a-cycles, β b-cycles, etc.

order	element in A_{12}	element in G
1	1	1_A
2	2^2	2_A
	2^4	2_B
	2^6	2_A
3	3	3_A
	3^2	3_A
	3^3	3_B
	3^4	3_A
4	$4\cdot 2$	4_A
	$4\cdot 2^3$	4_A
	4^2	4_B
	$4^2\cdot 2^2$	4_B
5	5	5_A
	5^2	5_B
6	$3\cdot 2^2$	6_A
	$3\cdot 2^4$	6_B
	$3^2\cdot 2^2$	6_A
	$6\cdot 2$	6_B
	$6\cdot 2^3$	6_A
	$6\cdot 3\cdot 2$	6_C
	6^2	6_A
7	7	7_A

order	element in A_{12}	element in G
8	$8\cdot 2$	8_B
	$8\cdot 4$	8_B
9	9	9_A
	$9\cdot 3$	9_A
	$9\cdot 3$	9_A
10	$5\cdot 2^2$	10_A
	$10\cdot 2$	10_B
11	11	11_A
	11	11_A
12	$4\cdot 3\cdot 2$	12_A
	$4\cdot 3^2\cdot 2$	12_A
	$4^2\cdot 3$	12_B
	$6\cdot 4$	12_A
14	$7\cdot 2^2$	14_A
15	$5\cdot 3$	15_A
	$5\cdot 3^2$	15_A
20	$5\cdot 4\cdot 2$	20_A
21	$7\cdot 3$	21_A
30	$5\cdot 3\cdot 2^2$	30_A
35	$7\cdot 5$	35_A
	$7\cdot 5$	35_B

By Lemma 2.14, we know the fusion of involutions and elements of order 3 of A_{12} in G, as is shown above.

As $H = C_G(t)$ has only one conjugacy class of elements whose squares are t, $4 \cdot 2 \sim 4 \cdot 2^3$. In the proof of Lemma 2.16, we have shown that $R_1 \subseteq O_2(M)$. As

$$(13)(24)(57)(68)(34)(56) = (1423)(5768) \in R_1$$

and

$$(1423)(5768)(9,\ 10)(11,\ 12) \in R_1 ,$$

we conclude by Lemma 2.9(1), $4^2 \sim 4^2 \cdot 2^2$ in G.

As $C_{A_{12}}(5) \cong Z_5 \times A_7$, $5 \sim 5_1$ of (H) by Lemma 2.17. Thus we obtain 5_A. $5^2 \in C_{A_{12}}(2^6)$ and $C_{A_{12}}(2^6)$ is an extension of E_{32} by S_6. Such an extension is uniquely determined in $C_G(t) = C_G(2_A)$ up to conjugacy [13]. Moreover, the S_6 acts on E_{64} of $C_G(2_A)$. Hence the element 5^2 is found in $C_G(< v_2, t >)$. Hence by (V), $5^2 \sim 5_2$. This is 5_B.

As H has only one conjugacy class of elements of order 3, $3 \cdot 2^2 \sim 3^2 \cdot 2^2 \sim 6 \cdot 2^3 \sim 6^2$ in G. By the fusion of elements of order 3 in M, we easily see $3 \cdot 2^4 \sim 6 \cdot 2 \not\sim 6 \cdot 3 \cdot 2$ in G.

$(8 \cdot 2)^2 = 4^2$ and $(8 \cdot 4)^2 = 4^2 \cdot 2^2$. We know that $4^2 , 4^2 \cdot 2^2 \in O_2(M)$. By the character table of $C_G(2_B)$, we see that $C_G(2_B)$ has only one class of elements of order 8 whose squares are in $O_2(M)$. Hence $8 \cdot 2 \sim 8 \cdot 4$.

The character table of $C_G(3_B)$ shows that there is only one class of cyclic groups of order 9 in $C_G(3_B)$. As the element 9 of A_{12} is rational in A_{12}, $9 \sim 9 \cdot 3 \sim 9 \cdot 3$ in G.

Clearly $5 \cdot 2^2 \not\sim 10 \cdot 2$, as $5 \not\sim 5^2$. By Lemmas 2.17 and 2.18, we know the conjugacy of 7-elements and 11-elements of G.

By (H), there is only conjugacy class of elements of order 12 whose 2-part is $4_4 = x$. Hence $4 \cdot 3 \cdot 2 \sim 4 \cdot 3^2 \cdot 2 \sim 6 \cdot 4$ in G. Obviously $4 \cdot 3 \cdot 2 \not\sim 4^2 \cdot 3$ in G.

Since $3 \sim 3^2$ and $C_G(3_A) \cong Z_3 \times A_9$, $5 \cdot 3 \sim 5 \cdot 3^2$ must hold. As $C_G(7_A) \cong Z_7 \times A_5$, two classes of elements of order 35 of A_{12} remain unfused in G.

Thus we have obtained 27 classes of G: 1_A, 2_A, 2_B, 3_A, 3_B, 4_A, 4_B, 5_A, 5_B, 6_A, 6_B, 6_C, 7_A, 8_B, 9_A, 10_A, 10_B, 11_A, 12_A, 12_B, 14_A, 15_A, 20_A, 21_A, 30_A, 35_A and 35_B.

We next investigate the fusion of elements of

$$H = C_G(t) = C_G(2_A)$$

in G. By Lemma 2.17(2), 5_3 of H is not conjugate to $5_1 = 5_A$ or $5_2 = 5_B$. Hence we have $5_3 = 5_C$. As $8_5^4 = 2_1 = 2_A$, we obtain $8_5 = 8_A$. Clearly $10_3 = 2_1 \cdot 5_3$ and $10_4 = 2_2 \cdot 5_3$ are new classes. We set $10_3 = 10_C$ and $10_4 = 10_D$.

As $C_M(5_B) \cong C_H(5_B) \cong Z_5 \times Z_2 \times D_{10}$ and 5_B is not rational in $M = C_G(z) = C_G(2_B)$, 10_7 and 10_8 of (H) are two new classes of G . We set $10_7 = 10_E$ and $10_8 = 10_F$.

We know that $4^2 = 4_B \in O_2(M)$ and all 240 elements of order 4 of $O_2(M)$ are conjugate in M . From this we obtain that $C_M(4_B)$ is an extension of $C_{O_2(M)}(4_B)$ by A_5. Thus 5-elements of $C_M(4_B)$ are not rational. Hence $20_2 = 20_B$ and $20_3 = 20_C$ are two new classes of G .

As $C_G(5_A) \cong Z_5 \times U_3(5)$ by Lemma 2.17(2) and $U_3(5)$ has two conjugacy classes of elements of order 8 (which are represented in $Z_5 \times SU^{\pm}(2, 5)$) , we conclude that $40 = 40_A$ and $40_2 = 40_B$ are new classes. 22 of (H) is obviously a new class.

Thus we have obtained eleven new classes: 5_C, 8_A, 10_C, 10_D, 10_E, 10_F, 20_B, 20_C, 22_A, 40_A, and 40_B .

We next investigate $M = C_G(2_B)$ to find out more classes. The element of order 4 in $C_M(\sigma_1)$ of Lemma 2.15 is clearly not contained in H or in any conjugate of it. Thus we have a new class 4_C . As $C_G(4_C)$ involves $SL(2, 5)$, we obtain 12_C, 20_D and 20_E . By Lemma 2.17(3), $5_4 = 5_D$ is a new class. Since a Sylow 2-subgroup of $N_G(< 5_D >)$ must be contained in M and $5_D \not\sim 5_D^2$ in M , $5_D \not\sim 5_D^2$ in G . Hence we obtain $5_D^2 = 5_E$. This gives rise to six more new classes: $10_G = 2_B \cdot 5_D$, $10_H = 2_B \cdot 5_E$, $15_B = 3_B \cdot 5_D$, and $15_C = 3_B \cdot 5_E$, $30_B = 2_B \cdot 3_B \cdot 5_D$ and $30_C = 2_B \cdot 3_B \cdot 5_E$.

Thus we have obtained 12 new classes: 4_C, 5_D, 5_E, 10_G, 10_H, 12_C, 15_B, 15_C, 20_D, 20_E, 30_B, and 30_C .

By Lemma 3.2, we have classes 19_A and 19_B. Finally, from the character table of $N_G(< 5_C >)$ (Table III), one finds two new classes, 25_A and 25_B. We have thus obtained 54 classes of G.

For all 54 classes obtained so far, we can get, without too much effort, the order of the centralizer of a representative of each class. Calculating the total number of elements obtained so far, we find that we, in fact, have found all conjugacy classes of G.

The Sylow 5-subgroups of G are isomorphic to those of the Lyons group. Hence the Lyons group contains elements of order 25, contrary to earlier belief.

The following table includes the names and orders of the representative, the decomposition into p-parts, the power map, the order of the centralizer, and some of the structure.

x		$\lvert C_G(x)\rvert$	$C_G(x)$
1_A		$2^{14}\cdot 3^6\cdot 5^6\cdot 7\cdot 11\cdot 19$	
2_A		$2^{11}\cdot 3^2\cdot 5^3\cdot 7\cdot 11$	$2HS\cdot 2 \cong \hat{Aut}(HS)$
2_B		$2^{14}\cdot 3^2\cdot 5^2$	$2^{1+8}\ A_5 \wr Z_2$
3_A		$2^6\cdot 3^5\cdot 5\cdot 7$	$3 \times A_9$
3_B		$2^3\cdot 3^6\cdot 5$	$3^{1+4}\ SL(2, 5)$
4_A	$4_A^2 = 2_A$	$2^7\cdot 3^2\cdot 5$	
4_B	$4_B^2 = 2_B$	$2^{10}\cdot 3\cdot 5$	
4_C	$4_C^2 = 2_B$	$2^8\cdot 3\cdot 5$	
5_A		$2^4\cdot 3^2\cdot 5^4\cdot 7$	$5 \times U_3(5)$
5_B		$2^2\cdot 5^4$	
5_C		$2^5\cdot 5^6$	$5^{1+4}\cdot 2^{1+4}\cdot 5$
5_D		$2^3\cdot 3\cdot 5^4$	$5^3\cdot SL(2, 5)$
5_E	5_D^2	$2^3\cdot 3\cdot 5^4$	
6_A	$2_A\cdot 3_A$	$2^5\cdot 3^2\cdot 5$	
6_B	$2_B\cdot 3_A$	$2^6\cdot 3^2$	
6_C	$2_B\cdot 3_B$	$2^3\cdot 3^2\cdot 5$	
7_A		$2^2\cdot 3\cdot 5\cdot 7$	$7 \times A_5$
8_A	$8_A^2 = 4_A$	$2^4\cdot 5$	
8_B	$8_B^2 = 4_B$	2^6	
9_A		3^3	
10_A	$2_A\cdot 5_A$	$2^4\cdot 3\cdot 5^2$	
10_B	$2_A\cdot 5_B$	$2^2\cdot 5^2$	

(Cont.)

x		$\lvert C_G(x) \rvert$	$C_G(x)$
10_C	$2_A \cdot 5_C$	$2^4 \cdot 5^3$	
10_D	$2_B \cdot 5_C$	$2^5 \cdot 5^2$	
10_E	$2_B \cdot 5_B$	$2^2 \cdot 5^2$	
10_F	$2_B \cdot 5_B^2$	$2^2 \cdot 5^2$	
10_G	$2_B \cdot 5_D$	$2^3 \cdot 3 \cdot 5^2$	
10_H	$2_B \cdot 5_E$	$2^3 \cdot 3 \cdot 5^2$	
11_A		$2 \cdot 11$	
12_A	$4_A \cdot 3_A$	$2^3 \cdot 3^2$	
12_B	$4_B \cdot 3_A$	$2^4 \cdot 3$	
12_C	$4_C \cdot 3_B$	$2^2 \cdot 3$	
14_A	$2_A \cdot 7_A$	$2^2 \cdot 7$	
15_A	$3_A \cdot 5_A$	$2^2 \cdot 3^2 \cdot 5$	
15_B	$3_B \cdot 5_D$	$2 \cdot 3 \cdot 5$	
15_C	$3_B \cdot 5_E$	$2 \cdot 3 \cdot 5$	
19_A		19	
19_B		19	
20_A	$4_A \cdot 5_A$	$2^3 \cdot 5$	
20_B	$4_B \cdot 5_C$	$2^4 \cdot 5$	
20_C	$4_B \cdot 5_C^2$	$2^4 \cdot 5$	
20_D	$4_C \cdot 5_D$	$2^2 \cdot 5$	
20_E	$4_C \cdot 5_E$	$2^2 \cdot 5$	
21_A	$3_A \cdot 7_A$	$3 \cdot 7$	
		(Cont.)	

x		$\lvert C_G(x)\rvert$	$C_G(x)$
22_A	$2_A \cdot 11_A$	$2 \cdot 11$	
25_A		5^2	
25_B	25_A^2	5^2	
30_A	$2_A \cdot 3_A \cdot 5_A$	$2^2 \cdot 3 \cdot 5$	
30_B	$2_B \cdot 3_B \cdot 5_D$	$2 \cdot 3 \cdot 5$	
30_C	$2_B \cdot 3_B \cdot 5_E$	$2 \cdot 3 \cdot 5$	
35_A	$5_A \cdot 7_A$	$5 \cdot 7$	
35_B	$5_A^2 \cdot 7_A$	$5 \cdot 7$	
40_A	$8_A \cdot 5_A$	$2^3 \cdot 5$	
40_B	$8_A^{-1} \cdot 5_A$	$2^3 \cdot 5$	

6. THE EXISTENCE OF THE CHARACTER OF DEGREE 133.

As seen in Section 5, it is important to obtain the degrees of all irreducible characters of G if we want to construct the character table of G. However, as is seen in the next section, it is impossible for us to find out the degrees unless we know a number of characters (irreducible or compound) beforehand.

We shall establish in this section the existence of an irreducible character of degree 133, 133 being the smallest possible degree of any nontrivial character of G, as shown below.

LEMMA 6.1. *Let* $\chi \neq 1$ *be an irreducible character of* G. *Then* $\chi(1) \geq 133$.

PROOF: Suppose $\chi(1) < 133$. Then $\chi|_H$ is a linear combination of the principal character and the characters of degrees 22 and 56:

$$\chi(1) = a \cdot 1 + b \cdot 22 + c \cdot 56, \qquad c \geq 1.$$

As $2_1 \sim 2_3$ in G (see (H)), we have, using the character table of H (Table I),

$$a + 22b - 56c = a + 6b + 8c.$$

Hence $b = 4c$ and so $\chi(1) = a + 144c \geq 144$, contrary to $\chi(1) < 133$.

LEMMA 6.2. *If* χ *is an irreducible character of degree* 133, *then*

$$\chi|_{G_0} = 1 + 132 \qquad (G_0 \cong A_{12})$$

and

$$\chi|_H = 77 + 56.$$

(Here 132 *etc. denotes an irreducible character of degree* 132 *etc. This convention applies through the rest of the paper.) Furthermore,* χ *is irrational. The values of* χ *on all*

classes of G *are uniquely determined up to the conjugacy in the field* $R(\sqrt{-5})$.

PROOF: The proof of the previous lemma shows that 77 and 56 must appear in $\chi|_H$.

As for $\chi|_{G_0}$, it suffices to observe that $\phi(1^9 \cdot 3) > \phi(1^6 \cdot 3^2)$ holds for any irreducible character of A_{12} of degree less than 133 except $\phi = 1$ or 132. As $1^9{\cdot}3 \sim 1^6{\cdot}3^2$ in G , $\chi|_{G_0} = 1 + 132$ must hold.

Thus we can fill in the values of χ (and χ') on 38 classes. It is easy to find the values of the remaining 16 classes of G (see Table V).

We have established the uniqueness of the pair of irreducible characters 133 and 133' if they exist. We shall next prove the existence.

First we state Brauer's famous theorem:

THEOREM. *(characterization of characters).*

Let ϕ *be a complex valued class function of* G . *Suppose further that:*

(1) *the restriction* $\phi|N$ *of* ϕ *to every nilpotent subgroup* N *of* G *is a generalized character of* N .

(2) $\phi(1) > 0$ *and* $< \phi, \phi >_G = 1$.

Then ϕ *is an irreducible character of* G .

We apply the theorem to show that the "class functions" 133 and 133' obtained in the previous lemma are actually characters.

We need to classify all nilpotent subgroups of G.

PROPOSITION 6.3. Every nilpotent subgroup of G is conjugate to a subgroup of one of the following groups:

$$C_G(19_A) \cong Z_{19},\quad C_G(11_A) \cong Z_2 \times Z_{11},\quad C_G(5_A) \cong Z_5 \times U_3(5)\ ,$$

$$N_G(< 5_C >),\quad C_G(3_B),\quad C_G(2_B),\quad \text{or}\quad G_0 \cong A_{12}\ .$$

PROOF: Let N be a nilpotent subgroup of G. If 11 or 19 divides $|N|$, then it is clear. As

$$C_G(7_A) \cong Z_7 \times A_5 \cong C_{G_0}(7_A)\ ,$$

$N \subset A_{12}$ if 7 divides $|N|$.

Suppose that 3 divides $|N|$. Then $N \subset Z_3 \times A_9 \subset A_{12}$ or $N \subset C_G(3_B)$. Hence we may assume that N is a 2, 5-group with $10 \mid |N|$.

If the center of a Sylow 5-subgroup of N contains 5_A, 5_C, 5_F or 5_G, our result is trivial. Therefore it suffices to show that any nilpotent 2, 5-group of $C_G(5_B)$ or $C_G(5_D)$ is contained in $N_G(< 5_C >)$ or $C_G(2_B)$.

We know that $N_G(< 5_C >)$ is an extension of an extra-special group P of order 5^5 by a group of order $2^5 \cdot 5 \cdot 4$. We also know (see Table III) that P contains 5_B and

$$C_G(5_B) \subset N_G(< 5_C >) .$$

As $C_G(5_D)$ is an extension of $Z_5 \times Z_5 \times Z_5$ by SL(2,5) and the involutions of $C_G(5_D)$ are of 2_B-type, $N \subset C_G(2_B)$.

PROPOSITION 6.4. The 133 and 133' are irreducible characters of G .

PROOF: One can readily check that $133|C_G(19_A)$ is 7 times the regular character of $C_G(19_A)$, that $133|C_G(11_A) = 7 \cdot \rho + 5\lambda \cdot \rho + \lambda \cdot 1$ where ρ is the regular character of 11_A , λ the nonprincipal character of Z_2 , and that

$$133|C_G(5_A) = 1 \cdot 21 + \lambda \cdot 28_1 + \lambda^2 \cdot 28_2 + \lambda^3 \cdot 28_2 + \lambda^4 \cdot 28_1$$

where λ is a nonprincipal character of Z_5 and 28_1 , 28_2 are suitable characters of $U_3(5)$ of degree 28 .

We already know that $133|G_0$ is a character.

Finally, it is easy to find the decompositions of $133|N_G(< 5_C >)$, $133|C_G(3_B)$ and $133|C_G(2_B)$, as given at the bottom of the corresponding character table. This completes the proof.

7. THE CHARACTER TABLE OF G.

(Computed by Conway, Curtis, Smith, Norton and the author).

In the preceding sections we have determined the conjugacy classes of G and have shown the existence of the characters of degree 133. We are now in a position to construct the character table of G . We shall omit the details of the construction.

For brevity, the characters (irreducible or compound) of G are denoted by their degrees as before. If a character is known to be irreducible, we place a bar underneath: $\underline{133}$. The character $X(x) \otimes Y(x)$ will be denoted simply by $X \cdot Y$.

The first question we ask ourselves in constructing the character table is this: Can we determine all degrees (54 degrees in our case) before we attempt to fill in the values? In some cases the work seems to be fairly easy (Lyons group, Rudvalis group, for example). Hoping that we can determine all degrees, we apply Brauer's theorem on modular characters.

By the structure of the Sylow 19- and 11-normalizer we obtain the following condition: If f is the degree of an irreducible character of G, then:

(1) $19 | f$ or $f \equiv \pm 1, \pm 9 \pmod{19}$, and

(2) $11 | f$ or $f \equiv \pm 1 \pmod{11}$.

As $C_G(7_A) \cong Z_7 \times A_5$ we have no condition for the prime 7. We of course have an obvious

(3) $f | \, |G|$ and $f^2 < |G|$.

A computer at Cambridge, U. K. punched out about 800 possible degrees. Apparently it is impossible to choose 54 "good" degrees by staring at those 800 numbers or even by working hard on each individual number. We may be able to eliminate some, but not 750. As a matter of fact, that is why we showed the existence of $\underline{133}$ in the previous section.

Failing to obtain all degrees, we try to make use of $\underline{133}$ (and $\underline{133'}$) as much as possible.

We see, by taking the norms, that $(133^2 - 133(x^2))/2 = \underline{8,778}$ is an irrational irreducible character and that

$$(133^2 + 133(x^2))/2$$

splits into $\underline{1} + \underline{8,910}$. The norm of $133 \cdot 133'$ is 2 and it does not contain any irreducible character obtained previously. As we go through all 800 eligible numbers, we find that 760 + 16,929, 4,389 + 13,300 and 5,985 + 11,704 are the possibilities. We know that $133 = 133' = \underline{1} + \underline{132}$ on A_{12} and so $133 \cdot 133'|A_{12} = \underline{1} + 2 \cdot \underline{132} + \underline{132}^2$. A formula for the square of an irreducible character of the symmetric groups is found in [14, p. 64]. By the formula, we can obtain the decomposition of $133 \cdot 133'$ on A_{12}. By a tedious calculation (mostly playing with numbers), we conclude that only 760 + 16,929 is possible. At the same time we obtain the values of $\underline{760}$ on A_{12}-classes. Using the character table of $H = C_G(2_A)$, we also obtain the values of $\underline{760}$ on H-classes. It is then easy to fill in all values of $\underline{760}$, hence of $\underline{16,929}$. Thus we have obtained eight characters $\underline{1}$, $\underline{133}$, $\underline{133'}$, $\underline{760}$, $\underline{8,778}$, $\underline{8,778'}$, $\underline{8,910}$ and $\underline{16,929}$.

To get more characters, we form $\underline{133} \cdot \underline{8,778}$ and $(\underline{133}^3 - \underline{133}(x^3))/3$. It turns out that the latter is fully contained in the former and that

$$\underline{133} \cdot \underline{8,778} - (\underline{133}^3 - \underline{133}(x^3))/3 - \underline{8,778} = \underline{374,528}$$

is irreducible. Also

$$\underline{8,778} \cdot \underline{133}' - (\underline{760} \cdot \underline{133} - \underline{133}') = \underline{1,066,527}$$

is irreducible. Furthermore,

$$(\underline{760}^2 - \underline{760}(x^2))/2 - (\underline{8,778} + \underline{8,778}') = \underline{270,864}$$

has turned out to be irreducible.

Those are just about all characters that we can obtain easily by inspection. It appears to be impossible to carry out this method to get all characters. The next thing we do is as follows. Even though we cannot get an irreducible character by forming the product of known characters, we can get a compound character of small norm. If we collect many compound characters of small norm and if an irreducible character appears in two distinct compound characters, then this will give us useful information.

For example, $760 \cdot 133 - 133' = a_2$ is a character of norm 2 and $(133^3 - 133(x^3))/3 - 133 = c_2$ is also a character of norm 2. The inner product $\langle a_2', c_2' \rangle$ of the conjugate of a_2 and c_2 is 1 . Hence a_2' and c_2' share a common irreducible character. Now go through the 800 possible numbers to find out the pairs of numbers which make up $\deg(a_2) = 100{,}947$ and $\deg(c_2) = 784{,}035$. As we know that there is a common number, the pairs are almost always determined uniquely. Thus we obtain three degrees. In the example above we have

$$a_2 = 35{,}112' + 65{,}835'$$

and

$$c_2 = 718{,}200 + 65{,}835' .$$

By such processes we obtain small degrees like 3,344 and 9,405. We then complete the character $\underline{3{,}344}$ and $\underline{9{,}405}$ by filling in all the values. This gives us other completed characters such as $\underline{267{,}520}$, since $\underline{3{,}344} + \underline{267{,}520}$ is a known compound character.

Having obtained those characters of small degree and hence many tensor products of small norm thereof, we attempt to get all degrees.

We use, at this final stage, thirteen "base" compound characters and their conjugates. Here "base" means "linearly independent." There will be no need to explain those base compound characters which are derived from the tensor product of known irreducible characters, as there will be many other ways to get such characters. We explain only the compound characters obtained from permutation representations.

As G contains a subgroup $G_0 \cong A_{12}$ we have a permutation character ϕ of G/G_0 of degree 1,140,000 . As we know the conjugacy of elements of G_0 in G , it is easy to work out the values of ϕ on all classes of G . We compute that $< \phi, \phi > = 12$ and that $\phi = \underline{1} + \underline{133} + \underline{133'} + \underline{760} + \underline{3,344} + \underline{8,910} + \underline{16,929} + \underline{267,520} + b_2 + f_2$ where b_2 is a base compound character of norm 2 derived from the tensor product. Thus we have obtained a norm 2 character f_2.

Similarly, if ψ_1 is the permutation character of G/H, then $\psi_1 = \underline{1} + \underline{3,344} + \underline{8,910} + \underline{9,405} + \underline{16,929} + \underline{267,520} + g_3$. As $< \psi_1, \psi_1 > = 9$, we have thus obtained a norm 3 character g_3 . If ψ_2 is the permutation character of G/H', then $\psi_2 = \psi_1 + \underline{760} + \underline{270,864} + j_2$. Hence we obtain j_2 .

One more character we used is the induced character 11^G of the irreducible character of $G_0 \cong A_{12}$ of degree 11 . We compute $< 11^G, 11^G > = 7$ and $11^G = \underline{9,405} + p_6$. Thus we obtain a norm 6 character p_6 .

In addition to those "genuine" characters, we use many generalized characters which are derived from a theory of modular characters.

By the structure of the Sylow 19 normalizers, there is a linear combination Δ_{19} of irreducible characters $X_1, \ldots, X_9$ and X_{10} of G such that

$$\Delta_{19} = \varepsilon_1 X_1 + \ldots + \varepsilon_9 X_9 + \varepsilon_{10} X_{10}$$

vanishes on all 19-regular elements and

$$\varepsilon_i = \pm 1, \qquad \chi_i \equiv \varepsilon_i \pmod{19}, \ 1 \leqslant i \leqslant 9$$

$$\chi_{10} \equiv \pm 9 \pmod{19}.$$

Furthermore any character whose degree is prime to 19 appears in $\{\chi_1, \ldots, \chi_9, \chi_{10}, \chi'_{10}\}$.

Similarly we have two generalized characters for the prime 11

$$\Delta_{11}^1 \quad \text{and} \quad \Delta_{11}^2$$

and four for the prime 7, Δ_7^1, Δ_7^2, Δ_7^3 and Δ_7^4. As

$$C_G(3_A) \cong Z_3 \times A_9$$

and A_9 has an irreducible character of degree 162 which is of defect 0 for the prime 3, G has a block of defect 1 for the prime 3 and there is a generalized character Δ_3 which vanishes on all 3_A-free elements.

Since $C_G(5_A) \cong Z_5 \times U_3(5)$ and $U_3(5)$ has an irreducible character of degree 125, G has a 5-block of defect 1 and a generalized character Δ_5 which vanishes on all 5_A-free elements.

With all the information given above we are able to determine all the degrees of irreducible characters of G. Among those degrees are $3{,}424{,}256 = 2^{14} \cdot 11 \cdot 19$, $2{,}375{,}000 = 5^6 \cdot 2^3 \cdot 19$, $4{,}156{,}250 = 5^6 \cdot 2 \cdot 7 \cdot 19$ and $656{,}250 = 5^6 \cdot 2 \cdot$

$3 \cdot 7$. The first is of 2-defect 0 and the last three are of 5-defect 0 . The values of such characters are very easy to fill in. Also, such characters are very useful to complete other characters, as they vanish on lots of places.

The character tables of $C_G(2_A)$, $C_G(3_B)$, $N_G(< 5_C >)$ and $C_G(2_B)$ and G are presented here, even though they have not yet been checked by computer.

REFERENCES

1. J. L. Alperin, R. Brauer, and D. Gorenstein, *Finite groups with quasi-dihedral and wreathed Sylow 2-subgroups*, Trans. Amer. Math. Soc., 151(1970), 1-261.
2. R. Brauer, *On simple groups of order* $5 \cdot 3^a \cdot 2^b$, Bull. Amer. Math. Soc., 74(1968), 900-903.
3. A. H. Clifford, *Representations induced in an invariant subgroup*, Ann. Math., 38(1937), 333-550.
4. J. S. Frame, *The characters of the Weyl group* E_8, Computational Problems in Abstract Algebra, 111-130, Pergamon, Oxford, 1970.
5. ___________, *Computation of characters of the Higman-Sims Group and its Automorphism Group*, J. Alg. 20(1972) 320-349.
6. D. Goldschmidt, 2-*Fusion in finite groups*, Ann. Math., 99(1974), 70-117.
7. D. Gorenstein and K. Harada, *Finite groups with Sylow 2-subgroups of type* PSp(4, q), q *odd*, J. Fac. Sci. Univ. of Tokyo, 20(1973), 341-372.
8. D. Gorenstein and K. Harada, *Finite groups whose 2-subgroups are generated by at most* 4 *elements*, Memoirs Amer. Math. Soc., No. 147 (1974).

9. D. Gorenstein and M. Harris, *A Characterization of the Higman-Sims simple group*, J. Alg., 24(1973), 565-590.

10. M. Hall, *Simple groups of order less than one million*, J. Alg., 20(1972), 98-102.

11. K. Harada, *On finite groups having self-centralizing* 2-*subgroups of small order*, J. Alg., 33(1975), 144-160.

12. B. Huppert, *Endliche Gruppen I*, Springer-Verlag,Berlin, Heidelberg, New York, 1967.

13. S. S. Magliveras, *The subgroup structure of the Higman-Sims simple group*, Bull. Amer. Math. Soc., 77(1971), 535-539.

14. G. Robinson, *Representation theory of the symmetric group*, Univ. of Toronto Press, Toronto, 1961.

15. A. Rudvalis, *The character table of* 2HS, (to appear).

16. J. G. Thompson, *Nonsolvable finite groups all of whose local subgroups are solvable*, Sections 1-6, Bull. Amer. Math. Soc., 74(1968), 383-438.

OHIO STATE UNIVERSITY
COLUMBUS, OHIO

TABLE I

THE CHARACTER TABLE OF 2HS·2

|G|	|G|	30,720	30,720	1440	512	128	7680	5760
1	2_1	2_2	2_3	3	4_1	4_2	4_3	4_4
1	1	1	1	1	1	1	1	1
1	1	1	1	1	1	1	1	1
22	22	6	6	4	2	2	-6	-2
22	22	6	6	4	2	2	-6	-2
77	77	13	13	5	5	1	5	1
77	77	13	13	5	5	1	5	1
175	175	15	15	4	-1	3	15	11
175	175	15	15	4	-1	3	15	11
231	231	7	7	6	-1	-1	15	-9
231	231	7	7	6	-1	-1	15	-9
1056	1056	32	32	-6	0	0	0	0
1056	1056	32	32	-6	0	0	0	0
825	825	25	25	6	1	1	-15	9
825	825	25	25	6	1	1	-15	9
770	770	34	34	5	2	-2	-14	-10
770	770	34	34	5	2	-2	-14	-10
1925	1925	5	5	-1	5	-3	5	-19
1925	1925	5	5	-1	5	-3	5	-19
1925	1925	5	5	-1	-3	1	-35	1
1925	1925	5	5	-1	-3	1	-35	1
3200	3200	0	0	-4	0	0	0	-16
3200	3200	0	0	-4	0	0	0	-16
1408	1408	0	0	4	0	0	0	16
1408	1408	0	0	4	0	0	0	16
2750	2750	-50	-50	5	2	2	10	-10
2750	2750	-50	-50	5	2	2	10	-10
1750	1750	-10	-10	-5	6	2	-10	10
1750	1750	-10	-10	-5	6	2	-10	10
693	693	21	21	0	5	1	21	9
693	693	21	21	0	5	1	21	9
154	154	10	10	1	6	-2	-2	10
154	154	10	10	1	6	-2	-2	10
1386	1386	-6	-6	0	-2	-2	6	18
1386	1386	-6	-6	0	-2	-2	6	18
2520	2520	24	24	0	-8	0	24	0
2520	2520	24	24	0	-8	0	24	0
308	308	20	20	2	-4	4	-20	-20

TABLE I (Cont.)

$\lvert G \rvert$ 1	$\lvert G \rvert$ 2_1	30,720 2_2	30,720 2_3	1440 3	512 4_1	128 4_2	7680 4_3	5760 4_4
1540	1540	-28	-28	10	-4	-4	-20	20
1792	1792	0	0	-8	0	0	0	32
56	-56	-8	8	2	0	0	0	0
56	-56	-8	8	2	0	0	0	0
352	-352	-32	32	10	0	0	0	0
1232	-1232	-48	48	8	0	0	0	0
1848	-1848	-8	8	12	0	0	0	0
1000	-1000	40	-40	10	0	0	0	0
1000	-1000	40	-40	10	0	0	0	0
2464	-2464	32	-32	-2	0	0	0	0
1792	-1792	0	0	-8	0	0	0	0
1792	-1792	0	0	-8	0	0	0	0
1848	-1848	-8	8	-6	0	0	0	0
1848	-1848	-8	8	-6	0	0	0	0
3960	-3960	-72	72	0	0	0	0	0
4608	-4608	0	0	0	0	0	0	0
2520	-2520	24	-24	0	0	0	0	0
2520	-2520	24	-24	0	0	0	0	0
2520	-2520	24	-24	0	0	0	0	0
2520	-2520	24	-24	0	0	0	0	0

(Cont.)

TABLE I (Cont.)

|G|	1200	100	2000	1400	96	96	28	32	16	1200	100
1	5_1	5_2	5_3	6_1	6_2	6_3	7	8_1	8_2	10_1	10_2
1	1	1	1	1	1	1	1	1	1	1	1
1	1	1	1	1	1	1	1	1	1	1	1
22	2	2	-3	4	0	0	1	0	0	2	2
22	2	2	-3	4	0	0	1	0	0	2	2
77	-3	2	2	5	1	1	0	1	-1	-3	2
77	-3	2	2	5	1	1	0	1	-1	-3	2
175	5	0	0	4	0	0	0	-1	1	5	0
175	5	0	0	4	0	0	0	-1	1	5	0
231	1	1	6	6	-2	-2	0	-1	-1	1	1
231	1	1	6	6	-2	-2	0	-1	-1	1	1
1056	-4	1	6	-6	2	2	-1	0	0	-4	1
1056	-4	1	6	-6	2	2	-1	0	0	-4	1
825	-5	0	0	6	-2	-2	-1	1	1	-5	0
825	-5	0	0	6	-2	-2	-1	1	1	-5	0
770	0	0	-5	5	1	1	0	-2	0	0	0
770	0	0	-5	5	1	1	0	-2	0	0	0
1925	5	0	0	-1	-1	-1	0	1	1	5	0
1925	5	0	0	-1	-1	-1	0	1	1	5	0
1925	5	0	0	-1	-1	-1	0	1	-1	5	0
1925	5	0	0	-1	-1	-1	0	1	-1	5	0
3200	-5	0	0	-4	0	0	1	0	0	-5	0
3200	-5	0	0	-4	0	0	1	0	0	-5	0
1408	-7	-2	8	4	0	0	1	0	0	-7	-2
1408	-7	-2	8	4	0	0	1	0	0	-7	-2
2750	0	0	0	5	1	1	-1	0	0	0	0
2750	0	0	0	5	1	1	-1	0	0	0	0
1750	0	0	0	-5	-1	-1	0	-2	0	0	0
1750	0	0	0	-5	-1	-1	0	-2	0	0	0
693	3	-2	-7	0	0	0	0	1	-1	3	-2
693	3	-2	-7	0	0	0	0	1	-1	3	-2
154	4	-1	4	1	1	1	0	0	0	4	-1
154	4	-1	4	1	1	1	0	0	0	4	-1
1386	6	1	11	0	0	0	0	0	0	6	1
1386	6	1	11	0	0	0	0	0	0	6	1
2520	0	0	-5	0	0	0	0	0	0	0	0
2520	0	0	-5	0	0	0	0	0	0	0	0
308	8	-2	8	2	2	2	0	0	0	8	-2

(Cont.)

TABLE I (Cont.)

|G|	1200	100	2000	1400	96	96	28	32	16	1200	100
1	5_1	5_2	5_3	6_1	6_2	6_3	7	8_1	8_2	10_1	10_2
1540	0	0	-10	10	2	2	0	0	0	0	0
1792	2	2	-8	-8	0	0	0	0	0	2	2
56	-4	1	6	-2	-2	2	0	0	0	4	-1
56	-4	1	6	-2	-2	2	0	0	0	4	-1
352	12	2	2	-10	-2	2	2	0	0	-12	-2
1232	-8	2	-18	-8	0	0	0	0	0	8	-2
1848	8	-2	-2	-12	4	-4	0	0	0	-8	2
1000	0	0	0	-10	-2	2	-1	0	0	0	0
1000	0	0	0	-10	-2	2	-1	0	0	0	0
2464	4	4	14	2	2	-2	0	0	0	-4	-4
1792	2	2	-8	8	0	0	0	0	0	-2	-2
1792	2	2	-8	8	0	0	0	0	0	-2	-2
1848	8	-2	-2	6	-2	2	0	0	0	-8	2
1848	8	-2	-2	6	-2	2	0	0	0	-8	2
3960	0	0	10	0	0	0	-2	0	0	0	0
4608	-12	-2	8	0	0	0	2	0	0	12	2
2520	0	0	-5	0	0	0	0	0	0	0	0
2520	0	0	-5	0	0	0	0	0	0	0	0
2520	0	0	-5	0	0	0	0	0	0	0	0
2520	0	0	-5	0	0	0	0	0	0	0	0

(Cont.)

TABLE I (Cont.)

$\|G\|$	2000	80	80	22	72	24	28	60	40	40	40
1	10_3	10_4	10_5	11	12_1	12_2	14_1	15	20_1	20_2	20_3
1	1	1	1	1	1	1	1	1	1	1	1
1	1	1	1	1	1	1	1	1	1	1	1
22	-3	1	1	0	-2	0	1	-1	-2	-1	-1
22	-3	1	1	0	-2	0	1	-1	-2	-1	-1
77	2	-2	-2	0	1	-1	0	0	1	0	0
77	2	-2	-2	0	1	-1	0	0	1	0	0
175	0	0	0	-1	2	0	0	-1	1	0	0
175	0	0	0	-1	2	0	0	-1	1	0	0
231	6	2	2	0	0	0	0	1	1	0	0
231	6	2	2	0	0	0	0	1	1	0	0
1056	6	2	2	0	0	0	-1	-1	0	0	0
1056	6	2	2	0	0	0	-1	-1	0	0	0
825	0	0	0	0	0	0	-1	1	-1	0	0
825	0	0	0	0	0	0	-1	1	-1	0	0
770	-5	-1	-1	0	-1	1	0	0	0	1	1
770	-5	-1	-1	0	-1	1	0	0	0	1	1
1925	0	0	0	0	-1	-1	0	-1	1	0	0
1925	0	0	0	0	-1	-1	0	-1	1	0	0
1925	0	0	0	0	1	1	0	-1	1	0	0
1925	0	0	0	0	1	1	0	-1	1	0	0
3200	0	0	0	-1	2	0	1	1	-1	0	0
3200	0	0	0	-1	2	0	1	1	-1	0	0
1408	8	0	0	0	-2	0	1	-1	1	0	0
1408	8	0	0	0	-2	0	1	-1	1	0	0
2750	0	0	0	0	-1	1	-1	0	0	0	0
2750	0	0	0	0	-1	1	-1	0	0	0	0
1750	0	0	0	1	1	-1	0	0	0	0	0
1750	0	0	0	1	1	-1	0	0	0	0	0
693	-7	1	1	0	0	0	0	0	-1	1	1
693	-7	1	1	0	0	0	0	0	-1	1	1
154	4	0	0	0	1	1	0	1	0	-2	-2
154	4	0	0	0	1	1	0	1	0	-2	-2
1386	11	-1	-1	0	0	0	0	0	-2	1	1
1386	11	-1	-1	0	0	0	0	0	-2	1	1
2520	-5	-1	-1	1	0	0	0	0	0	-1	-1
2520	-5	-1	-1	1	0	0	0	0	0	-1	-1
308	8	0	0	0	-2	-2	0	2	0	0	0

(Cont.)

TABLE I (Cont.)

$\|G\|$	2000	80	80	22	72	24	28	60	40	40	40
1	10_3	10_4	10_5	11	12_1	12_2	14_1	15	20_1	20_2	20_3
1540	-10	2	2	0	2	-2	0	0	0	0	0
1792	-8	0	0	-1	-4	0	0	2	2	0	0
56	-6	2	-2	1	0	0	0	2	0	0	0
56	-6	2	-2	1	0	0	0	2	0	0	0
352	-2	-2	2	0	0	0	-2	0	0	0	0
1232	18	2	-2	0	0	0	0	-2	0	0	0
1848	2	2	-2	0	0	0	0	2	0	0	0
1000	0	0	0	-1	0	0	1	0	0	0	0
1000	0	0	0	-1	0	0	1	0	0	0	0
2464	-14	2	-2	0	0	0	0	-2	0	0	0
1792	8	0	0	-1	0	0	0	2	0	0	0
1792	8	0	0	-1	0	0	0	2	0	0	0
1848	2	2	-2	0	0	0	0	-1	0	0	0
1848	2	2	-2	0	0	0	0	-1	0	0	0
3960	-10	-2	2	0	0	0	2	0	0	0	0
4608	-8	0	0	-1	0	0	-2	0	0	0	0
2520	5	-1	1	1	0	0	0	0	0	$\sqrt{5}$	$-\sqrt{5}$
2520	5	-1	1	1	0	0	0	0	0	$\sqrt{5}$	$-\sqrt{5}$
2520	5	-1	1	1	0	0	0	0	0	$-\sqrt{5}$	$\sqrt{5}$
2520	5	-1	1	1	0	0	0	0	0	$-\sqrt{5}$	$\sqrt{5}$

(Cont.)

TABLE I (Cont.)

|G|	22	60	80,640	3840	192	640	72	48	720	64	64
1	22	30_2	2_4	2_5	4_5	4_6	6_4	6_5	6_6	8_3	8_4
1	1	1	1	1	1	1	1	1	1	1	1
1	1	1	-1	-1	-1	-1	-1	-1	-1	-1	-1
22	0	-1	8	0	0	4	2	0	-4	2	-2
22	0	-1	-8	0	0	-4	-2	0	4	-2	2
77	0	0	21	5	1	5	3	-1	3	1	1
77	0	0	-21	-5	-1	-5	-3	1	-3	-1	-1
175	-1	-1	21	5	1	5	0	2	6	1	1
175	-1	-1	-21	-5	-1	-5	0	-2	-6	-1	-1
231	0	1	21	-11	-3	5	0	-2	6	1	1
231	0	1	-21	11	3	-5	0	2	-6	-1	-1
1056	0	-1	48	16	0	0	0	-2	6	0	0
1056	0	-1	-48	-16	0	0	0	2	-6	0	0
825	0	1	69	5	-3	5	0	2	-6	1	1
825	0	1	-69	-5	3	-5	0	-2	6	-1	-1
770	0	0	70	-10	2	6	1	-1	-5	-2	-2
770	0	0	-70	10	-2	-6	-1	1	5	2	2
1925	0	-1	91	-5	-5	-5	1	1	1	-1	-1
1925	0	-1	-91	5	5	5	-1	-1	-1	1	1
1925	0	-1	21	5	1	5	-3	-1	-9	1	1
1925	0	-1	-21	-5	-1	-5	3	1	9	-1	-1
3200	-1	1	64	0	0	0	-2	0	4	0	0
3200	-1	1	-64	0	0	0	2	0	-4	0	0
1408	0	-1	64	0	0	0	-2	0	4	0	0
1408	0	-1	-64	0	0	0	2	0	-4	0	0
2750	0	0	20	-20	4	0	-1	1	5	2	-2
2750	0	0	-20	20	-4	0	1	-1	-5	-2	2
1750	1	0	70	-10	2	-10	1	-1	-5	2	2
1750	1	0	-70	10	-2	10	-1	1	5	-2	-2
693	0	0	63	15	3	-1	0	0	0	-1	-1
693	0	0	-63	-15	-3	1	0	0	0	1	1
154	0	1	28	4	4	0	1	1	1	-2	2
154	0	1	-28	-4	-4	0	-1	-1	-1	2	-2
1386	0	0	0	-24	0	4	0	0	0	-2	2
1386	0	0	0	24	0	-4	0	0	0	2	-2
2520	1	0	0	0	0	0	0	0	0	0	0
2520	1	0	0	0	0	0	0	0	0	0	0
308	0	2	0	0	0	0	0	0	0	0	0

(Cont.)

TABLE I (Cont.)

$\lvert G \rvert$	22	60	80,640	3840	192	640	72	48	720	64	64
1	22	30_2	2_4	2_5	4_5	4_6	6_4	6_5	6_6	8_3	8_4
1540	0	0	0	0	0	0	0	0	0	0	0
1792	-1	2	0	0	0	0	0	0	0	0	0
56	-1	-2	0	0	0	0	0	0	0	0	0
56	-1	-2	0	0	0	0	0	0	0	0	0
352	0	0	0	0	0	0	0	0	0	0	0
1232	0	2	0	0	0	0	0	0	0	0	0
1848	0	-2	0	0	0	0	0	0	0	0	0
1000	1	0	0	0	0	0	0	0	0	0	0
1000	1	0	0	0	0	0	0	0	0	0	0
2464	0	2	0	0	0	0	0	0	0	0	0
1792	1	-2	0	0	0	0	0	0	0	0	0
1792	1	-2	0	0	0	0	0	0	0	0	0
1848	0	1	0	0	0	0	0	0	0	0	0
1848	0	1	0	0	0	0	0	0	0	0	0
3960	0	0	0	0	0	0	0	0	0	0	0
4608	1	0	0	0	0	0	0	0	0	0	0
2520	-1	0	0	0	0	0	0	0	0	0	0
2520	-1	0	0	0	0	0	0	0	0	0	0
2520	-1	0	0	0	0	0	0	0	0	0	0
2520	-1	0	0	0	0	0	0	0	0	0	0

(Cont.)

TABLE I (Cont.)

|G|	80	60	20	20	24	28	28	40
1	8_5	10_6	10_7	10_8	12_3	14_2	14_3	20_4
1	1	1	1	1	1	1	1	1
1	-1	-1	-1	-1	-1	-1	-1	-1
22	0	-2	0	0	0	1	1	-1
22	0	2	0	0	0	-1	-1	1
77	-1	1	0	0	1	0	0	0
77	1	-1	0	0	-1	0	0	0
175	-1	1	0	0	-2	0	0	0
175	1	-1	0	0	2	0	0	0
231	1	1	-1	-1	0	0	0	0
231	-1	-1	1	1	0	0	0	0
1056	0	-2	1	1	0	-1	-1	0
1056	0	2	-1	-1	0	1	1	0
825	1	-1	0	0	0	-1	-1	0
825	-1	1	0	0	0	1	1	0
770	0	0	0	0	-1	0	0	1
770	0	0	0	0	1	0	0	-1
1925	-1	1	0	0	1	0	0	0
1925	1	-1	0	0	-1	0	0	0
1925	-1	1	0	0	1	0	0	0
1925	1	-1	0	0	-1	0	0	0
3200	-4	-1	0	0	0	1	1	0
3200	4	1	0	0	0	-1	-1	0
1408	4	-1	0	0	0	1	1	0
1408	-4	1	0	0	0	-1	-1	0
2750	0	0	0	0	1	-1	-1	0
2750	0	0	0	0	-1	1	1	0
1750	0	0	0	0	-1	0	0	0
1750	0	0	0	0	1	0	0	0
693	1	3	0	0	0	0	0	-1
693	-1	-3	0	0	0	0	0	1
154	0	-2	-1	-1	1	0	0	0
154	0	2	1	1	-1	0	0	0
1386	0	0	1	1	0	0	0	-1
1386	0	0	-1	-1	0	0	0	1
2520	0	0	0	0	0	0	0	$-\sqrt{5}$
2520	0	0	0	0	0	0	0	$--\sqrt{5}$
308	0	0	0	0	0	0	0	0

(Cont.)

TABLE I (Cont.)

$\lvert G\rvert$ 1	80 8_5	60 10_6	20 10_7	20 10_8	24 12_3	28 14_2	28 14_3	40 20_4
1540	0	0	0	0	0	0	0	0
1792	0	0	0	0	0	0	0	0
56	0	0	$\sqrt{5}$	$-\sqrt{5}$	0	0	0	0
56	0	0	$-\sqrt{5}$	$\sqrt{5}$	0	0	0	0
352	0	0	0	0	0	0	0	0
1232	0	0	0	0	0	0	0	0
1848	0	0	0	0	0	0	0	0
1000	0	0	0	0	0	$\sqrt{7}\,i$	$-\sqrt{7}\,i$	0
1000	0	0	0	0	0	$-\sqrt{7}\,i$	$\sqrt{7}\,i$	0
2464	0	0	0	0	0	0	0	0
1792	0	0	0	0	0	0	0	0
1792	0	0	0	0	0	0	0	0
1848	0	0	0	0	0	0	0	0
1848	0	0	0	0	0	0	0	0
3960	0	0	0	0	0	0	0	0
4608	0	0	0	0	0	0	0	0
2520	0	0	0	0	0	0	0	αi
2520	0	0	0	0	0	0	0	$-\alpha i$
2520	0	0	0	0	0	0	0	βi
2520	0	0	0	0	0	0	0	$-\beta i$

(Cont.)

TABLE I (Cont.)

|G|	40	40	40	60	60	40	40
1	20_5	20_6	20_7	30_2	30_3	40_1	40_2
1	1	1	1	1	1	1	1
1	-1	-1	-1	-1	-1	-1	-1
22	-1	-1	-1	1	1	0	0
22	1	1	1	-1	-1	0	0
77	0	0	0	-2	-2	-1	-1
77	0	0	0	2	2	1	1
175	0	0	0	1	1	-1	-1
175	0	0	0	-1	-1	1	1
231	0	0	0	1	1	1	1
231	0	0	0	-1	-1	-1	-1
1056	0	0	0	1	1	0	0
1056	0	0	0	-1	-1	0	0
825	0	0	0	-1	-1	1	1
825	0	0	0	1	1	-1	-1
770	1	1	1	0	0	0	0
770	-1	-1	-1	0	0	0	0
1925	0	0	0	1	1	-1	-1
1925	0	0	0	-1	-1	1	1
1925	0	0	0	1	1	-1	-1
1925	0	0	0	-1	-1	1	1
3200	0	0	0	-1	-1	1	1
3200	0	0	0	1	1	-1	-1
1408	0	0	0	-1	-1	-1	-1
1408	0	0	0	1	1	1	1
2750	0	0	0	0	0	0	0
2750	0	0	0	0	0	0	0
1750	0	0	0	0	0	0	0
1750	0	0	0	0	0	0	0
693	-1	-1	-1	0	0	1	1
693	1	1	1	0	0	-1	-1
154	0	0	0	1	1	0	0
154	0	0	0	-1	-1	0	0
1386	-1	-1	-1	0	0	0	0
1386	1	1	1	0	0	0	0
2520	$\sqrt{5}$	$-\sqrt{5}$	$-\sqrt{5}$	0	0	0	0
2520	$-\sqrt{5}$	$\sqrt{5}$	$\sqrt{5}$	0	0	0	0
308	0	0	0	0	0	0	0

(Cont.)

TABLE I (Cont.)

$\lvert G \rvert$ 1	40 20_5	40 20_6	40 20_7	60 30_2	60 30_3	40 40_1	40 40_2
1540	0	0	0	0	0	0	0
1792	0	0	0	0	0	0	0
56	0	0	0	0	0	0	0
56	0	0	0	0	0	0	0
352	0	0	0	0	0	0	0
1232	0	0	0	0	0	0	0
1848	0	0	0	0	0	0	0
1000	0	0	0	0	0	0	0
1000	0	0	0	0	0	0	0
2464	0	0	0	0	0	0	0
1792	0	0	0	0	0	$\sqrt{10}\,i$	$-\sqrt{10}\,i$
1792	0	0	0	0	0	$-\sqrt{10}\,i$	$\sqrt{10}\,i$
1848	0	0	0	$\sqrt{15}$	$-\sqrt{15}$	0	0
1848	0	0	0	$-\sqrt{15}$	$\sqrt{15}$	0	0
3960	0	0	0	0	0	0	0
4608	0	0	0	0	0	0	0
2520	$-\alpha i$	βi	$-\beta i$	0	0	0	0
2520	αi	$-\beta i$	βi	0	0	0	0
2520	$-\beta i$	αi	$-\alpha i$	0	0	0	0
2520	βi	$-\alpha i$	αi	0	0	0	0

$\alpha = \sqrt{5 + 2\sqrt{5}}$ $\qquad$ $\beta = \sqrt{5 - 2\sqrt{5}}$

TABLE II

CHARACTER TABLE OF $3^{1+4} \cdot SL(2, 5)$

x	1	3_1	3_1^2	3_2	3_3	2	3_4	3_5
$\lvert C_G(x)\rvert$	29,160	29,160	29,160	243	243	360	162	162
	1	1	1	1	1	1	1	1
	3	3	3	3	3	3	0	0
	3	3	3	3	3	3	0	0
	4	4	4	4	4	4	1	1
	5	5	5	5	5	5	-1	-1
	2	2	2	2	2	-2	-1	-1
	2	2	2	2	2	-2	-1	-1
	4	4	4	4	4	-4	1	1
	6	6	6	6	6	-6	0	0
	40	40	40	4	-5	0	4	4
	40	40	40	4	-5	0	-2	-2
	40	40	40	4	-5	0	-2	-2
	40	40	40	-5	4	0	4	4
	40	40	40	-5	4	0	-2	-2
	40	40	40	-5	4	0	-2	-2
	9	$9w$	$9w^2$	0	0	1	3	$3w$
	9	$9w^2$	$9w$	0	0	1	3	$3w^2$
	27	$27w$	$27w^2$	0	0	3	0	0
	27	$27w^2$	$27w$	0	0	3	0	0
	27	$27w$	$27w^2$	0	0	3	0	0
	27	$27w^2$	$27w$	0	0	3	0	0
	36	$36w$	$36w^2$	0	0	4	3	$3w$
	36	$36w^2$	$36w$	0	0	4	3	$3w^2$

(Cont.)

TABLE II (Cont.)

x	1	3_1	3_1^2	3_2	3_3	2	3_4	3_5
$\lvert C_G(x)\rvert$	29,160	29,160	29,160	243	243	360	162	162
	45	$45w$	$45w^2$	0	0	5	−3	$-3w$
	45	$45w^2$	$45w$	0	0	5	−3	$-3w^2$
	18	$18w$	$18w^2$	0	0	−2	−3	$-3w$
	18	$18w^2$	$18w$	0	0	−2	−3	$-3w^2$
	18	$18w$	$18w^2$	0	0	−2	−3	$-3w$
	18	$18w^2$	$18w$	0	0	−2	−3	$-3w^2$
	36	$36w$	$36w^2$	0	0	−4	3	$3w$
	36	$36w^2$	$36w$	0	0	−4	3	$3w^2$
	54	$54w$	$54w^2$	0	0	−6	0	0
	54	$54w^2$	$54w$	0	0	−6	0	0

(Cont.)

TABLE II (Cont.)

1	3_6	3_7	3_8	4	5_1	5_1^2	6_1	6_2	6_3
29,160	162	27	27	12	30	30	360	360	18
1	1	1	1	1	1	1	1	1	1
3	0	0	0	-1	α	β	3	3	0
3	0	0	0	-1	β	α	3	3	0
4	1	1	1	0	-1	-1	4	4	1
5	-1	-1	-1	1	0	0	5	5	-1
2	-1	-1	-1	0	$-\beta$	$-\alpha$	-2	-2	1
2	-1	-1	-1	0	$-\alpha$	$-\beta$	-2	-2	1
4	1	1	1	0	-1	-1	-4	-4	-1
6	0	0	0	0	1	1	-6	-6	0
40	4	-2	-2	0	0	0	0	0	0
40	-2	1	1	0	0	0	0	0	0
40	-2	1	1	0	0	0	0	0	0
40	4	1	1	0	0	0	0	0	0
40	-2	$1+3w$	$1+3w^2$	0	0	0	0	0	0
40	-2	$1+3w^2$	$1+3w$	0	0	0	0	0	0
9	$3w^2$	0	0	1	-1	-1	w	w^2	1
9	$3w$	0	0	1	-1	-1	w^2	w	1
27	0	0	0	-1	$-\alpha$	$-\beta$	$3w$	$3w^2$	0
27	0	0	0	-1	$-\alpha$	$-\beta$	$3w^2$	$3w$	0
27	0	0	0	-1	$-\beta$	$-\alpha$	$3w$	$3w^2$	0
27	0	0	0	-1	$-\beta$	$-\alpha$	$3w^2$	$3w$	0
36	$3w^2$	0	0	0	1	1	$4w$	$4w^2$	1
36	$3w$	0	0	0	1	1	$4w^2$	$4w$	1

(Cont.)

TABLE II (Cont.)

1	3_6	3_7	3_8	4	5_1	5_1^2	6_1	6_2	6_3
29,160	162	27	27	12	30	30	360	360	18
45	$-3w^2$	0	0	1	0	0	$5w$	$5w^2$	-1
45	$-3w$	0	0	1	0	0	$5w^2$	$5w$	-1
18	$-3w^2$	0	0	0	β	α	$-2w$	$-2w^2$	1
18	$-3w$	0	0	0	β	α	$-2w^2$	$-2w$	1
18	$-3w^2$	0	0	0	α	β	$-2w$	$-2w^2$	1
18	$-3w$	0	0	0	α	β	$-2w^2$	$-2w$	1
36	$3w^2$	0	0	0	1	1	$-4w$	$-4w^2$	-1
36	$3w$	0	0	0	1	1	$-4w^2$	$-4w$	-1
54	0	0	0	0	-1	-1	$-6w$	$-6w^2$	0
54	0	0	0	0	-1	-1	$-6w^2$	$-6w$	0

(Cont.)

TABLE II (Cont.)

1	6_4	6_5	9_1	9_2	10_1	10_2	12_1	12_2
29,160	18	18	27	27	30	30	12	12
1	1	1	1	1	1	1	1	1
3	0	0	0	0	α	β	-1	-1
3	0	0	0	0	β	α	-1	-1
4	1	1	1	1	-1	-1	0	0
5	-1	-1	-1	-1	0	0	1	1
2	1	1	-1	-1	β	α	0	0
2	1	1	-1	-1	α	β	0	0
4	-1	-1	1	1	1	1	0	0
6	0	0	0	0	-1	-1	0	0
40	0	0	1	1	0	0	0	0
40	0	0	$1+3w$	$1+3w^2$	0	0	0	0
40	0	0	$1+3w^2$	$1+3w$	0	0	0	0
40	0	0	-2	-2	0	0	0	0
40	0	0	1	1	0	0	0	0
40	0	0	1	1	0	0	0	0
9	w	w^2	0	0	1	1	w	w^2
9	w^2	w	0	0	1	1	w^2	w
27	0	0	0	0	α	β	$-w$	$-w^2$
27	0	0	0	0	α	β	$-w^2$	$-w$
27	0	0	0	0	β	α	$-w$	$-w^2$
27	0	0	0	0	β	α	$-w^2$	$-w$
36	w	w^2	0	0	-1	-1	0	0
36	w^2	w	0	0	-1	-1	0	0

(Cont.)

TABLE II (Cont.)

1	6_4	6_5	9_1	9_2	10_1	10_2	12_1	12_2
29,160	18	18	27	27	30	30	12	12
45	$-w$	$-w^2$	0	0	0	0	w	w^2
45	$-w^2$	$-w$	0	0	0	0	w^2	w
18	w	w^2	0	0	β	α	0	0
18	w^2	w	0	0	β	α	0	0
18	w	w^2	0	0	α	β	0	0
18	w^2	w	0	0	α	β	0	0
36	$-w$	$-w^2$	0	0	1	1	0	0
36	$-w^2$	$-w$	0	0	1	1	0	0
54	0	0	0	0	-1	-1	0	0
54	0	0	0	0	-1	-1	0	0

(Cont.)

TABLE II (Cont.)

1	15_1	15_2	15_3	15_4	30_1	30_2	30_3	30_4
29,160	30	30	30	30	30	30	30	30
1	1	1	1	1	1	1	1	1
3	α	α	β	β	α	α	β	β
3	β	β	α	α	β	β	α	α
4	-1	-1	-1	-1	-1	-1	-1	-1
5	0	0	0	0	0	0	0	0
2	$-\beta$	$-\beta$	$-\alpha$	$-\alpha$	β	β	α	α
2	$-\alpha$	$-\alpha$	$-\beta$	$-\beta$	α	α	β	β
4	-1	-1	-1	-1	1	1	1	1
6	1	1	1	1	-1	-1	-1	-1
40	0	0	0	0	0	0	0	0
40	0	0	0	0	0	0	0	0
40	0	0	0	0	0	0	0	0
40	0	0	0	0	0	0	0	0
40	0	0	0	0	0	0	0	0
40	0	0	0	0	0	0	0	0
9	$-w$	$-w^2$	$-w$	$-w^2$	w	w^2	w	w^2
9	$-w^2$	$-w$	$-w^2$	$-w$	w^2	w	w^2	w
27	$-\alpha w$	$-\alpha w^2$	$-\beta w$	$-\beta w^2$	αw	αw^2	βw	βw^2
27	$-\alpha w^2$	$-\alpha w$	$-\beta w^2$	$-\beta w$	αw^2	αw	βw^2	βw
27	$-\beta w$	$-\beta w^2$	$-\alpha w$	$-\alpha w^2$	βw	βw^2	αw	αw^2
27	$-\beta w^2$	$-\beta w$	$-\alpha w^2$	$-\alpha w$	βw^2	βw	αw^2	αw
36	w	w^2	w	w^2	$-w$	$-w^2$	$-w$	$-w^2$
36	w^2	w	w^2	w	$-w^2$	$-w$	$-w^2$	$-w$

(Cont.)

TABLE II (Cont.)

1	15_1	15_2	15_3	15_4	30_1	30_2	30_3	30_4
29,160	30	30	30	30	30	30	30	30
45	0	0	0	0	0	0	0	0
45	0	0	0	0	0	0	0	0
18	βw	βw^2	αw	αw^2	βw	βw^2	αw	αw^2
18	βw^2	βw	αw^2	αw	βw^2	βw	αw^2	αw
18	αw	αw^2	βw	βw^2	αw	αw^2	βw	βw^2
18	αw^2	αw	βw^2	βw	αw^2	αw	βw^2	βw
36	w	w^2	w	w^2	w	w^2	w	w^2
36	w^2	w	w^2	w	w^2	w	w^2	w
54	$-w$	$-w^2$	$-w$	$-w^2$	$-w$	$-w^2$	$-w$	$-w^2$
54	$-w^2$	$-w$	$-w^2$	$-w$	$-w^2$	$-w$	$-w^2$	$-w$

$$1 + w + w^2 = 0 ,$$

$$\alpha = (1 + \sqrt{5})/2 ,$$

$$\beta = (1 - \sqrt{5})/2 = 1 - \alpha .$$

REMARK. There are two extensions of the extra-special group of order 3^5 by SL(2, 5). In the other extension, 3_7 and 3_8 become elements of order 9. This is the only difference.

TABLE III

THE CHARACTER TABLE OF $5^{1+4} \cdot 2^{1+4} \cdot 5 \cdot 4$

$\lvert C_G(x)\rvert$	2×10^6	5×10^5	5000	2500	2500	2500	6250	3200	800
x	1	5_1	5_2	5_3	5_4	5_5	5_6	2_1	10
	1	1	1	1	1	1	1	1	1
	1	1	1	1	1	1	1	1	1
	1	1	1	1	1	1	1	1	1
	1	1	1	1	1	1	1	1	1
	4	4	4	4	4	4	4	4	4
	5	5	5	5	5	5	5	5	5
	5	5	5	5	5	5	5	5	5
	5	5	5	5	5	5	5	5	5
	5	5	5	5	5	5	5	5	5
	10	10	10	10	10	10	10	10	10
	10	10	10	10	10	10	10	10	10
	4	4	4	4	4	4	4	-4	-4
	4	4	4	4	4	4	4	-4	-4
	4	4	4	4	4	4	4	-4	-4
	4	4	4	4	4	4	4	-4	-4
	16	16	16	16	16	16	16	-16	-4
	80	80	-5	10	0	0	-20	0	0
	80	80	-5	10	0	0	-20	0	0
	80	80	-5	10	0	0	-20	0	0

(Cont.)

TABLE III (Cont.)

$\lvert C_G(x)\rvert$	2×10^6	5×10^5	5000	2500	2500	2500	6250	3200	800
x	1	5_1	5_2	5_3	5_4	5_5	5_6	2_1	10
	80	80	-5	10	0	0	-20	0	0
	80	80	-5	10	0	0	-20	0	0
	80	80	-5	10	0	0	-20	0	0
	80	80	-5	10	0	0	-20	0	0
	80	80	-5	10	0	0	-20	0	0
	160	160	20	5	-10	-10	10	0	0
	160	160	20	5	-10	-10	10	0	0
	160	160	20	5	-10	-10	10	0	0
	160	160	20	5	-10	-10	10	0	0
	320	320	0	-20	10	10	-5	0	0
	320	320	0	-20	10	10	-5	0	0
	32	32	-8	2	$\frac{-1+5\sqrt{5}}{2}$	$\frac{-1-5\sqrt{5}}{2}$	7	0	0
	32	32	-8	2	$\frac{-1+5\sqrt{5}}{2}$	$\frac{-1-5\sqrt{5}}{2}$	7	0	0
	32	32	-8	2	$\frac{-1+5\sqrt{5}}{2}$	$\frac{-1-5\sqrt{5}}{2}$	7	0	0
	32	32	-8	2	$\frac{-1+5\sqrt{5}}{2}$	$\frac{-1-5\sqrt{5}}{2}$	7	0	0
	128	128	-32	8	$-2+10\sqrt{5}$	$-2-10\sqrt{5}$	28	0	0
	32	32	-8	2	$\frac{-1-5\sqrt{5}}{2}$	$\frac{-1+5\sqrt{5}}{2}$	7	0	0
	32	32	-8	2	$\frac{-1-5\sqrt{5}}{2}$	$\frac{-1+5\sqrt{5}}{2}$	7	0	0
	32	32	-8	2	$\frac{-1-5\sqrt{5}}{2}$	$\frac{-1+5\sqrt{5}}{2}$	7	0	0

(Cont.)

TABLE III (Cont.)

$\lvert C_G(x)\rvert$	2×10^6	5×10^5	5000	2500	2500	2500	6250	3200	800
x	1	5_1	5_2	5_3	5_4	5_5	5_6	2_1	10
	32	32	-8	2	$\frac{-1-5\sqrt{5}}{2}$	$\frac{-1+5\sqrt{5}}{2}$	7	0	0
	128	128	-32	8	$-2-10\sqrt{5}$	$-2+10\sqrt{5}$	28	0	0
	100	-25	0	0	0	0	0	4	-1
	100	-25	0	0	0	0	0	4	-1
	100	-25	0	0	0	0	0	4	-1
	100	-25	0	0	0	0	0	4	-1
	100	-25	0	0	0	0	0	4	-1
	500	-125	0	0	0	0	0	20	-5
	500	-125	0	0	0	0	0	20	-5
	500	-125	0	0	0	0	0	20	-5
	400	-100	0	0	0	0	0	-16	4
	400	-100	0	0	0	0	0	-16	4
	400	-100	0	0	0	0	0	-16	4
	400	-100	0	0	0	0	0	-16	4
	400	-100	0	0	0	0	0	-16	4

(Cont.)

TABLE III (Cont.)

2×10^6 1	8000 2_2	160 4_1	2000 10	200 10	100 10	80 20	80 20	250 5_7	250 5_8
1	1	1	1	1	1	1	1	1	1
1	1	1	1	1	1	1	1	1	1
1	1	1	1	1	1	1	1	1	1
1	1	1	1	1	1	1	1	1	1
4	4	4	4	4	4	4	4	-1	-1
5	-3	1	-3	-3	-3	1	1	0	0
5	-3	1	-3	-3	-3	1	1	0	0
5	-3	1	-3	-3	-3	1	1	0	0
5	-3	1	-3	-3	-3	1	1	0	0
10	2	-2	2	2	2	-2	-2	0	0
10	2	-2	2	2	2	-2	-2	0	0
4	0	0	0	0	0	0	0	-1	-1
4	0	0	0	0	0	0	0	-1	-1
4	0	0	0	0	0	0	0	-1	-1
4	0	0	0	0	0	0	0	-1	-1
16	0	0	0	0	0	0	0	1	1
80	8	0	8	3	-2	0	0	0	0
80	8	0	8	3	-2	0	0	0	0
80	8	0	8	3	-2	0	0	0	0

(Cont.)

TABLE III (Cont.)

2×10^6	8000	160	2000	200	100	80	80	250	250
1	2_2	4_1	10	10	10	20	20	5_7	5_8
80	8	0	8	3	-2	0	0	0	0
80	-8	0	-8	-3	2	0	0	0	0
80	-8	0	-8	-3	2	0	0	0	0
80	-8	0	-8	-3	2	0	0	0	0
80	-8	0	-8	-3	2	0	0	0	0
160	16	0	16	-4	1	0	0	0	0
160	16	0	16	-4	1	0	0	0	0
160	-16	0	-16	4	-1	0	0	0	0
160	-16	0	-16	4	-1	0	0	0	0
320	0	0	0	0	0	0	0	0	0
320	0	0	0	0	0	0	0	0	0
32	0	0	0	0	0	0	0	2	2
32	0	0	0	0	0	0	0	2	2
32	0	0	0	0	0	0	0	2	2
32	0	0	0	0	0	0	0	2	2
128	0	0	0	0	0	0	0	-2	-2
32	0	0	0	0	0	0	0	2	2
32	0	0	0	0	0	0	0	2	2
32	0	0	0	0	0	0	0	2	2

(Cont.)

TABLE III (Cont.)

2×10^6	8000	160	2000	200	100	80	80	250	250
1	2_2	4_1	10	10	10	20	20	5_7	5_8
32	0	0	0	0	0	0	0	2	2
128	0	0	0	0	0	0	0	-2	-2
100	20	4	-5	0	0	-1	-1	0	5
100	20	4	-5	0	0	-1	-1	5	$\frac{-5-5\sqrt{5}}{2}$
100	20	4	-5	0	0	-1	-1	-5	0
100	20	4	-5	0	0	-1	-1	-5	0
100	20	4	-5	0	0	-1	-1	5	$\frac{-5+5\sqrt{5}}{2}$
500	-60	4	15	0	0	-1	-1	0	0
500	20	-4	-5	0	0	$1-2\sqrt{5}$	$1+2\sqrt{5}$	0	0
500	20	-4	-5	0	0	$1+2\sqrt{5}$	$1-2\sqrt{5}$	0	0
400	0	0	0	0	0	0	0	0	-5
400	0	0	0	0	0	0	0	-5	$\frac{5+5\sqrt{5}}{2}$
400	0	0	0	0	0	0	0	5	0
400	0	0	0	0	0	0	0	5	0
400	0	0	0	0	0	0	0	-5	$\frac{5-5\sqrt{5}}{2}$

(Cont.)

TABLE III (Cont.)

2×10^6 1	250 5_9	250 5_{10}	250 5_{11}	25 25_1	25 25_2	50 10	50 10	50 10	50 10
1	1	1	1	1	1	1	1	1	1
1	1	1	1	1	1	1	1	1	1
1	1	1	1	1	1	1	1	1	1
1	1	1	1	1	1	1	1	1	1
4	-1	-1	-1	-1	-1	-1	-1	-1	-1
5	0	0	0	0	0	0	0	0	0
5	0	0	0	0	0	0	0	0	0
5	0	0	0	0	0	0	0	0	0
5	0	0	0	0	0	0	0	0	0
10	0	0	0	0	0	0	0	0	0
10	0	0	0	0	0	0	0	0	0
4	-1	-1	-1	-1	-1	1	1	1	1
4	-1	-1	-1	-1	-1	1	1	1	1
4	-1	-1	-1	-1	-1	1	1	1	1
4	-1	-1	-1	-1	-1	1	1	1	1
16	1	1	1	1	1	-1	-1	-1	-1
80	0	0	0	0	0	0	0	0	0
80	0	0	0	0	0	0	0	0	0
80	0	0	0	0	0	0	0	0	0

(Cont.)

TABLE III (Cont.)

2×10^6	250	250	250	25	25	50	50	50	50
1	5_9	5_{10}	5_{11}	25_1	25_2	10	10	10	10
80	0	0	0	0	0	0	0	0	0
80	0	0	0	0	0	0	0	0	0
80	0	0	0	0	0	0	0	0	0
80	0	0	0	0	0	0	0	0	0
80	0	0	0	0	0	0	0	0	0
160	0	0	0	0	0	0	0	0	0
160	0	0	0	0	0	0	0	0	0
160	0	0	0	0	0	0	0	0	0
160	0	0	0	0	0	0	0	0	0
320	0	0	0	0	0	0	0	0	0
320	0	0	0	0	0	0	0	0	0
32	2	2	2	$\frac{-1+\sqrt{5}}{2}$	$\frac{-1-\sqrt{5}}{2}$	0	0	0	0
32	2	2	2	$\frac{-1+\sqrt{5}}{2}$	$\frac{-1-\sqrt{5}}{2}$	0	0	0	0
32	2	2	2	$\frac{-1+\sqrt{5}}{2}$	$\frac{-1-\sqrt{5}}{2}$	0	0	0	0
32	2	2	2	$\frac{-1+\sqrt{5}}{2}$	$\frac{-1-\sqrt{5}}{2}$	0	0	0	0
128	-2	-2	-2	$\frac{1-\sqrt{5}}{2}$	$\frac{1+\sqrt{5}}{2}$	0	0	0	0
32	2	2	2	$\frac{-1-\sqrt{5}}{2}$	$\frac{-1+\sqrt{5}}{2}$	0	0	0	0
32	2	2	2	$\frac{-1-\sqrt{5}}{2}$	$\frac{-1+\sqrt{5}}{2}$	0	0	0	0
32	2	2	2	$\frac{-1-\sqrt{5}}{2}$	$\frac{-1+\sqrt{5}}{2}$	0	0	0	0

(Cont.)

TABLE III (Cont.)

2×10^6	250	250	250	25	25	50	50	50	50
1	5_9	5_{10}	5_{11}	25_1	25_2	10	10	10	10
32	2	2	2	$\frac{-1-\sqrt{5}}{2}$	$\frac{-1+\sqrt{5}}{2}$	0	0	0	0
128	-2	-2	-2	$\frac{1+\sqrt{5}}{2}$	$\frac{1-\sqrt{5}}{2}$	0	0	0	0
100	5	-5	-5	0	0	4	-1	-1	-1
100	$\frac{-5+5\sqrt{5}}{2}$	0	0	0	0	-1	$\frac{3-\sqrt{5}}{2}$	$\frac{3+\sqrt{5}}{2}$	$-1-\sqrt{5}$
100	0	$\frac{5-5\sqrt{5}}{2}$	$\frac{5+5\sqrt{5}}{2}$	0	0	-1	$-1-\sqrt{5}$	$-1+\sqrt{5}$	$\frac{3+\sqrt{5}}{2}$
100	0	$\frac{5+5\sqrt{5}}{2}$	$\frac{5-5\sqrt{5}}{2}$	0	0	-1	$-1+\sqrt{5}$	$-1-\sqrt{5}$	$\frac{3-\sqrt{5}}{2}$
100	$\frac{-5-5\sqrt{5}}{2}$	0	0	0	0	-1	$\frac{3+\sqrt{5}}{2}$	$\frac{3-\sqrt{5}}{2}$	$-1+\sqrt{5}$
500	0	0	0	0	0	0	0	0	0
500	0	0	0	0	0	0	0	0	0
500	0	0	0	0	0	0	0	0	0
400	-5	5	5	0	0	4	-1	-1	-1
400	$\frac{5-5\sqrt{5}}{2}$	0	0	0	0	-1	$\frac{3-\sqrt{5}}{2}$	$\frac{3+\sqrt{5}}{2}$	$-1-\sqrt{5}$
400	0	$\frac{-5+5\sqrt{5}}{2}$	$\frac{-5-5\sqrt{5}}{2}$	0	0	-1	$-1-\sqrt{5}$	$-1+\sqrt{5}$	$\frac{3+\sqrt{5}}{2}$
400	0	$\frac{-5-5\sqrt{5}}{2}$	$\frac{-5+5\sqrt{5}}{2}$	0	0	-1	$-1+\sqrt{5}$	$-1-\sqrt{5}$	$\frac{3-\sqrt{5}}{2}$
400	$\frac{5+5\sqrt{5}}{2}$	0	0	0	0	-1	$\frac{3+\sqrt{5}}{2}$	$\frac{3-\sqrt{5}}{2}$	$-1+\sqrt{5}$

(Cont.)

TABLE III (Cont.)

2×10^6	50	400	160	160	8	100	100	100	100
1	10	2_3	4_2	4_2^{-1}	8_1	10	10	10	10
1	1	1	1	1	1	1	1	1	1
1	1	1	1	1	1	1	1	1	1
1	1	-1	-1	-1	-1	-1	-1	-1	-1
1	1	-1	-1	-1	-1	-1	-1	-1	-1
4	-1	0	0	0	0	0	0	0	0
5	0	1	1	1	-1	1	1	1	1
5	0	1	1	1	-1	1	1	1	1
5	0	-1	-1	-1	1	-1	-1	-1	-1
5	0	-1	-1	-1	1	-1	-1	-1	-1
10	0	2	-2	-2	0	2	2	2	2
10	0	-2	2	2	0	-2	-2	-2	-2
4	1	0	2i	-2i	0	0	0	0	0
4	1	0	2i	-2i	0	0	0	0	0
4	1	0	-2i	2i	0	0	0	0	0
4	1	0	-2i	2i	0	0	0	0	0
16	-1	0	0	0	0	0	0	0	0
80	0	0	4	4	0	0	0	0	0
80	0	0	4	4	0	0	0	0	0
80	0	0	-4	-4	0	0	0	0	0

(Cont.)

TABLE III (Cont.)

2×10^6	50	400	160	160	8	100	100	100	100
1	10	2_3	4_2	4_2^{-1}	8_1	10	10	10	10
80	0	0	-4	-4	0	0	0	0	0
80	0	0	4i	-4i	0	0	0	0	0
80	0	0	-4i	4i	0	0	0	0	0
80	0	0	4i	-4i	0	0	0	0	0
80	0	0	-4i	4i	0	0	0	0	0
160	0	8	0	0	0	3	3	-2	-2
160	0	-8	0	0	0	-3	-3	2	2
160	0	0	0	0	0	$\sqrt{5}$	$-\sqrt{5}$	$-2\sqrt{5}$	$2\sqrt{5}$
160	0	0	0	0	0	$-\sqrt{5}$	$\sqrt{5}$	$2\sqrt{5}$	$-2\sqrt{5}$
320	0	8	0	0	0	-2	-2	-2	-2
320	0	-8	0	0	0	2	2	2	2
32	0	4	0	0	0	$-1+\sqrt{5}$	$-1-\sqrt{5}$	$\frac{3+\sqrt{5}}{2}$	$\frac{3-\sqrt{5}}{2}$
32	0	4	0	0	0	$-1+\sqrt{5}$	$-1-\sqrt{5}$	$\frac{3+\sqrt{5}}{2}$	$\frac{3-\sqrt{5}}{2}$
32	0	-4	0	0	0	$1-\sqrt{5}$	$1+\sqrt{5}$	$\frac{-3-\sqrt{5}}{2}$	$\frac{-3+\sqrt{5}}{2}$
32	0	-4	0	0	0	$1-\sqrt{5}$	$1+\sqrt{5}$	$\frac{-3-\sqrt{5}}{2}$	$\frac{-3+\sqrt{5}}{2}$
128	0	0	0	0	0	0	0	0	0
32	0	4	0	0	0	$-1-\sqrt{5}$	$-1+\sqrt{5}$	$\frac{3-\sqrt{5}}{2}$	$\frac{3+\sqrt{5}}{2}$
32	0	4	0	0	0	$-1-\sqrt{5}$	$-1+\sqrt{5}$	$\frac{3-\sqrt{5}}{2}$	$\frac{3+\sqrt{5}}{2}$
32	0	-4	0	0	0	$1+\sqrt{5}$	$1-\sqrt{5}$	$\frac{-3+\sqrt{5}}{2}$	$\frac{-3-\sqrt{5}}{2}$

(Cont.)

TABLE III (Cont.)

2×10^6	50	400	160	160	8	100	100	100	100
1	10	2_3	4_2	4_2^{-1}	8_1	10	10	10	10
32	0	-4	0	0	0	$1+\sqrt{5}$	$1-\sqrt{5}$	$\frac{-3+\sqrt{5}}{2}$	$\frac{-3-\sqrt{5}}{2}$
128	0	0	0	0	0	0	0	0	0
100	-1	0	0	0	0	0	0	0	0
100	$-1+\sqrt{5}$	0	0	0	0	0	0	0	0
100	$\frac{3-\sqrt{5}}{2}$	0	0	0	0	0	0	0	0
100	$\frac{3+\sqrt{5}}{2}$	0	0	0	0	0	0	0	0
100	$-1-\sqrt{5}$	0	0	0	0	0	0	0	0
500	0	0	0	0	0	0	0	0	0
500	0	0	0	0	0	0	0	0	0
500	0	0	0	0	0	0	0	0	0
400	-1	0	0	0	0	0	0	0	0
400	$-1+\sqrt{5}$	0	0	0	0	0	0	0	0
400	$\frac{3-\sqrt{5}}{2}$	0	0	0	0	0	0	0	0
400	$\frac{3+\sqrt{5}}{2}$	0	0	0	0	0	0	0	0
400	$-1-\sqrt{5}$	0	0	0	0	0	0	0	0

(Cont.)

TABLE III (Cont.)

2×10^6	50	40	40	40	40	80	80	16	16
1	10	20	20	4_3	4_3^{-1}	8_2	8_2^{-1}	8_3	8_3^{-1}
1	1	1	1	1	1	1	1	1	1
1	1	1	1	-1	-1	-1	-1	-1	-1
1	-1	-1	-1	i	-i	i	-i	i	-i
1	-1	-1	-1	-i	i	-i	i	-i	i
4	0	0	0	0	0	0	0	0	0
5	1	1	1	1	1	-1	-1	-1	-1
5	1	1	1	-1	-1	1	1	1	1
5	-1	-1	-1	i	-i	-i	i	i	-i
5	-1	-1	-1	-i	i	i	-i	-i	i
10	2	-2	-2	0	0	0	0	0	0
10	-2	2	2	0	0	0	0	0	0
4	0	2i	-2i	0	0	1-i	1+i	-1+i	-1-i
4	0	2i	-2i	0	0	-1+i	-1-i	1-i	1+i
4	0	-2i	2i	0	0	1+i	1-i	-1-i	-1+i
4	0	-2i	2i	0	0	-1-i	-1+i	1+i	1-i
16	0	0	0	0	0	0	0	0	0
80	0	-1	-1	0	0	4	4	0	0
80	0	-1	-1	0	0	-4	-4	0	0
80	0	1	1	0	0	4i	-4i	0	0

(Cont.)

TABLE III (Cont.)

2×10^6 1	50 10	40 20	40 20	40 4_3	40 4_3^{-1}	80 8_2	80 8_2^{-1}	16 8_3	16 8_3^{-1}
80	0	1	1	0	0	-4i	4i	0	0
80	0	-i	i	0	0	0	0	0	0
80	0	i	-i	0	0	0	0	0	0
80	0	-i	i	0	0	0	0	0	0
80	0	i	-i	0	0	0	0	0	0
160	-2	0	0	0	0	0	0	0	0
160	2	0	0	0	0	0	0	0	0
160	0	0	0	0	0	0	0	0	0
160	0	0	0	0	0	0	0	0	0
320	3	0	0	0	0	0	0	0	0
320	-3	0	0	0	0	0	0	0	0
32	-1	0	0	2	2	0	0	0	0
32	-1	0	0	-2	-2	0	0	0	0
32	1	0	0	2i	-2i	0	0	0	0
32	1	0	0	-2i	2i	0	0	0	0
128	0	0	0	0	0	0	0	0	0
32	-1	0	0	2	2	0	0	0	0
32	-1	0	0	-2	-2	0	0	0	0
32	1	0	0	2i	-2i	0	0	0	0

(Cont.)

TABLE III (Cont.)

2×10^6 1	50 10	40 20	40 20	40 4_3	40 4_3^{-1}	80 8_2	80 8_2^{-1}	16 8_3	16 8_3^{-1}
32	1	0	0	-2i	2i	0	0	0	0
128	0	0	0	0	0	0	0	0	0
100	0	0	0	0	0	0	0	0	0
100	0	0	0	0	0	0	0	0	0
100	0	0	0	0	0	0	0	0	0
100	0	0	0	0	0	0	0	0	0
100	0	0	0	0	0	0	0	0	0
500	0	0	0	0	0	0	0	0	0
500	0	0	0	0	0	0	0	0	0
500	0	0	0	0	0	0	0	0	0
400	0	0	0	0	0	0	0	0	0
400	0	0	0	0	0	0	0	0	0
400	0	0	0	0	0	0	0	0	0
400	0	0	0	0	0	0	0	0	0
400	0	0	0	0	0	0	0	0	0

(Cont.)

TABLE III (Cont.)

2×10^6	20	20	20	20	40	40	40	40
1	20	20	20	20	40	40	40	40
1	1	1	1	1	1	1	1	1
1	-1	-1	-1	-1	-1	-1	-1	-1
1	i	i	-i	-i	i	-i	i	-i
1	-i	-i	i	i	-i	i	-i	i
4	0	0	0	0	0	0	0	0
5	1	1	1	1	-1	-1	-1	-1
5	-1	-1	-1	-1	1	1	1	1
5	i	i	-i	-i	-i	i	-i	i
5	-i	-i	i	i	i	-i	i	-i
10	0	0	0	0	0	0	0	0
10	0	0	0	0	0	0	0	0
4	0	0	0	0	1-i	1+i	1-i	1+i
4	0	0	0	0	-1+i	-1-i	-1+i	-1-i
4	0	0	0	0	1+i	1-i	1+i	1-i
4	0	0	0	0	-1-i	-1+i	-1-i	-1+i
16	0	0	0	0	0	0	0	0
80	0	0	0	0	-1	-1	-1	-1
80	0	0	0	0	1	1	1	1
80	0	0	0	0	i	-i	i	-i

(Cont.)

TABLE III (Cont.)

2×10^6	20	20	20	20	40	40	40	40
1	20	20	20	20	40	40	40	40
80	0	0	0	0	$-i$	i	$-i$	i
80	0	0	0	0	$\sqrt{5}\rho$	$\sqrt{5}\rho^3$	$\sqrt{5}\rho$	$-\sqrt{5}\rho^3$
80	0	0	0	0	$\sqrt{5}\rho^3$	$\sqrt{5}\rho$	$-\sqrt{5}\rho^3$	$-\sqrt{5}\rho$
80	0	0	0	0	$-\sqrt{5}\rho$	$-\sqrt{5}\rho^3$	$\sqrt{5}\rho$	$\sqrt{5}\rho^3$
80	0	0	0	0	$-\sqrt{5}\rho^3$	$-\sqrt{5}\rho$	$\sqrt{5}\rho^3$	$\sqrt{5}\rho$
160	0	0	0	0	0	0	0	0
160	0	0	0	0	0	0	0	0
160	0	0	0	0	0	0	0	0
160	0	0	0	0	0	0	0	0
320	0	0	0	0	0	0	0	0
320	0	0	0	0	0	0	0	0
32	$\frac{-1+\sqrt{5}}{2}$	$\frac{1-\sqrt{5}}{2}$	$\frac{-1+\sqrt{5}}{2}$	$\frac{-1+\sqrt{5}}{2}$	0	0	0	0
32	$\frac{1-\sqrt{5}}{2}$	$\frac{1+\sqrt{5}}{2}$	$\frac{1-\sqrt{5}}{2}$	$\frac{1-\sqrt{5}}{2}$	0	0	0	0
32	$\frac{-1+\sqrt{5}}{2}i$	$\frac{1-\sqrt{5}}{2}i$	$\frac{1-\sqrt{5}}{2}i$	$\frac{-1+\sqrt{5}}{2}i$	0	0	0	0
32	$\frac{1-\sqrt{5}}{2}i$	$\frac{1+\sqrt{5}}{2}i$	$\frac{-1+\sqrt{5}}{2}i$	$\frac{-1-\sqrt{5}}{2}i$	0	0	0	0
128	0	0	0	0	0	0	0	0
32	$\frac{-1-\sqrt{5}}{2}$	$\frac{1+\sqrt{5}}{2}$	$\frac{-1-\sqrt{5}}{2}$	$\frac{-1-\sqrt{5}}{2}$	0	0	0	0
32	$\frac{1+\sqrt{5}}{2}$	$\frac{1-\sqrt{5}}{2}$	$\frac{1+\sqrt{5}}{2}$	$\frac{-1-\sqrt{5}}{2}$	0	0	0	0
32	$\frac{-1-\sqrt{5}}{2}i$	$\frac{-1+\sqrt{5}}{2}i$	$\frac{1+\sqrt{5}}{2}i$	$\frac{-1-\sqrt{5}}{2}i$	0	0	0	0

(Cont.)

TABLE III (Cont.)

2×10^6	20	20	20	20	40	40	40	40
1	20	20	20	20	40	40	40	40
32	$\frac{1+\sqrt{5}}{2}i$	$\frac{1-\sqrt{5}}{2}i$	$\frac{-1-\sqrt{5}}{2}i$	$\frac{1+\sqrt{5}}{2}i$	0	0	0	0
128	0	0	0	0	0	0	0	0
100	0	0	0	0	0	0	0	0
100	0	0	0	0	0	0	0	0
100	0	0	0	0	0	0	0	0
100	0	0	0	0	0	0	0	0
100	0	0	0	0	0	0	0	0
500	0	0	0	0	0	0	0	0
500	0	0	0	0	0	0	0	0
500	0	0	0	0	0	0	0	0
400	0	0	0	0	0	0	0	0
400	0	0	0	0	0	0	0	0
400	0	0	0	0	0	0	0	0
400	0	0	0	0	0	0	0	0
400	0	0	0	0	0	0	0	0

$$i = \sqrt{-1}$$

ρ = the primitive 8th root of unity

133 = 1* + 32* + 100

TABLE IV

THE CHARACTER TABLE OF $2^{1+8}A_5 \wr Z_2$

$\lvert C_G(x) \rvert$	3686400	3686400	24,516	30,720	15,360	576	360	800
x	1	2_1	2_2	2_2	4_1	3_1	3_2	5_1
	1	1	1	1	1	1	1	1
	1	1	1	1	1	1	1	1
	16	16	16	16	16	1	4	1
	16	16	16	16	16	1	4	1
	25	25	25	25	25	1	-5	0
	25	25	25	25	25	1	-5	0
	9	9	9	9	9	0	0	-1
	9	9	9	9	9	0	0	-1
	9	9	9	9	9	0	0	-1
	9	9	9	9	9	0	0	-1
	8	8	8	8	8	2	5	-2
	10	10	10	10	10	-2	4	0
	6	6	6	6	6	0	3	1
	6	6	6	6	6	0	3	1
	40	40	40	40	40	-2	1	0
	24	24	24	24	24	0	3	-1
	24	24	24	24	24	0	3	-1
	30	30	30	30	30	0	-3	0
	30	30	30	30	30	0	-3	0

(Cont.)

TABLE IV (Cont.)

$\|C_G(x)\|$	3686400	3686400	24,516	30,720	15,360	576	360	800
x	1	2_1	2_2	2_2	4_1	3_1	3_2	5_1
	18	18	18	18	18	0	0	3
	60	60	-4	12	-4	3	0	5
	60	60	-4	12	-4	3	0	5
	240	240	-16	48	-16	3	0	-5
	240	240	-16	48	-16	3	0	-5
	300	300	-20	60	-20	-3	0	0
	300	300	-20	60	-20	-3	0	0
	360	360	-24	72	-24	0	0	5
	120	120	-8	-8	8	6	0	10
	480	480	-32	-32	32	6	0	-10
	600	600	-40	-40	40	-6	0	0
	360	360	-24	-24	24	0	0	5
	360	360	-24	-24	24	0	0	5
	75	75	11	-5	-5	6	0	0
	75	75	11	-5	-5	6	0	0
	75	75	11	-5	-5	-3	0	0
	75	75	11	-5	-5	-3	0	0
	75	75	11	-5	-5	-3	0	0
	75	75	11	-5	-5	-3	0	0

(Cont.)

TABLE IV (Cont.)

$\|C_G(x)\|$	3686400	3686400	24,516	30,720	15,360	576	360	800
x	1	2_1	2_2	2_2	4_1	3_1	3_2	5_1
	225	225	33	-15	-15	0	0	0
	225	225	33	-15	-15	0	0	0
	450	450	66	-30	-30	0	0	0
	450	450	66	-30	-30	0	0	0
	384	-384	0	0	0	0	-6	4
	384	-384	0	0	0	0	-6	4
	768	-768	0	0	0	0	6	8
	576	-576	0	0	0	0	0	-4
	576	-576	0	0	0	0	0	-4
	64	-64	0	0	0	4	-2	4
	64	-64	0	0	0	4	-2	4
	64	-64	0	0	0	4	-2	4
	64	-64	0	0	0	4	-2	4
	128	-128	0	0	0	8	-4	-12
	256	-256	0	0	0	4	4	-4
	256	-256	0	0	0	4	4	-4
	256	-256	0	0	0	-8	-2	-4
	256	-256	0	0	0	-8	-2	-4

(Cont.)

TABLE IV (Cont.)

3686400 1	100 5_2	100 5_3	600 5_4	600 5_5	30 15_1	30 15_2	576 6_1	360 6_2
1	1	1	1	1	1	1	1	1
1	1	1	1	1	1	1	1	1
16	1	1	-4	-4	-1	-1	1	4
16	1	1	-4	-4	-1	-1	1	4
25	0	0	0	0	0	0	1	-5
25	0	0	0	0	0	0	1	-5
9	$\frac{3-\sqrt{5}}{2}$	$\frac{3+\sqrt{5}}{2}$	$\frac{3-3\sqrt{5}}{2}$	$\frac{3+3\sqrt{5}}{2}$	0	0	0	0
9	$\frac{3+\sqrt{5}}{2}$	$\frac{3-\sqrt{5}}{2}$	$\frac{3+3\sqrt{5}}{2}$	$\frac{3-3\sqrt{5}}{2}$	0	0	0	0
9	$\frac{3+\sqrt{5}}{2}$	$\frac{3-\sqrt{5}}{2}$	$\frac{3+3\sqrt{5}}{2}$	$\frac{3-3\sqrt{5}}{2}$	0	0	0	0
9	$\frac{3-\sqrt{5}}{2}$	$\frac{3+\sqrt{5}}{2}$	$\frac{3-3\sqrt{5}}{2}$	$\frac{3+3\sqrt{5}}{2}$	0	0	0	0
8	-2	-2	3	3	0	0	2	5
10	0	0	5	5	-1	-1	-2	4
6	$1-\sqrt{5}$	$1+\sqrt{5}$	$\frac{7-\sqrt{5}}{2}$	$\frac{7+\sqrt{5}}{2}$	$\frac{1-\sqrt{5}}{2}$	$\frac{1+\sqrt{5}}{2}$	0	3
6	$1+\sqrt{5}$	$1-\sqrt{5}$	$\frac{7+\sqrt{5}}{2}$	$\frac{7-\sqrt{5}}{2}$	$\frac{1+\sqrt{5}}{2}$	$\frac{1-\sqrt{5}}{2}$	0	3
40	0	0	-5	-5	1	1	-2	1
24	$-1+\sqrt{5}$	$-1-\sqrt{5}$	$-1-2\sqrt{5}$	$-1+2\sqrt{5}$	$\frac{1-\sqrt{5}}{2}$	$\frac{1+\sqrt{5}}{2}$	0	3
24	$-1-\sqrt{5}$	$-1+\sqrt{5}$	$-1+2\sqrt{5}$	$-1-2\sqrt{5}$	$\frac{1+\sqrt{5}}{2}$	$\frac{1-\sqrt{5}}{2}$	0	3
30	0	0	$\frac{5-5\sqrt{5}}{2}$	$\frac{5+5\sqrt{5}}{2}$	$\frac{-1+\sqrt{5}}{2}$	$\frac{-1-\sqrt{5}}{2}$	0	-3
30	0	0	$\frac{5+5\sqrt{5}}{2}$	$\frac{5-5\sqrt{5}}{2}$	$\frac{-1-\sqrt{5}}{2}$	$\frac{-1+\sqrt{5}}{2}$	0	-3

(Cont.)

TABLE IV (Cont.)

3686400 1	100 5_2	100 5_3	600 5_4	600 5_5	30 15_1	30 15_2	576 6_1	360 6_2
18	-2	-2	3	3	0	0	0	0
60	0	0	0	0	0	0	3	0
60	0	0	0	0	0	0	3	0
240	0	0	0	0	0	0	3	0
240	0	0	0	0	0	0	3	0
300	0	0	0	0	0	0	-3	0
300	0	0	0	0	0	0	-3	0
360	0	0	0	0	0	0	0	0
120	0	0	0	0	0	0	6	0
480	0	0	0	0	0	0	6	0
600	0	0	0	0	0	0	-6	0
360	0	0	0	0	0	0	0	0
360	0	0	0	0	0	0	0	0
75	0	0	0	0	0	0	6	0
75	0	0	0	0	0	0	6	0
75	0	0	0	0	0	0	-3	0
75	0	0	0	0	0	0	-3	0
75	0	0	0	0	0	0	-3	0
75	0	0	0	0	0	0	-3	0

(Cont.)

TABLE IV (Cont.)

3686400	100	100	600	600	30	30	576	360
1	5_2	5_3	5_4	5_5	15_1	15_2	6_1	6_2
225	0	0	0	0	0	0	0	0
225	0	0	0	0	0	0	0	0
450	0	0	0	0	0	0	0	0
450	0	0	0	0	0	0	0	0
384	$-1+\sqrt{5}$	$-1-\sqrt{5}$	$-1+3\sqrt{5}$	$-1-3\sqrt{5}$	-1	-1	0	6
384	$-1-\sqrt{5}$	$-1+\sqrt{5}$	$-1-3\sqrt{5}$	$-1+3\sqrt{5}$	-1	-1	0	6
768	-2	-2	-2	-2	1	1	0	-6
576	1	1	6	6	0	0	0	0
576	1	1	6	6	0	0	0	0
64	$\frac{3+\sqrt{5}}{2}$	$\frac{3-\sqrt{5}}{2}$	$-1-\sqrt{5}$	$-1+\sqrt{5}$	$\frac{1+\sqrt{5}}{2}$	$\frac{1-\sqrt{5}}{2}$	-4	2
64	$\frac{3-\sqrt{5}}{2}$	$\frac{3+\sqrt{5}}{2}$	$-1+\sqrt{5}$	$-1-\sqrt{5}$	$\frac{1-\sqrt{5}}{2}$	$\frac{1+\sqrt{5}}{2}$	-4	2
64	$\frac{3+\sqrt{5}}{2}$	$\frac{3-\sqrt{5}}{2}$	$-1-\sqrt{5}$	$-1+\sqrt{5}$	$\frac{1+\sqrt{5}}{2}$	$\frac{1-\sqrt{5}}{2}$	-4	2
64	$\frac{3-\sqrt{5}}{2}$	$\frac{3+\sqrt{5}}{2}$	$-1+\sqrt{5}$	$-1-\sqrt{5}$	$\frac{1-\sqrt{5}}{2}$	$\frac{1+\sqrt{5}}{2}$	-4	2
128	-2	-2	-2	-2	1	1	-8	4
256	1	1	-4	-4	-1	-1	-4	-4
256	1	1	-4	-4	-1	-1	-4	-4
256	$1+\sqrt{5}$	$1-\sqrt{5}$	$-4+2\sqrt{5}$	$-4-2\sqrt{5}$	$\frac{1+\sqrt{5}}{2}$	$\frac{1-\sqrt{5}}{2}$	8	2
256	$1-\sqrt{5}$	$1+\sqrt{5}$	$-4-2\sqrt{5}$	$-4+2\sqrt{5}$	$\frac{1-\sqrt{5}}{2}$	$\frac{1+\sqrt{5}}{2}$	8	2

(Cont.)

TABLE IV (Cont.)

3686400	800	100	100	600	600	30	30
1	10_1	10_2	10_3	10_4	10_5	30_1	30_2
1	1	1	1	1	1	1	1
1	1	1	1	1	1	1	1
16	1	1	1	-4	-4	-1	-1
16	1	1	1	-4	-4	-1	-1
25	0	0	0	0	0	0	0
25	0	0	0	0	0	0	0
9	-1	$\frac{3-\sqrt{5}}{2}$	$\frac{3+\sqrt{5}}{2}$	$\frac{3-3\sqrt{5}}{2}$	$\frac{3+3\sqrt{5}}{2}$	0	0
9	-1	$\frac{3+\sqrt{5}}{2}$	$\frac{3-\sqrt{5}}{2}$	$\frac{3+3\sqrt{5}}{2}$	$\frac{3-3\sqrt{5}}{2}$	0	0
9	-1	$\frac{3+\sqrt{5}}{2}$	$\frac{3-\sqrt{5}}{2}$	$\frac{3+3\sqrt{5}}{2}$	$\frac{3-3\sqrt{5}}{2}$	0	0
9	-1	$\frac{3-\sqrt{5}}{2}$	$\frac{3+\sqrt{5}}{2}$	$\frac{3-3\sqrt{5}}{2}$	$\frac{3+3\sqrt{5}}{2}$	0	0
8	-2	-2	-2	3	3	0	0
10	0	0	0	5	5	-1	-1
6	1	$1-\sqrt{5}$	$1+\sqrt{5}$	$\frac{7-\sqrt{5}}{2}$	$\frac{7+\sqrt{5}}{2}$	$\frac{1-\sqrt{5}}{2}$	$\frac{1+\sqrt{5}}{2}$
6	1	$1+\sqrt{5}$	$1-\sqrt{5}$	$\frac{7+\sqrt{5}}{2}$	$\frac{7-\sqrt{5}}{2}$	$\frac{1+\sqrt{5}}{2}$	$\frac{1-\sqrt{5}}{2}$
40	0	0	0	-5	-5	1	1
24	-1	$-1+\sqrt{5}$	$-1-\sqrt{5}$	$-1-2\sqrt{5}$	$-1+2\sqrt{5}$	$\frac{1-\sqrt{5}}{2}$	$\frac{1+\sqrt{5}}{2}$
24	-1	$-1-\sqrt{5}$	$-1+\sqrt{5}$	$-1+2\sqrt{5}$	$-1-2\sqrt{5}$	$\frac{1+\sqrt{5}}{2}$	$\frac{1-\sqrt{5}}{2}$
30	0	0	0	$\frac{5-5\sqrt{5}}{2}$	$\frac{5+5\sqrt{5}}{2}$	$\frac{-1+\sqrt{5}}{2}$	$\frac{-1-\sqrt{5}}{2}$
30	0	0	0	$\frac{5+5\sqrt{5}}{2}$	$\frac{5-5\sqrt{5}}{2}$	$\frac{-1-\sqrt{5}}{2}$	$\frac{-1+\sqrt{5}}{2}$

(Cont.)

TABLE IV (Cont.)

3686400 1	800 10_1	100 10_2	100 10_3	600 10_4	600 10_5	30 30_1	30 30_2
18	3	-2	-2	3	3	0	0
60	5	0	0	0	0	0	0
60	5	0	0	0	0	0	0
240	-5	0	0	0	0	0	0
240	-5	0	0	0	0	0	0
300	0	0	0	0	0	0	0
300	0	0	0	0	0	0	0
360	5	0	0	0	0	0	0
120	10	0	0	0	0	0	0
480	-10	0	0	0	0	0	0
600	0	0	0	0	0	0	0
360	5	0	0	0	0	0	0
360	5	0	0	0	0	0	0
75	0	0	0	0	0	0	0
75	0	0	0	0	0	0	0
75	0	0	0	0	0	0	0
75	0	0	0	0	0	0	0
75	0	0	0	0	0	0	0
75	0	0	0	0	0	0	0

(Cont.)

TABLE IV (Cont.)

3686400	800	100	100	600	600	30	30
1	10_1	10_2	10_3	10_4	10_5	30_1	30_2
225	0	0	0	0	0	0	0
225	0	0	0	0	0	0	0
450	0	0	0	0	0	0	0
450	0	0	0	0	0	0	0
384	-4	$1-\sqrt{5}$	$1+\sqrt{5}$	$1-3\sqrt{5}$	$1+3\sqrt{5}$	1	1
384	-4	$1+\sqrt{5}$	$1-\sqrt{5}$	$1+3\sqrt{5}$	$1-3\sqrt{5}$	1	1
768	-8	2	2	2	2	-1	-1
576	4	-1	-1	-6	-6	0	0
576	4	-1	-1	-6	-6	0	0
64	-4	$\frac{-3-\sqrt{5}}{2}$	$\frac{-3+\sqrt{5}}{2}$	$1+\sqrt{5}$	$1-\sqrt{5}$	$\frac{-1-\sqrt{5}}{2}$	$\frac{-1+\sqrt{5}}{2}$
64	-4	$\frac{-3+\sqrt{5}}{2}$	$\frac{-3-\sqrt{5}}{2}$	$1-\sqrt{5}$	$1+\sqrt{5}$	$\frac{-1+\sqrt{5}}{2}$	$\frac{-1-\sqrt{5}}{2}$
64	-4	$\frac{-3-\sqrt{5}}{2}$	$\frac{-3+\sqrt{5}}{2}$	$1+\sqrt{5}$	$1-\sqrt{5}$	$\frac{-1-\sqrt{5}}{2}$	$\frac{-1+\sqrt{5}}{2}$
64	-4	$\frac{-3+\sqrt{5}}{2}$	$\frac{-3-\sqrt{5}}{2}$	$1-\sqrt{5}$	$1+\sqrt{5}$	$\frac{-1+\sqrt{5}}{2}$	$\frac{-1-\sqrt{5}}{2}$
128	12	2	2	2	2	-1	-1
256	4	-1	-1	4	4	1	1
256	4	-1	-1	4	4	1	1
256	4	$-1-\sqrt{5}$	$-1+\sqrt{5}$	$4-2\sqrt{5}$	$4+2\sqrt{5}$	$\frac{-1-\sqrt{5}}{2}$	$\frac{-1+\sqrt{5}}{2}$
256	4	$-1+\sqrt{5}$	$-1-\sqrt{5}$	$4+2\sqrt{5}$	$4-2\sqrt{5}$	$\frac{-1+\sqrt{5}}{2}$	$\frac{-1-\sqrt{5}}{2}$

(Cont.)

TABLE IV (Cont.)

3686400 1	96 6_3	96 6_4	96 6_5	80 10_6	48 12_1	80 20_1	80 20_2	512 2_4	3840 4_2
1	1	1	1	1	1	1	1	1	1
1	1	1	1	1	1	1	1	1	1
16	1	1	1	1	1	1	1	0	0
16	1	1	1	1	1	1	1	0	0
25	1	1	1	0	1	0	0	1	5
25	1	1	1	0	1	0	0	1	5
9	0	0	0	-1	0	-1	-1	1	-3
9	0	0	0	-1	0	-1	-1	1	-3
9	0	0	0	-1	0	-1	-1	1	-3
9	0	0	0	-1	0	-1	-1	1	-3
8	2	2	2	-2	2	-2	-2	0	4
10	-2	-2	-2	0	-2	0	0	2	6
6	0	0	0	1	0	1	1	-2	2
6	0	0	0	1	0	1	1	-2	2
40	-2	-2	-2	0	-2	0	0	0	4
24	0	0	0	-1	0	-1	-1	0	-4
24	0	0	0	-1	0	-1	-1	0	-4
30	0	0	0	0	0	0	0	-2	-2
30	0	0	0	0	0	0	0	-2	-2

(Cont.)

TABLE IV (Cont.)

3686400	96	96	96	80	48	80	80	512	3840
1	6_3	6_4	6_5	10_6	12_1	20_1	20_2	2_4	4_2
18	0	0	0	3	0	3	3	2	-6
60	-1	-1	3	-3	-1	1	1	4	0
60	-1	-1	3	-3	-1	1	1	4	0
240	-1	-1	3	3	-1	-1	-1	0	0
240	-1	-1	3	3	-1	-1	-1	0	0
300	1	1	-3	0	1	0	0	4	0
300	1	1	-3	0	1	0	0	4	0
360	0	0	0	-3	0	1	1	-8	0
120	-2	-2	-2	2	2	-2	-2	-8	0
480	-2	-2	-2	-2	2	2	2	0	0
600	2	2	2	0	-2	0	0	-8	0
360	0	0	0	1	0	$-1+2\sqrt{5}$	$-1-2\sqrt{5}$	8	0
360	0	0	0	1	0	$-1-2\sqrt{5}$	$-1+2\sqrt{5}$	8	0
75	2	2	-2	0	-2	0	0	-1	15
75	2	2	-2	0	-2	0	0	-1	15
75	$-1+2\sqrt{-3}$	$-1-2\sqrt{-3}$	1	0	1	0	0	-1	15
75	$-1-2\sqrt{-3}$	$-1+2\sqrt{-3}$	1	0	1	0	0	-1	15
75	$-1+2\sqrt{-3}$	$-1-2\sqrt{-3}$	1	0	1	0	0	-1	15
75	$-1-2\sqrt{-3}$	$-1+2\sqrt{-3}$	1	0	1	0	0	-1	15

(Cont.)

TABLE IV (Cont.)

3686400 1	96 6_3	96 6_4	96 6_5	80 10_6	48 12_1	80 20_1	80 20_2	512 2_4	3840 4_2
225	0	0	0	0	0	0	0	5	-15
225	0	0	0	0	0	0	0	5	-15
450	0	0	0	0	0	0	0	-6	-30
450	0	0	0	0	0	0	0	2	30
384	0	0	0	0	0	0	0	0	0
384	0	0	0	0	0	0	0	0	0
768	0	0	0	0	0	0	0	0	0
576	0	0	0	0	0	0	0	0	0
576	0	0	0	0	0	0	0	0	0
64	0	0	0	0	0	0	0	0	0
64	0	0	0	0	0	0	0	0	0
64	0	0	0	0	0	0	0	0	0
64	0	0	0	0	0	0	0	0	0
128	0	0	0	0	0	0	0	0	0
256	0	0	0	0	0	0	0	0	0
256	0	0	0	0	0	0	0	0	0
256	0	0	0	0	0	0	0	0	0
256	0	0	0	0	0	0	0	0	0

(Cont.)

TABLE IV (Cont.)

3686400	256	128	256	512	64	12	20	20
1	4_3	4_4	4_5	4_6	8_1	12_2	20_3	20_4
1	1	1	1	1	1	1	1	1
1	1	1	1	1	1	1	1	1
16	0	0	0	0	0	0	0	0
16	0	0	0	0	0	0	0	0
25	5	1	1	1	1	-1	0	0
25	5	1	1	1	1	-1	0	0
9	-3	1	1	1	1	0	$\frac{-1+\sqrt{5}}{2}$	$\frac{-1-\sqrt{5}}{2}$
9	-3	1	1	1	1	0	$\frac{-1-\sqrt{5}}{2}$	$\frac{-1+\sqrt{5}}{2}$
9	-3	1	1	1	1	0	$\frac{-1-\sqrt{5}}{2}$	$\frac{-1+\sqrt{5}}{2}$
9	-3	1	1	1	1	0	$\frac{-1+\sqrt{5}}{2}$	$\frac{-1-\sqrt{5}}{2}$
8	4	0	0	0	0	1	-1	-1
10	6	2	2	2	2	0	1	1
6	2	-2	-2	-2	-2	-1	$\frac{-1-\sqrt{5}}{2}$	$\frac{-1+\sqrt{5}}{2}$
6	2	-2	-2	-2	-2	-1	$\frac{-1+\sqrt{5}}{2}$	$\frac{-1-\sqrt{5}}{2}$
40	4	0	0	0	0	1	-1	-1
24	-4	0	0	0	0	-1	1	1
24	-4	0	0	0	0	-1	1	1
30	-2	-2	-2	-2	-2	1	$\frac{1-\sqrt{5}}{2}$	$\frac{1+\sqrt{5}}{2}$
30	-2	-2	-2	-2	-2	1	$\frac{1+\sqrt{5}}{2}$	$\frac{1-\sqrt{5}}{2}$

(Cont.)

TABLE IV (Cont.)

3686400	256	128	256	512	64	12	20	20
1	4_3	4_4	4_5	4_6	8_1	12_2	20_3	20_4
18	-6	2	2	2	2	0	-1	-1
60	0	0	-4	4	0	0	0	0
60	0	0	-4	4	0	0	0	0
240	0	0	0	0	0	0	0	0
240	0	0	0	0	0	0	0	0
300	0	0	-4	4	0	0	0	0
300	0	0	-4	4	0	0	0	0
360	0	0	8	-8	0	0	0	0
120	0	0	0	8	0	0	0	0
480	0	0	0	0	0	0	0	0
600	0	0	0	8	0	0	0	0
360	0	0	0	-8	0	0	0	0
360	0	0	0	-8	0	0	0	0
75	-1	3	-1	-1	-1	0	0	0
75	-1	3	-1	-1	-1	0	0	0
75	-1	3	-1	-1	-1	0	0	0
75	-1	3	-1	-1	-1	0	0	0
75	-1	3	-1	-1	-1	0	0	0
75	-1	3	-1	-1	-1	0	0	0

(Cont.)

TABLE IV (Cont.)

3686400 1	256 4_3	128 4_4	256 4_5	512 4_6	64 8_1	12 12_2	20 20_3	20 20_4
225	1	1	5	5	-3	0	0	0
225	1	1	5	5	-3	0	0	0
450	2	2	-6	-6	2	0	0	0
450	-2	-6	2	2	2	0	0	0
384	0	0	0	0	0	0	0	0
384	0	0	0	0	0	0	0	0
768	0	0	0	0	0	0	0	0
576	0	0	0	0	0	0	0	0
576	0	0	0	0	0	0	0	0
64	0	0	0	0	0	0	0	0
64	0	0	0	0	0	0	0	0
64	0	0	0	0	0	0	0	0
64	0	0	0	0	0	0	0	0
128	0	0	0	0	0	0	0	0
256	0	0	0	0	0	0	0	0
256	0	0	0	0	0	0	0	0
256	0	0	0	0	0	0	0	0
256	0	0	0	0	0	0	0	0

(Cont.)

TABLE IV (Cont.)

3686400	3840	3840	48	48	20	20	20	20
1	2_5	2_6	6_6	6_7	10_6	10_7	10_8	10_9
1	1	1	1	1	1	1	1	1
1	-1	-1	-1	-1	-1	-1	-1	-1
16	4	4	1	1	-1	-1	-1	-1
16	-4	-4	-1	-1	1	1	1	1
25	-5	-5	1	1	0	0	0	0
25	5	5	-1	-1	0	0	0	0
9	3	3	0	0	$\frac{1+\sqrt{5}}{2}$	$\frac{1-\sqrt{5}}{2}$	$\frac{1+\sqrt{5}}{2}$	$\frac{1-\sqrt{5}}{2}$
9	-3	-3	0	0	$\frac{1-\sqrt{5}}{2}$	$\frac{1+\sqrt{5}}{2}$	$\frac{1-\sqrt{5}}{2}$	$\frac{1+\sqrt{5}}{2}$
9	3	3	0	0	$\frac{-1-\sqrt{5}}{2}$	$\frac{-1+\sqrt{5}}{2}$	$\frac{-1-\sqrt{5}}{2}$	$\frac{-1+\sqrt{5}}{2}$
9	-3	-3	0	0	$\frac{-1+\sqrt{5}}{2}$	$\frac{-1-\sqrt{5}}{2}$	$\frac{-1+\sqrt{5}}{2}$	$\frac{-1-\sqrt{5}}{2}$
8	0	0	0	0	0	0	0	0
10	0	0	0	0	0	0	0	0
6	0	0	0	0	0	0	0	0
6	0	0	0	0	0	0	0	0
40	0	0	0	0	0	0	0	0
24	0	0	0	0	0	0	0	0
24	0	0	0	0	0	0	0	0
30	0	0	0	0	0	0	0	0
30	0	0	0	0	0	0	0	0

(Cont.)

TABLE IV (Cont.)

3686400	3840	3840	48	48	20	20	20	20
1	2_5	2_6	6_6	6_7	10_6	10_7	10_8	10_9
18	0	0	0	0	0	0	0	0
60	10	10	1	1	0	0	0	0
60	-10	-10	-1	-1	0	0	0	0
240	20	20	-1	-1	0	0	0	0
240	-20	-20	1	1	0	0	0	0
300	10	10	1	1	0	0	0	0
300	-10	-10	-1	-1	0	0	0	0
360	0	0	0	0	0	0	0	0
120	0	0	0	0	0	0	0	0
480	0	0	0	0	0	0	0	0
600	0	0	0	0	0	0	0	0
360	0	0	0	0	0	0	0	0
360	0	0	0	0	0	0	0	0
75	5	5	2	2	0	0	0	0
75	-5	-5	-2	-2	0	0	0	0
75	5	5	-1	-1	0	0	0	0
75	5	5	-1	-1	0	0	0	0
75	-5	-5	1	1	0	0	0	0
75	-5	-5	1	1	0	0	0	0

(Cont.)

TABLE IV (Cont.)

3686400 1	3840 2_5	3840 2_6	48 6_6	48 6_7	20 10_6	20 10_7	20 10_8	20 10_9
225	15	15	0	0	0	0	0	0
225	-15	-15	0	0	0	0	0	0
450	0	0	0	0	0	0	0	0
450	0	0	0	0	0	0	0	0
384	0	0	0	0	0	0	0	0
384	0	0	0	0	0	0	0	0
768	0	0	0	0	0	0	0	0
576	24	-24	0	0	-1	-1	1	1
576	-24	24	0	0	1	1	-1	-1
64	8	-8	2	-2	$\frac{1-\sqrt{5}}{2}$	$\frac{1+\sqrt{5}}{2}$	$\frac{-1+\sqrt{5}}{2}$	$\frac{-1-\sqrt{5}}{2}$
64	8	-8	2	-2	$\frac{1+\sqrt{5}}{2}$	$\frac{1-\sqrt{5}}{2}$	$\frac{-1-\sqrt{5}}{2}$	$\frac{-1+\sqrt{5}}{2}$
64	-8	8	-2	2	$\frac{-1+\sqrt{5}}{2}$	$\frac{-1-\sqrt{5}}{2}$	$\frac{1-\sqrt{5}}{2}$	$\frac{1+\sqrt{5}}{2}$
64	-8	8	-2	2	$\frac{-1-\sqrt{5}}{2}$	$\frac{-1+\sqrt{5}}{2}$	$\frac{1+\sqrt{5}}{2}$	$\frac{1-\sqrt{5}}{2}$
128	0	0	0	0	0	0	0	0
256	16	-16	-2	2	1	1	-1	-1
256	-16	16	2	-2	-1	-1	1	1
256	0	0	0	0	0	0	0	0
256	0	0	0	0	0	0	0	0

(Cont.)

TABLE IV (Cont.)

3686400	192	384	24	24	24	16	32	32
1	4_7	4_8	12_3	12_4	12_5	8_2	4_9	8_3
1	1	1	1	1	1	1	1	1
1	-1	-1	-1	-1	-1	-1	-1	-1
16	4	4	1	1	1	0	0	0
16	-4	-4	-1	-1	-1	0	0	0
25	-5	-5	1	1	1	-1	-1	-1
25	5	5	-1	-1	-1	1	1	1
9	3	3	0	0	0	-1	-1	-1
9	-3	-3	0	0	0	1	1	1
9	3	3	0	0	0	-1	-1	-1
9	-3	-3	0	0	0	1	1	1
8	0	0	0	0	0	0	0	0
10	0	0	0	0	0	0	0	0
6	0	0	0	0	0	0	0	0
6	0	0	0	0	0	0	0	0
40	0	0	0	0	0	0	0	0
24	0	0	0	0	0	0	0	0
24	0	0	0	0	0	0	0	0
30	0	0	0	0	0	0	0	0
30	0	0	0	0	0	0	0	0

(Cont.)

TABLE IV (Cont.)

3686400 1	192 4_7	384 4_8	24 12_3	24 12_4	24 12_5	16 8_2	32 4_9	32 8_3
18	0	0	0	0	0	0	0	0
60	-2	2	1	-1	-1	0	-2	2
60	2	-2	-1	1	1	0	2	-2
240	-4	4	-1	1	1	0	0	0
240	4	-4	1	-1	-1	0	0	0
300	-2	2	1	-1	-1	0	-2	2
300	2	-2	-1	1	1	0	2	-2
360	0	0	0	0	0	0	0	0
120	0	0	0	0	0	0	0	0
480	0	0	0	0	0	0	0	0
600	0	0	0	0	0	0	0	0
360	0	0	0	0	0	0	0	0
360	0	0	0	0	0	0	0	0
75	1	-3	-2	0	0	-1	1	1
75	-1	3	2	0	0	1	-1	-1
75	1	-3	1	$\sqrt{-3}$	$-\sqrt{-3}$	-1	1	1
75	1	-3	1	$-\sqrt{-3}$	$\sqrt{-3}$	-1	1	1
75	-1	3	-1	$-\sqrt{-3}$	$\sqrt{-3}$	1	-1	-1
75	-1	3	-1	$\sqrt{-3}$	$-\sqrt{-3}$	1	-1	-1

(Cont.)

TABLE IV (Cont.)

3686400 1	192 4_7	384 4_8	24 12_3	24 12_4	24 12_5	16 8_2	32 4_9	32 8_3
225	3	-9	0	0	0	1	-1	-1
225	-3	9	0	0	0	-1	1	1
450	0	0	0	0	0	0	0	0
450	0	0	0	0	0	0	0	0
384	0	0	0	0	0	0	0	0
384	0	0	0	0	0	0	0	0
768	0	0	0	0	0	0	0	0
576	0	0	0	0	0	0	0	0
576	0	0	0	0	0	0	0	0
64	0	0	0	0	0	0	0	0
64	0	0	0	0	0	0	0	0
64	0	0	0	0	0	0	0	0
64	0	0	0	0	0	0	0	0
128	0	0	0	0	0	0	0	0
256	0	0	0	0	0	0	0	0
256	0	0	0	0	0	0	0	0
256	0	0	0	0	0	0	0	0
256	0	0	0	0	0	0	0	0

133 = 9* + 60* + 64*

TABLE V

THE CHARACTER TABLE OF THE SIMPLE GROUP OF ORDER $2^{14} \cdot 3^6 \cdot 5^6 \cdot 7 \cdot 11 \cdot 19$

1	2A	2B	3A	3B	4A	4B	4C	5A	5B	5C
1	1	1	1	1	1	1	1	1	1	1
133	21	5	7	-2	1	5	-3	-7	3	8
133	21	5	7	-2	1	5	-3	-7	3	8
760	56	-8	22	4	0	8	0	20	5	10
3344	176	16	41	-4	8	16	0	14	9	-31
8778	154	-54	21	3	-10	10	2	28	3	28
8778	154	-54	21	3	-10	10	2	28	3	28
8910	286	78	27	0	10	14	6	20	5	35
9405	77	61	36	9	9	-3	-11	55	5	30
16929	385	33	27	0	1	17	9	29	4	54
35112	616	40	84	12	0	24	0	-28	7	-13
35112	616	40	84	12	0	24	0	-28	7	-13
65835	539	-85	63	-18	-1	11	3	-105	5	85
65835	539	-85	63	-18	-1	11	3	-105	5	85
69255	-153	135	0	0	-9	-9	15	45	10	5
69255	-153	135	0	0	-9	-9	15	45	10	5
214016	1408	0	104	-40	16	0	0	161	6	16
267520	1408	256	148	4	0	0	0	120	-5	20
270864	880	-240	189	0	-8	16	0	134	4	-11

(Cont.)

TABLE V (Cont.)

1	2A	2B	3A	3B	4A	4B	4C	5A	5B	5C
365750	2310	310	35	35	10	6	-10	0	0	125
374528	0	-256	56	38	0	0	0	-112	-2	28
374528	0	-256	56	38	0	0	0	-112	-2	28
406296	2904	24	162	0	0	40	0	-84	11	-204
653125	1925	325	-41	-5	21	5	5	125	0	0
656250	-1750	250	105	15	-10	10	10	0	0	0
656250	-1750	250	105	15	-10	10	10	0	0	0
718200	2520	120	0	-27	0	24	-8	0	0	75
718200	2520	120	0	-27	0	24	-8	0	0	75
1053360	1232	-592	63	36	8	-16	0	210	10	235
1066527	2079	-225	0	0	-9	15	-9	-63	-3	152
1066527	2079	-225	0	0	-9	15	-9	-63	-3	152
1185030	1078	390	189	0	10	-26	30	-280	10	155
1354320	2288	336	-27	0	8	16	0	70	-5	-55
1361920	1792	0	112	40	32	0	0	-70	-10	-80
1361920	-1792	0	112	40	0	0	0	-70	-10	-80
1361920	-1792	0	112	40	0	0	0	-70	-10	-80
1575936	3200	0	216	0	-16	0	0	91	-4	-64
1625184	-1056	96	162	0	0	-32	0	204	-1	-66
2031480	1848	120	-162	0	0	8	0	120	10	-270

(Cont.)

TABLE V (Cont.)

1	2A	2B	3A	3B	4A	4B	4C	5A	5B	5C
2375000	-1000	600	-190	80	0	40	0	0	0	0
2407680	-1408	256	36	36	0	0	0	280	5	180
2661120	0	768	0	-54	0	0	0	0	-20	120
2784375	3575	375	-189	0	-1	-25	15	-125	0	0
2985984	4608	0	0	0	0	0	0	-36	-6	-16
3200000	-3200	0	104	-40	16	0	0	125	0	0
3424256	0	0	-64	8	0	0	0	176	-4	256
3878280	-2520	648	0	0	0	-24	-24	0	20	-95
4156250	-1750	-550	-175	5	30	10	10	0	0	0
4561920	-1408	0	216	0	-16	0	0	55	-10	-80
4809375	175	-225	-216	0	-29	15	15	125	0	0
5103000	2520	-360	0	0	0	-24	0	0	0	-125
5103000	2520	-360	0	0	0	-24	0	0	0	-125
5332635	-693	27	0	0	-9	-21	-21	-315	10	135
5878125	-4675	-275	36	-45	9	45	5	-125	0	0

(Cont.)

TABLE V (Cont.)

1	5D	5E	6A	6B	6C	7	8A	8B	9	10A
1	1	1	1	1	1	1	1	1	1	1
133	$\frac{1-5\sqrt{5}}{2}$	$\frac{1+5\sqrt{5}}{2}$	3	-1	2	0	-1	1	1	1
133	$\frac{1+5\sqrt{5}}{2}$	$\frac{1-5\sqrt{5}}{2}$	3	-1	2	0	-1	1	1	1
760	-10	-10	2	-2	4	4	0	0	1	-4
3344	4	4	5	1	4	5	0	0	2	6
8778	$\frac{31-5\sqrt{5}}{2}$	$\frac{31+5\sqrt{5}}{2}$	1	-3	3	0	0	-2	0	4
8778	$\frac{31+5\sqrt{5}}{2}$	$\frac{31-5\sqrt{5}}{2}$	1	-3	3	0	0	-2	0	4
8910	15	15	7	3	0	-1	0	2	0	-4
9405	5	5	-4	4	1	4	-1	1	0	7
16929	-21	-21	7	3	0	-4	1	1	0	5
35112	$-8+10\sqrt{5}$	$-8-10\sqrt{5}$	4	4	4	0	0	0	0	-4
35112	$-8-10\sqrt{5}$	$-8+10\sqrt{5}$	4	4	4	0	0	0	0	-4
65835	$\frac{5+25\sqrt{5}}{2}$	$\frac{5-25\sqrt{5}}{2}$	-1	-1	2	0	1	-1	0	-1
65835	$\frac{5-25\sqrt{5}}{2}$	$\frac{5+25\sqrt{5}}{2}$	-1	-1	2	0	1	-1	0	-1
69255	0	0	0	0	0	-3	-1	-1	0	-3
69255	0	0	0	0	0	-3	-1	-1	0	-3
214016	-24	-24	4	0	0	5	4	0	-1	-7
267520	20	20	4	4	4	1	0	0	-2	8
270864	24	24	-11	-3	0	6	0	0	0	-10

(Cont.)

TABLE V (Cont.)

1	5D	5E	6A	6B	6C	7	8A	8B	9	10A
365750	0	0	15	-5	-5	0	0	2	-1	0
374528	$8-10\sqrt{5}$	$8+10\sqrt{5}$	0	8	2	0	0	0	-1	0
374528	$8+10\sqrt{5}$	$8-10\sqrt{5}$	0	8	2	0	0	0	-1	0
406296	6	6	6	-6	0	-5	0	0	0	4
653125	0	0	-1	7	-5	4	1	-3	1	5
656250	0	0	5	1	-5	0	0	-2	0	0
656250	0	0	5	1	-5	0	0	-2	0	0
718200	$\frac{25+25\sqrt{5}}{2}$	$\frac{25-25\sqrt{5}}{2}$	0	0	-3	0	0	0	0	0
718200	$\frac{25-25\sqrt{5}}{2}$	$\frac{25+25\sqrt{5}}{2}$	0	0	-3	0	0	0	0	0
1053360	10	10	-1	-1	-4	0	0	0	0	2
1066527	$-18+15\sqrt{5}$	$-18-15\sqrt{5}$	0	0	0	0	-1	-1	0	9
1066527	$-18-15\sqrt{5}$	$-18+15\sqrt{5}$	0	0	0	0	-1	-1	0	9
1185030	0	0	-11	-3	0	0	0	2	0	8
1354320	-30	-30	-7	-3	0	-5	0	0	0	-2
1361920	0	0	8	0	0	0	0	0	1	2
1361920	0	0	8	0	0	0	0	0	1	-2
1361920	0	0	8	0	0	0	0	0	1	-2
1575936	-24	-24	-4	0	0	5	-4	0	0	-5
1625184	-6	-6	6	-6	0	1	0	0	0	4
2031480	0	0	-6	6	0	-4	0	0	0	8

(Cont.)

TABLE V (Cont.)

1	5D	5E	6A	6B	6C	7	8A	8B	9	10A
2375000	0	0	-10	-6	0	5	0	0	-1	0
2407680	-20	-20	-4	4	4	-5	0	0	0	-8
2661120	10	10	0	0	6	0	0	0	0	0
2784375	0	0	11	3	0	6	-1	-1	0	-5
2985984	24	24	0	0	0	-6	0	0	0	-12
3200000	0	0	4	0	0	-1	-4	0	-1	5
3424256	16	16	0	0	0	-4	0	0	2	0
3878280	15	15	0	0	0	0	0	0	0	0
4156250	0	0	5	-7	5	0	0	-2	-1	0
4561920	0	0	-4	0	0	-1	4	0	0	7
4809375	0	0	4	0	0	4	1	3	0	5
5103000	0	0	0	0	0	0	0	0	0	0
5103000	0	0	0	0	0	0	0	0	0	0
5332635	-15	-15	0	0	0	0	1	-1	0	-3
5878125	0	0	-4	4	-5	1	-1	1	0	-5

(Cont.)

TABLE V (Cont.)

1	10B	10C	10D	10E	10F	10G	10H	11	12A	12B
1	1	1	1	1	1	1	1	1	1	1
133	1	-4	0	$\sqrt{5}$	$-\sqrt{5}$	$\frac{5-\sqrt{5}}{2}$	$\frac{5+\sqrt{5}}{2}$	1	1	-1
133	1	-4	0	$-\sqrt{5}$	$\sqrt{5}$	$\frac{5+\sqrt{5}}{2}$	$\frac{5-\sqrt{5}}{2}$	1	1	-1
760	1	6	2	-3	-3	2	2	1	0	2
3344	1	1	1	1	1	-4	-4	0	-1	1
8778	-1	4	-4	1	1	$\frac{7-5\sqrt{5}}{2}$	$\frac{7+5\sqrt{5}}{2}$	0	-1	1
8778	-1	4	-4	1	1	$\frac{7+5\sqrt{5}}{2}$	$\frac{7-5\sqrt{5}}{2}$	0	-1	1
8910	1	11	3	3	3	3	3	0	1	-1
9405	-3	2	6	1	1	1	1	0	0	0
16929	0	10	-2	-2	-2	3	3	0	1	-1
35112	1	-9	-5	$-\sqrt{5}$	$\sqrt{5}$	$-2\sqrt{5}$	$2\sqrt{5}$	0	0	0
35112	1	-9	-5	$\sqrt{5}$	$-\sqrt{5}$	$2\sqrt{5}$	$-2\sqrt{5}$	0	0	0
65835	-1	-11	5	$-\sqrt{5}$	$\sqrt{5}$	$\frac{5+\sqrt{5}}{2}$	$\frac{5-\sqrt{5}}{2}$	0	-1	-1
65835	-1	-11	5	$\sqrt{5}$	$-\sqrt{5}$	$\frac{5-\sqrt{5}}{2}$	$\frac{5+\sqrt{5}}{2}$	0	-1	-1
69255	2	-3	5	0	0	0	0	-1	0	0
69255	2	-3	5	0	0	0	0	-1	0	0
214016	-2	8	0	0	0	0	0	0	-2	0
267520	3	8	-4	1	1	-4	-4	0	0	0
270864	0	5	5	0	0	0	0	0	1	1

(Cont.)

TABLE V (Cont.)

1	10B	10C	10D	10E	10F	10G	10H	11	12A	12B
365750	0	-15	5	0	0	0	0	0	1	3
374528	0	0	4	$-1-\sqrt{5}$	$-1+\sqrt{5}$	$4-2\sqrt{5}$	$4+2\sqrt{5}$	0	0	0
374528	0	0	4	$-1+\sqrt{5}$	$-1-\sqrt{5}$	$4+2\sqrt{5}$	$4-2\sqrt{5}$	0	0	0
406296	-1	4	4	-1	-1	-6	-6	0	0	-2
653125	0	0	0	0	0	0	0	0	3	-1
656250	0	0	0	0	0	0	0	1	-1	1
656250	0	0	0	0	0	0	0	1	-1	1
718200	0	-5	-5	0	0	$\frac{5+5\sqrt{5}}{2}$	$\frac{5-5\sqrt{5}}{2}$	-1	0	0
718200	0	-5	-5	0	0	$\frac{5-5\sqrt{5}}{2}$	$\frac{5+5\sqrt{5}}{2}$	-1	0	0
1053360	2	7	3	-2	-2	-2	-2	0	-1	-1
1066527	-1	4	0	$\sqrt{5}$	$-\sqrt{5}$	$-3\sqrt{5}$	$3\sqrt{5}$	0	0	0
1066527	-1	4	0	$-\sqrt{5}$	$\sqrt{5}$	$3\sqrt{5}$	$-3\sqrt{5}$	0	0	0
1185030	-2	3	-5	0	0	0	0	0	1	1
1354320	3	13	1	1	1	6	6	0	-1	1
1361920	2	-8	0	0	0	0	0	-1	-4	0
1361920	-2	8	0	0	0	0	0	-1	0	0
1361920	-2	8	0	0	0	0	0	-1	0	0
1575936	0	0	0	0	0	0	0	-1	2	0
1625184	-1	-6	6	1	1	6	6	0	0	-2
2031480	-2	-2	10	0	0	0	0	0	0	2

(Cont.)

TABLE V (Cont.)

1	10B	10C	10D	10E	10F	10G	10H	11	12A	12B
2375000	0	0	0	0	0	0	0	1	0	-2
2407680	-3	-8	-4	1	1	-4	-4	0	0	0
2661120	0	0	8	-2	-2	-2	-2	0	0	0
2784375	0	0	0	0	0	0	0	0	-1	-1
2985984	-2	8	0	0	0	0	0	1	0	0
3200000	0	0	0	0	0	0	0	1	-2	0
3424256	0	0	0	0	0	0	0	0	0	0
3878280	0	5	-7	-2	-2	3	3	-1	0	0
4156250	0	0	0	0	0	0	0	-1	3	1
4561920	2	-8	0	0	0	0	0	0	2	0
4809375	0	0	0	0	0	0	0	-1	-2	0
5103000	0	-5	-5	0	0	0	0	1	0	0
5103000	0	-5	-5	0	0	0	0	1	0	0
5332635	2	7	7	2	2	-3	-3	0	0	0
5878125	0	0	0	0	0	0	0	0	0	0

(Cont.)

TABLE V (Cont.)

1	12C	14A	15A	15B	15C	19A	19B	20A
1	1	1	1	1	1	1	1	1
133	0	0	2	$\frac{1+\sqrt{5}}{2}$	$\frac{1-\sqrt{5}}{2}$	0	0	1
133	0	0	2	$\frac{1-\sqrt{5}}{2}$	$\frac{1+\sqrt{5}}{2}$	0	0	1
760	0	0	2	-1	-1	0	0	0
3344	0	1	-4	1	1	0	0	-2
8778	-1	0	1	$\frac{1+\sqrt{5}}{2}$	$\frac{1-\sqrt{5}}{2}$	0	0	0
8778	-1	0	1	$\frac{1-\sqrt{5}}{2}$	$\frac{1+\sqrt{5}}{2}$	0	0	0
8910	0	-1	2	0	0	-1	-1	0
9405	1	0	1	-1	-1	0	0	-1
16929	0	0	2	0	0	0	0	1
35112	0	0	-1	$\frac{-1-\sqrt{5}}{2}$	$\frac{-1+\sqrt{5}}{2}$	0	0	0
35112	0	0	-1	$\frac{-1+\sqrt{5}}{2}$	$\frac{-1-\sqrt{5}}{2}$	0	0	0
65835	0	0	3	$\frac{-1-\sqrt{5}}{2}$	$\frac{-1+\sqrt{5}}{2}$	0	0	-1
65835	0	0	3	$\frac{-1+\sqrt{5}}{2}$	$\frac{-1-\sqrt{5}}{2}$	0	0	-1
69255	0	1	0	0	0	0	0	1
69255	0	1	0	0	0	0	0	1
214016	0	1	-1	0	0	0	0	1
267520	0	1	3	-1	-1	0	0	0
270864	0	-2	-1	0	0	0	0	2

(Cont.)

TABLE V (Cont.)

1	12C	14A	15A	15B	15C	19A	19B	20A
365750	-1	0	0	0	0	0	0	0
374528	0	0	-4	$\frac{1+\sqrt{5}}{2}$	$\frac{1-\sqrt{5}}{2}$	0	0	0
374528	0	0	-4	$\frac{1-\sqrt{5}}{2}$	$\frac{1+\sqrt{5}}{2}$	0	0	0
406296	0	-1	-3	0	0	0	0	0
653125	-1	0	-1	0	0	0	0	1
656250	1	0	0	0	0	$\frac{-1+\sqrt{-19}}{2}$	$\frac{-1-\sqrt{-19}}{2}$	0
656250	1	0	0	0	0	$\frac{-1-\sqrt{-19}}{2}$	$\frac{-1+\sqrt{-19}}{2}$	0
718200	1	0	0	$\frac{1+\sqrt{5}}{2}$	$\frac{1-\sqrt{5}}{2}$	0	0	0
718200	1	0	0	$\frac{1-\sqrt{5}}{2}$	$\frac{1+\sqrt{5}}{2}$	0	0	0
1053360	0	0	3	1	1	0	0	-2
1066527	0	0	0	0	0	0	0	1
1066527	0	0	0	0	0	0	0	1
1185030	0	0	-1	0	0	0	0	0
1354320	0	-1	-2	0	0	0	0	-2
1361920	0	0	2	0	0	0	0	2
1361920	0	0	2	0	0	0	0	0
1361920	0	0	2	0	0	0	0	0
1575936	0	1	1	0	0	0	0	-1
1625184	0	1	-3	0	0	0	0	0
2031480	0	0	3	0	0	0	0	0

(Cont.)

TABLE V (Cont.)

1	12C	14A	15A	15B	15C	19A	19B	20A
2375000	0	1	0	0	0	0	0	0
2407680	0	-1	1	1	1	0	0	0
2661120	0	0	0	1	1	-1	-1	0
2784375	0	-2	1	0	0	1	1	-1
2985984	0	2	0	0	0	1	1	0
3200000	0	-1	-1	0	0	1	1	1
3424256	0	0	-4	-2	-2	0	0	0
3878280	0	0	0	0	0	0	0	0
4156250	1	0	0	0	0	0	0	0
4561920	0	-1	1	0	0	1	1	-1
4809375	0	0	-1	0	0	0	0	1
5103000	0	0	0	0	0	-1	-1	0
5103000	0	0	0	0	0	-1	-1	0
5332635	0	0	0	0	0	0	0	1
5878125	-1	1	1	0	0	0	0	-1

(Cont.)

TABLE V (Cont.)

1	20B	20C	20D	20E	21	22	25A	25B
1	1	1	1	1	1	1	1	1
133	0	0	$\frac{-1+\sqrt{5}}{2}$	$\frac{-1-\sqrt{5}}{2}$	0	-1	$\frac{1+\sqrt{5}}{2}$	$\frac{1-\sqrt{5}}{2}$
133	0	0	$\frac{-1-\sqrt{5}}{2}$	$\frac{-1+\sqrt{5}}{2}$	0	-1	$\frac{1-\sqrt{5}}{2}$	$\frac{1+\sqrt{5}}{2}$
760	-2	-2	0	0	1	1	0	0
3344	1	1	0	0	-1	0	-1	-1
8778	0	0	$\frac{-1-\sqrt{5}}{2}$	$\frac{-1+\sqrt{5}}{2}$	0	0	$\frac{1+\sqrt{5}}{2}$	$\frac{1-\sqrt{5}}{2}$
8778	0	0	$\frac{-1+\sqrt{5}}{2}$	$\frac{-1-\sqrt{5}}{2}$	0	0	$\frac{1-\sqrt{5}}{2}$	$\frac{1+\sqrt{5}}{2}$
8910	-1	-1	1	1	-1	0	0	0
9405	2	2	-1	-1	1	0	0	0
16929	2	2	-1	-1	-1	0	-1	-1
35112	-1	-1	0	0	0	0	$\frac{-1+\sqrt{5}}{2}$	$\frac{-1-\sqrt{5}}{2}$
35112	-1	-1	0	0	0	0	$\frac{-1-\sqrt{5}}{2}$	$\frac{-1+\sqrt{5}}{2}$
65835	1	1	$\frac{1+\sqrt{5}}{2}$	$\frac{1-\sqrt{5}}{2}$	0	0	0	0
65835	1	1	$\frac{1-\sqrt{5}}{2}$	$\frac{1+\sqrt{5}}{2}$	0	0	0	0
69255	1	1	0	0	0	1	0	0
69255	1	1	0	0	0	1	0	0
214016	0	0	0	0	-1	0	1	1
267520	0	0	0	0	1	0	0	0
270864	1	1	0	0	0	0	-1	-1

(Cont.)

TABLE V (Cont.)

1	20B	20C	20D	20E	21	22	25A	25B
365750	1	1	0	0	0	0	0	0
374528	0	0	0	0	0	0	$\frac{1-\sqrt{5}}{2}$	$\frac{1+\sqrt{5}}{2}$
374528	0	0	0	0	0	0	$\frac{1+\sqrt{5}}{2}$	$\frac{1-\sqrt{5}}{2}$
406296	0	0	0	0	1	0	1	1
653125	0	0	0	0	1	0	0	0
656250	0	0	0	0	0	-1	0	0
656250	0	0	0	0	0	-1	0	0
718200	-1	-1	$\frac{-1-\sqrt{5}}{2}$	$\frac{-1+\sqrt{5}}{2}$	0	1	0	0
718200	-1	-1	$\frac{-1+\sqrt{5}}{2}$	$\frac{-1-\sqrt{5}}{2}$	0	1	0	0
1053360	-1	-1	0	0	0	0	0	0
1066527	0	0	1	1	0	0	$\frac{-1-\sqrt{5}}{2}$	$\frac{-1+\sqrt{5}}{2}$
1066527	0	0	1	1	0	0	$\frac{-1+\sqrt{5}}{2}$	$\frac{-1-\sqrt{5}}{2}$
1185030	-1	-1	0	0	0	0	0	0
1354320	1	1	0	0	1	0	0	0
1361920	0	0	0	0	0	-1	0	0
1361920	0	0	0	0	0	1	0	0
1361920	0	0	0	0	0	1	0	0
1575936	0	0	0	0	-1	-1	1	1
1625184	-2	-2	0	0	1	0	-1	-1
2031480	-2	-2	0	0	-1	0	0	0

(Cont.)

TABLE V (Cont.)

1	20B	20C	20D	20E	21	22	25A	25B
2375000	0	0	0	0	-1	1	0	0
2407680	0	0	0	0	1	0	0	0
2661120	0	0	0	0	0	0	0	0
2784375	0	0	0	0	0	0	0	0
2985984	0	0	0	0	0	-1	-1	-1
3200000	0	0	0	0	-1	1	0	0
3424256	0	0	0	0	-1	0	1	1
3878280	1	1	1	1	0	-1	0	0
4156250	0	0	0	0	0	-1	0	0
4561920	0	0	0	0	-1	0	0	0
4809375	0	0	0	0	1	-1	0	0
5103000	$1+2\sqrt{5}$	$1-2\sqrt{5}$	0	0	0	1	0	0
5103000	$1-2\sqrt{5}$	$1+2\sqrt{5}$	0	0	0	1	0	0
5332635	-1	-1	-1	-1	0	0	0	0
5878125	0	0	0	0	1	0	0	0

(Cont.)

TABLE V (Cont.)

1	30A	30B	30C	35A	35B	40A	40B
1	1	1	1	1	1	1	1
133	-2	$\frac{-1-\sqrt{5}}{2}$	$\frac{-1+\sqrt{5}}{2}$	0	0	-1	-1
133	-2	$\frac{-1+\sqrt{5}}{2}$	$\frac{-1-\sqrt{5}}{2}$	0	0	-1	-1
760	2	-1	-1	-1	-1	0	0
3344	0	-1	-1	0	0	0	0
8778	1	$\frac{1+\sqrt{5}}{2}$	$\frac{1-\sqrt{5}}{2}$	0	0	0	0
8778	1	$\frac{1-\sqrt{5}}{2}$	$\frac{1+\sqrt{5}}{2}$	0	0	0	0
8910	2	0	0	-1	-1	0	0
9405	1	1	1	-1	-1	-1	-1
16929	2	0	0	1	1	1	1
35112	-1	$\frac{3-\sqrt{5}}{2}$	$\frac{3+\sqrt{5}}{2}$	0	0	0	0
35112	-1	$\frac{3+\sqrt{5}}{2}$	$\frac{3-\sqrt{5}}{2}$	0	0	0	0
65835	-1	$\frac{-1+\sqrt{5}}{2}$	$\frac{-1-\sqrt{5}}{2}$	0	0	1	1
65835	-1	$\frac{-1-\sqrt{5}}{2}$	$\frac{-1+\sqrt{5}}{2}$	0	0	1	1
69255	0	0	0	$\frac{-1+\sqrt{-35}}{2}$	$\frac{-1-\sqrt{-35}}{2}$	-1	-1
69255	0	0	0	$\frac{-1-\sqrt{-35}}{2}$	$\frac{-1+\sqrt{-35}}{2}$	-1	-1
214016	-1	0	0	0	0	-1	-1
267520	-1	-1	-1	1	1	0	0
270864	-1	0	0	1	1	0	0

(Cont.)

TABLE V (Cont.)

1	30A	30B	30C	35A	35B	40A	40B
365750	0	0	0	0	0	0	0
374528	0	$\frac{-1-\sqrt{5}}{2}$	$\frac{-1+\sqrt{5}}{2}$	0	0	0	0
374528	0	$\frac{-1+\sqrt{5}}{2}$	$\frac{-1-\sqrt{5}}{2}$	0	0	0	0
406296	1	0	0	0	0	0	0
653125	-1	0	0	-1	-1	1	1
656250	0	0	0	0	0	0	0
656250	0	0	0	0	0	0	0
718200	0	$\frac{-1-\sqrt{5}}{2}$	$\frac{-1+\sqrt{5}}{2}$	0	0	0	0
718200	0	$\frac{-1+\sqrt{5}}{2}$	$\frac{-1-\sqrt{5}}{2}$	0	0	0	0
1053360	-1	1	1	0	0	0	0
1066527	0	0	0	0	0	-1	-1
1066527	0	0	0	0	0	-1	-1
1185030	-1	0	0	0	0	0	0
1354320	-2	0	0	0	0	0	0
1361920	2	0	0	0	0	0	0
1361920	-2	0	0	0	0	$\sqrt{-10}$	$-\sqrt{-10}$
1361920	-2	0	0	0	0	$-\sqrt{-10}$	$\sqrt{-10}$
1575936	1	0	0	0	0	1	1
1625184	1	0	0	1	1	0	0
2031480	-1	0	0	1	1	0	0

(Cont.)

TABLE V (Cont.)

1	30A	30B	30C	35A	35B	40A	40B
2375000	0	0	0	0	0	0	0
2407680	1	-1	-1	0	0	0	0
2661120	0	1	1	0	0	0	0
2784375	1	0	0	1	1	-1	-1
2985984	0	0	0	-1	-1	0	0
3200000	-1	0	0	-1	-1	1	1
3424256	0	0	0	1	1	0	0
3878280	0	0	0	0	0	0	0
4156250	0	0	0	0	0	0	0
4561920	1	0	0	-1	-1	-1	-1
4809375	-1	0	0	-1	-1	1	1
5103000	0	0	0	0	0	0	0
5103000	0	0	0	0	0	0	0
5332635	0	0	0	0	0	1	1
5878125	1	0	0	1	1	-1	-1

A MONOMIAL CHARACTER OF FISCHER'S BABY MONSTER

BY

D. G. HIGMAN*

In [1], [2], [3] we view monomial representations of a finite group as modified permutation representations and ask about the extent to which the theory of permutation representations extends to monomial representations. Here we illustrate this point of view by determining the degrees of the irreducible constituents of a certain monomial character of the group whose existence has been predicted by B. Fischer, now referred to as the "baby monster". These degrees have been determined by J. Leon [4] by a completely different method. The necessary preliminaries are outlined in Section 1 (see the above mentioned papers for details and further development) and the application to the baby monster is indicated in Section 2.

---

*Research supported in part by the National Science Foundation.

1. Monomial Spaces and Weights

A *monomial space* (X,u) for a finite group G consists of a G-space X and a map $u : G \times X \to$ complex roots of unity, $(\sigma,x) \to u_\sigma(x)$, such that

$$u_{\sigma\tau}(x) = u_\sigma(x)\, u_\tau\,(x^\sigma) \quad \text{and} \quad u_1(x) = 1$$

for all $\sigma,\tau \in G$, $x \in X$. The *value group* is

$$U = \langle u_\sigma(x) \mid \sigma \in G\,,\ x \in X\rangle$$

and the *support* is

$$\text{spt}\,(X,u) = \{(x,y) \in X^2 \mid u_\sigma(x) = u_\sigma(y) \quad \text{for all} \quad \sigma \in G_{x,y}\}.$$

We have $\text{spt}(X,u) = \bigcup_{f\in\mathcal{O}_{(X,u)}} f$ where $\mathcal{O}_{(X,u)}$ is a subset of the set $\mathcal{O}$ of all orbits for G in X^2. The *rank* of (X,u) is $r_{(X,u)} = |\,\mathcal{O}_{(X,u)}\,|$.

CX has the structure of a CG-module via

$$x\sigma = u_\sigma(x)\, x^\sigma \qquad (x \in X\,,\ \sigma \in G)\,.$$

We call $V(X,u) := \text{Hom}_{CG}\,(CX,CX)$ the *centralizer algebra* of (X,u); $V(X) := V(X,1)$ is the centralizer algebra of the G-space X. When convenient we write $\mathcal{O}_\chi$ for $\mathcal{O}_{(X,u)}$ and r_χ for $r_{(X,u)}$, where χ is the monomial character of G afforded by (X,u).

We put $U^* = \langle\zeta^{1/2}\rangle$ if $U = \langle\zeta\rangle$. There is a matrix $w: X^2 \to C$ with $\text{spt}\, w = \text{spt}\,(X,u)$ such that (i) w is hermitian, (ii) $w(x,y) \in U^* \cup \{0\}$ and $w(x,x) = 1$ for all $x,y \in X$, and (iii) $\{w_f | f \in \mathcal{O}_{(X,u)}\}$ is a basis of $V(X,u)$, where

$$w_f(x,y) = \begin{cases} w(x,y) & \text{if} \quad (x,y) \in f \\ 0 & \text{if} \quad (x,y) \in X^2 - f \end{cases}$$

We call a matrix w satisfying (i), (ii) and (iii) a *weight* afforded by (X,u) and refer to $\{w_f\}$ as the w-*basis*.

A triangle $(x,y,z) \in X^3$ has *type* (f,g,h) if $(x,y) \in f$, $(y,z) \in g$ and $(x,z) \in h$ with $f,g,h \in \mathcal{O}$, and weight $(\delta w)(x,y,z) = w(y,z)\,\overline{w(x,z)}\,w(x,y)$. In its action on X, G preserves types and weights of triangles. Put

$\beta_{fgh}(\alpha)$: = the number of triangles $(x,y,z) \in X^3$ of type (f,g,h) and weight α for given $(x,z) \in h$.

Then the structure constants for $V(x,u)$ with respect to the w-basis are

$$b_{fgh} = \sum_{\alpha \in U^*} \beta_{fgh}(\alpha)\,\alpha \qquad (f,g,h \in \mathcal{O}_{(X,u)})$$

and the intersection numbers for the G-space X; i.e., the structure constants for the Φ-basis of $V(X)$ where Φ is the "all 1" weight are

$$a_{fgh} = \sum_{\alpha \in U^* \cup \{0\}} \beta_{fgh}(\alpha) \qquad (f,g,h \in \mathcal{O}) .$$

Let ζ be the standard character of $V(X,u)$, so that $\zeta(\phi) = \text{trace}\ \phi$ for $\phi \in V(X,u)$, and let $\zeta_1, \ldots, \zeta_m$ be the distinct irreducible characters of $V(X,u)$, $\zeta_i(1) = e_i$. Then

$$\zeta = \sum_{i=1} z_i \zeta_i$$

with $z_1, \ldots, z_m$ positive integers and

$$\chi = \sum_{i=1}^{m} e_i \chi_i$$

where $\chi_1, \ldots, \chi_m$ are the distinct irreducible constituents of χ and $\chi_i(1) = z_i$.

The regular representation provides an isomorphism $V(X,u) \to \mathrm{Hom}_C(C\mathcal{O}_{(X,u)}, C\mathcal{O}_{(X,u)})$ such that $w_f \to \hat{w}_f$ for $f \in \mathcal{O}_{(X,u)}$, where $\hat{w}_f(g,h) = b_{gfh}$. The character multiplicity table

$$(\zeta_i\ (w_\alpha))\left|\begin{matrix} z_1 \\ \vdots \\ z_m \end{matrix}\right.$$

for $V(X,u)$ is determined by the $\hat{w}_f$ just as the corresponding table for $V(X)$ is determined by the intersection matrices $\hat{\phi}_f$, $f \in \mathcal{O}$.

2. Fischer's Baby Monster

B. Fischer has predicted the existence of a group G generated by a conjugacy class D of $\{3,4\}$-transpositions such that for $\tau \in D$, $H := C_G(\tau) \cong 2 \cdot {}^2E_6(2) \cdot 2$. According to Fischer, (i) $|D| = 13571955000$, (ii) G is rank 5 in its action on D , (iii) the subdegrees are $|i(\tau)| = 1$, $|\alpha(\tau)| = 3968055$, $|\beta(\tau)| = 23113728$, $|\gamma(\tau)| = 2370830336$

and $|\delta(\tau) = 11174042880$ where $\mathcal{O} = \{i, \alpha, \beta, \gamma, \delta\}$ is the set of G-orbits in D^2, and (iv) the intersection matrix $\hat{\Phi}_\alpha$ is

i	α	β	γ	δ
0	1	0	0	0
3968055	1782 + 44352	0	3510	648
0	0	69615	0	8064
0	2097152	0	142155 + 694980	663552
0	1824768	3898440	3127410	3295791

J. Conway has computed the character-multiplicity table:

1	3968055	23113728	2370830336	11174042880	1
1	228735	-709632	14483156	-14002560	92655
1	50895	133056	124928	-308880	9458750
1	1935	-4032	-31744	33840	4275362520
1	-945	1782	14336	-15120	9287037474

Now consider the monomial character χ of G induced by the alternating character λ of H. For each $f \in \mathcal{O}$ there is an involution $d = d_f \in G$ such that $(\tau, \tau^d) \in f$, and we find that $\lambda(\alpha) = \lambda(d\alpha d)$ for all $\alpha \in G_{\tau,\tau^d}$ if and only if $f \in \{i, \alpha, \gamma\}$. This means that $\mathcal{O}_\chi = \{i, \alpha, \gamma\}$ and hence that $r_\chi = 3$ and χ is the sum of three distinct irreducible characters. Since H has 2 orbits on the set of triangles

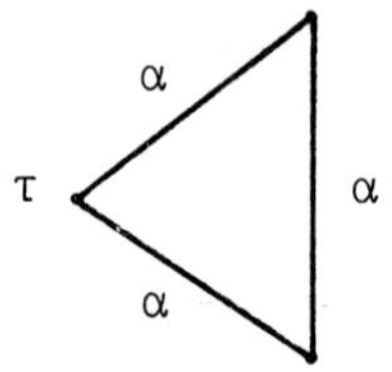

and the length of these orbits are as indicated in the corresponding entry in $\hat{\Phi}_\alpha$, we must have

$$b_{\alpha\alpha\alpha} \in \{\pm(1782 \pm 44352)\ ,\ 0\}\ .$$

Similarly

$$b_{\gamma\alpha\gamma} \in \{\pm(142155 \pm 694980)\ ,\ 0\}\ .$$

Since H is transitive on the set of triangles

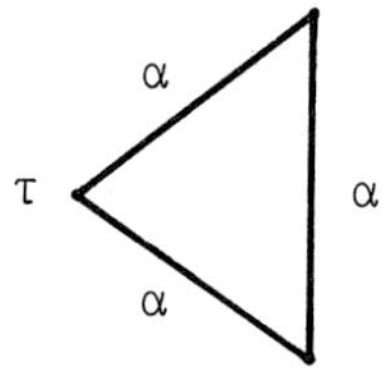

we may take $b_{\alpha\alpha\gamma} = a_{\alpha\alpha\gamma}$ and $b_{\gamma\alpha\alpha} = a_{\gamma\alpha\alpha}$. Then we must have $\hat{w}_\alpha$ equal to

$$\begin{array}{ccc} 0 & 1 & 0 \\ 3968055 & 42570 & 3510 \\ 0 & 2097152 & 552825. \end{array}$$

This gives the character-multiplicity table

1	566865	84672512	4371
1	28665	-114688	63532485
1	-135	512	13508418144

REFERENCES

1. D. G. Higman, *Coherent configurations*. Part II. Weights. Geometriae Dedicata, to appear.

2. __________, *Monomial characters*. Proceedings of the Symposium on finite groups, Sapporo, 1974, to appear.

3. __________, *Modular theory of monomial characters*, to appear.

4. J. Leon, Personal communication.

UNIVERSITY OF MICHIGAN
ANN ARBOR, MICHIGAN

ON THE IRREDUCIBLE CHARACTERS OF A SIMPLE GROUP OF ORDER $2^{41}\cdot3^{13}\cdot5^{6}\cdot7^{2}\cdot11\cdot13\cdot17\cdot19\cdot23\cdot31\cdot47$

BY

JEFFREY S. LEON*

Recently Fischer produced evidence for the existence of a simple group F_2 of order $2^{41}\cdot3^{13}\cdot5^{6}\cdot7^{2}\cdot11\cdot13\cdot17\cdot19\cdot23\cdot31\cdot47$. The group F_2 would be generated by a class of involutions, any two of whose members have product of order at most four. If d is an involution in this class,

$$C_{F_2}(d) \cong Z_2\backslash {}^2E_6(2)\backslash Z_2 .$$

As of this date, the existence and uniqueness of F_2 remains unproven. I shall assume that such a group does exist and shall present several results on its complex irreducible characters.

---

*Computing services used in this research were provided by the Computer Center of the University of Illinois at Chicago Circle. Their assistance is gratefully acknowledged.

Let E denote $C_{F_2}(d)$; then $|E:E'| = 2$, and E has two linear characters θ_1 and θ_2; here θ_1 is the identity character of E.

At present, the degrees of at least twenty irreducible characters of F_2 are known. I shall denote these characters by $\chi_1, \chi_2, \ldots, \chi_{20}$. For at least six characters, the values are essentially known everywhere. The known character degrees arise as follows:

(1) (Conway) $\theta_1{}^* = \chi_1 + \chi_2 + \chi_3 + \chi_4 + \chi_5$,

$$\chi_1(1) = 1,$$
$$\chi_2(1) = 3^3 \cdot 5 \cdot 23 \cdot 31,$$
$$\chi_3(1) = 2 \cdot 5^4 \cdot 7 \cdot 23 \cdot 47,$$
$$\chi_4(1) = 2^3 \cdot 3^3 \cdot 5 \cdot 11 \cdot 13 \cdot 19 \cdot 31 \cdot 47,$$
$$\chi_5(1) = 2 \cdot 3 \cdot 11 \cdot 13 \cdot 17 \cdot 19 \cdot 23 \cdot 31 \cdot 47.$$

(2) (D. Higman, Leon)

$$\theta_2{}^* = \chi_6 + \chi_7 + \chi_8,$$
$$\chi_6(1) = 3 \cdot 31 \cdot 47,$$
$$\chi_7(1) = 3^3 \cdot 5 \cdot 17 \cdot 19 \cdot 31 \cdot 47,$$
$$\chi_8(1) = 2^5 \cdot 3 \cdot 13 \cdot 17 \cdot 19 \cdot 23 \cdot 31 \cdot 47.$$

(3)
$$\chi_6{}^2 = \chi_1 + \phi_2 + \phi_3 + \chi_9,$$
$$\chi_9(1) = 3 \cdot 5 \cdot 19 \cdot 23 \cdot 31 \cdot 47,$$
$$\phi_2(1) = \chi_2(1),\ \phi_3(1) = \chi_3(1).$$

(4) (Landrock) The group F_2 has a 2-block $\{\chi_{10}, \chi_{11}, \chi_{12}, \chi_{13}, \chi_{14}\}$ with dihedral defect group.

$\chi_{10}(1) = \chi_{11}(1) = 2^{38}\cdot 3\cdot 5^{4}\cdot 23$,

$\chi_{12}(1) = \chi_{13}(1) = 2^{38}\cdot 23\cdot 31\cdot 47$,

$\chi_{14}(1) = 2^{39}\cdot 11\cdot 19\cdot 23$.

In [3] Landrock mentions that the values of these characters on elements of even order are determined by Steinberg characters of ${}^{2}E_{6}(2)$, $F_4(2)$, and ${}^{2}F_{4}(2)$. I have computed the values on elements of odd order (see Table 3). Thus the values of these five characters can be determined everywhere.

(5) The group F_2 has two 3-blocks of defect one. The degrees of the characters in the two 3-blocks are identical; namely,

$\chi_{15}(1) = \chi_{18}(1) = 2\cdot 3^{12}\cdot 5\cdot 7^{2}\cdot 13\cdot 17\cdot 23\cdot 31\cdot 47$,

$\chi_{16}(1) = \chi_{19}(1) = 2^{3}\cdot 3^{12}\cdot 5\cdot 13\cdot 17\cdot 19\cdot 23\cdot 31\cdot 47$,

$\chi_{17}(1) = \chi_{20}(1) = 2\cdot 3^{12}\cdot 5^{4}\cdot 13\cdot 17\cdot 23\cdot 31\cdot 47$.

Result (5) required extensive calculation. I have checked these calculations to a certain degree; however, more remains to be done.

Result (2) was obtained independently by D. Higman and the author using different methods. Of particular interest is the character X_6 of degree 4371. No other non-trivial character of F_2 has such low degree; in fact, the only irreducible character degrees less than one million are 1, 4371, and 96255. It is relatively easy to restrict X_6 to large subgroups of F_2. It can be shown that X_6 is the unique character of degree 4371.

I shall sketch my proof of result (2). Omitted details may be found in a more detailed write up available from the author. I make critical use of (a) Brauer's theory of blocks of defect one (see [1] and [2]), (b) a theorem of L. Scott on permutation characters (see [4], Cor A), (c) a theorem of Frame (see [5], Th 30.1), and (d) information about F_2 provided me by Fischer. Fischer's information included a partial list of conjugacy classes of F_2.

The subgroup E permutes the conjugates of d in five orbits of lengths $n_1 = 1$, $n_2 = 3^3\cdot5\cdot7\cdot13\cdot17\cdot19$, $n_3 = 2^{12}\cdot3^3\cdot11\cdot19$, $n_4 = 2^{20}\cdot7\cdot17\cdot19$, $n_5 = 2^7\cdot3^3\cdot5\cdot7\cdot11\cdot13\cdot17\cdot19$. The element d is fused in F_2 to elements of E outside E'; consequently, $\|\theta_2{}^*\| < \|\theta_1{}^*\| = 5$.

For $p \geq 7$, let g_p denote an element of F_2 of order p. Then $C_{F_2}(g_p) \cong \langle g_p \rangle \times C_p$, where C_p is $Z_2\backslash L_3(4)\backslash Z_2$,

S_5, S_4, $Z_2 \times Z_2$, Z_2, Z_2, 1, and 1 for $p = 7, 11, 13, 17, 19, 23, 31$, and 47 respectively. Also $N(\langle g_p \rangle)/C(g_p)$ has order $(p-1)/2$ for $p = 23, 31$, and 47 and order $p-1$ for all other p. Fischer labeled the four classes of involutions by z, θ, d, and f; correspondence between involutions of F_2 and those of C_p can be deduced from Fischer's partial list of conjugacy classes (see Table 1). The symbols α and a denote certain classes of order 3 and 5 respectively.

The p-modular character tables of $C_{F_2}(g_p)$ coincide with those of C_p and, except for $p = 7$, with the complex character tables for C_p. Several of these are given in Table 1. The modular table for C_7 was obtained easily from a complex character table of C_7 constructed by Peter Landrock. Distinct p-modular characters of C_p are fused in $N(\langle g_p \rangle)$ only for $p = 17$.

To each character ϕ of C_p of defect 0, there corresponds (except for $p = 17$) a unique p-block B of defect 1 in F_2; if X in B is nonexceptional, $X(g_p t) = e_X \phi(t)$ for p-regular elements t of C_p $(e_X = \pm 1)$. Data for the p-blocks of defect 1 is given in the table below. Row p, column i contains an entry x_{pi} and, in some cases, a second entry y_{pi}. If X is a nonexceptional character of the i'th p-block $B_i(p)$, then $X(1) \equiv X(g_p) = e_X x_{pi} \pmod p$; if X is exceptional, then $X(1) \equiv e_X y_{pi} \pmod p$. For $p = 7$, the congruences actually hold modulo p^2 as there is only one class of elements of order 7.

p \ i	1	2	3	4	5	6	7
7	35	35	70	126	56	70	70
11	1	1	4	4	5	5	6
13	1	1	2	3	3		
17	1	1	2,1				
19	1	1					
23	1,11	1,11					
31	1,15	1,15					
47	1,23	1,23					

Note that, for $p \geqslant 11$, this table yields congruences for the degrees of *all* irreducible characters of F_2. For example, $X(1) \equiv 0, \pm 1$, or $\pm 11 \pmod{23}$ for any irreducible character X; the alternatives ± 11 occur only if X is irrational.

Note also, for $p \geqslant 11$, $X(1)$ determines $X(g_p)$ uniquely via the congruence $X(g_p) \equiv X(1) \pmod{p}$ with one exception: if $p = 11$ and $X(1) \equiv 5,6 \pmod{11}$, $X(g_{11})$ could be 5 or -6, -5 or 6. Also $X(g_p)$ determines the value of X on p-singular elements, in some cases uniquely and in others up to a sign; these values may be read from Table 1.

Table 1 also gives the values of $\theta_1{}^*(g_p x)$ and $\theta_2{}^*(g_p x)$ for p-regular elements x of C_p' ($p = 7, 11, 13$). The values of $\theta_1{*}$ were easy to compute since $\theta_1{*}(g_p x)$ is the number of involutions of $C_{C_p}(x)$ conjugate to d in F_2. Also, $|\theta_2{*}(g_p x)| \leqslant \theta_1{*}(g_p x)$, and $\theta_2{*}(g_p x) = \theta_1{*}(g_p x)$ if x has odd order. Several values of $\theta_2{*}$ were obtained by local arguments and others by restriction to C_7: write

$$(\theta_1{*} + \theta_2{*})(g_7 x) = \sum_{i=1}^{11} (a_i^+ \phi_i^+(x) + a_i^- \phi_i^-(x))$$

and note that the a_i^+, a_i^- are integral and, for $i \geqslant 7$, nonnegative (nonnegativity follows from a theorem of L. Scott applied to the permutation character $\theta_1{*} + \theta_2{*}$).

From the values of $\theta_2{*}$ one obtains easily

$$\theta_2{*}(g_{13}x) = \phi_3^{\pm}(x) + \phi_3^{\pm}(x),$$

$$\theta_2{*}(g_{11}x) = \phi_1^{\pm}(x) + \phi_2^{\pm}(x) + \phi_3^{\pm}(x),$$

$$\theta_2{*}(g_7 x) = \sum_{i=1}^{11} (d_i^+ \phi_i^+(x) + d_i^- \phi_i^-(x)), \quad d_i^+ + d_i^- = 0 \text{ for } i \geqslant 7.$$

Write $\theta_2{*} = \sum_{i=6}^{k} X_i$, $k \leqslant 9$, and let $x_i = X_i(1)$. The Brauer theory for blocks of defect one tells us that $k \geqslant 8$ and

$$\text{(a)} \quad \{x_i : 6 \leqslant i \leqslant k\} \equiv \begin{cases} \{3,3,0\} \pmod{13} & \text{if } k = 8 \\ \{3,3,b,-b\} \pmod{13} & \text{if } k = 9, \end{cases}$$

$$\text{(b)} \quad \{x_i : 6 \leqslant i \leqslant k\} \equiv \begin{cases} \{1,4,5\} \pmod{11} & \text{if } k = 8 \\ \{1,4,5,0\} \pmod{11} & \text{if } k = 9, \end{cases}$$

(c) $\{x_i: 6 \leqslant i \leqslant k\}$ contains an even number of elements divisible by 7 but not by 7^2.

From (b) one deduces that all X_i are rational.

The induced character $\theta_1^* + \theta_2^*$ is the permutation character of F_2 on the cosets of E'; let $u_1 = 1$, $u_2 = 1$, $u_3, \ldots, u_k$ be the orbit lengths of a one-point stabilizer. A theorem of Frame [5, Theorem 30.1(C)] tells us that

$$|F_2:E'|^{k-2}\prod_j u_j / \prod_j x_j$$

is a square; in particular, the parities of the exponents of 7, 11, and 13 in $\prod_j u_j$ are the same as those in $\prod_j x_j$. From (a), (b), (c), and Conway's decomposition (see footnote) of θ_1^*, we know that the parities of the exponents of 7, 11, and 13 in $\prod_j x_j$ are odd, even, and odd respectively if $k = 8$ and odd, odd, and even respectively if $k = 9$. For $i = 2, 3, 4,$ and 5, there are either two values of j with $u_j = n_i$ or one value of j with $u_j = 2n_i$; the latter alternative occurs $10-k$ times. The values of the n_i were given earlier; if $k = 9$, it is easy to check that there is no way in which the parities of the exponents of 7, 11, and 13 in $\prod_j u_j$ can be correct. Thus $k = 8$.

Instead of using Conway's decomposition, we may apply Frame's Theorem to the permutation representation of F_2 on E.

As $|F_2:E'|$ is relatively prime to 7 and u_j is not divisible by 7^2 for any j, x_j is not divisible by 7^2 for any j [5, Theorem 30.4].

Let $T = \{n$: (i), (ii), (ii), (iv) hold$\}$.

(i) n divides $|F_2|$ and $n < |F_2:E|$,

(ii) 7^2 does not divide n,

(iii) $n \equiv 0,3 \pmod{13}$,

(iv) for $p = 17, 19, 23, 31, 47$, either $n \equiv 0$ (mod p) or n satisfies the congruences for nonexceptional characters in p-blocks of defect one given by the table.

Let $T_i = \{n \in T: n \equiv i \pmod{11}\}$. Renumbering the characters if necessary, we may assume $x_6 \in T_4$, $x_7 \in T_5$, and $x_8 \in T_1$. The sets T_4, T_5, and T_1 were constructed by computer; they contain 26, 16, and 21 elements respectively. There is only one way to choose x_6, x_7, and x_8 such that $x_6 + x_7 + x_8 = |F_2:E| = 13{,}571{,}955{,}000$, namely $x_6 = 4371$, $x_7 = 63532485$, $x_8 = 13508418144$. The sets T_4, T_5, and T_1 are shown in Table 2.

TABLE 1

C_7

$C(g)$	80640		256		16	64	64	36	36	10	10	
$\|g\|$	1	2	2	2	4	4	4	3	6	5	10	
g	1	d	z	θ	w			α	αd	a	ad	
$\phi_1^{\pm}$	1	1	1	1	1	1	1	1	1	1	1	
$\phi_2^{\pm}$	19	19	3	3	-1	-1	-1	1	1	-1	-1	
$\phi_3^{\pm}$	45	45	-3	-3	1	1	1	0	0	0	0	
$\phi_4^{\pm}$	10	-10	2	-2	0	2	-2	1	-1	0	0	
$\phi_5^{\pm}$	54	-54	-2	2	0	-2	2	0	0	-1	1	
$\phi_6^{\pm}$	36	-36	4	-4	0	0	0	0	0	1	-1	
$\phi_7^{\pm}$	35	35	3	3	-1	3	3	-1	-1	0	0	
ϕ_8^{+}	70	70	6	6	2	-2	-2	-2	-2	0	0	
ϕ_9^{+}	126	126	-2	-2	-2	-2	-2	0	0	1	1	defect zero
ϕ_{10}^{+}	56	-56	-8	8	0	0	0	2	-2	1	-1	
$\phi_{11}^{\pm}$	70	-70	-2	2	0	2	-2	-2	2	0	0	
$\theta_1{}^*$	121	121	9	9	1	5	5	4	4	1	1	
$\theta_2{}^*$	121	-119	9	-7	1	5	-3	4	-2	1	1	

(Cont.)

TABLE 1 (CONT.)

C_7(Cont.)

C(g)	672	672	32	32	16	16	12	12	
|g|	2	2	4	4	8	8	6	6	
g	d	f					αd	αf	
$\phi_1^{\pm}$	1	1	1	1	1	1	1	1	
$\phi_2^{\pm}$	5	5	1	1	-1	-1	-1	-1	
$\phi_3^{\pm}$	3	3	-1	-1	1	1	0	0	
$\phi_4^{\pm}$	-4	4	0	0	0	0	-1	1	
$\phi_5^{\pm}$	12	-12	0	0	0	0	0	0	
$\phi_6^{\pm}$	-6	6	-2	2	0	0	0	0	
$\phi_7^{\pm}$	-7	-7	1	1	1	1	-1	-1	defect zero
ϕ_8^{+}	0	0	0	0	0	0	0	0	defect zero
ϕ_9^{+}	0	0	0	0	0	0	0	0	defect zero
ϕ_{10}^{+}	0	0	0	0	0	0	0	0	defect zero
$\phi_{11}^{\pm}$	0	0	0	0	-2	2	0	0	defect zero

TABLE 1 (CONT.)

C_{11}

$C(g)$	120	8	6	5	12	4	6
$\|g\|$	1	2	3	5	2	4	6
g	1	z	α	a	d	q	αd
$\phi_1^{\pm}$	1	1	1	1	1	1	1
$\phi_2^{\pm}$	4	0	1	-1	-2	0	1
$\phi_3^{\pm}$	5	1	-1	0	1	-1	1
ϕ_4^{+}	6	-2	0	1	0	0	0
θ_1^*	10	2	1	0			
θ_2^*	10	2	1	0			

C_{13}

$C(g)$	24	8	3	4	4
$\|g\|$	1	2	3	2	4
g	1	f	α	d	v
$\phi_1^{\pm}$	1	1	1	1	1
ϕ_2^{+}	2	2	-1	0	0
$\phi_3^{\pm}$	3	-1	0	1	-1
θ_1^*	6	2	0		
θ_2^*	6	-2	0		

Note: ϕ_i^+ and ϕ_i^- differ by a sign on classes to the right of the vertical line.

TABLE 2

T_4	T_5	T_1
185148275	984988823	8864899407
4371	74494953	2999402055
128476803	12177059625	4221380670
2954966469	96255	725441850
5781456135	63532485	5716507446
10056818655	21782150	856512020
1407126890	347643114	91884
12327770	2532828402	9454380
4622913750	8398325754	799840548
217298250	32199700	95099940
7598485206	11819865876	1125701512
216481060	7449495300	3030734840
496633020	942740100	2120194440
2681818308	5628507560	335315344
10294721892	5348355600	3310886800
2317620760	5229065088	1525143984
10368303400		726259040
12515152104		13508418144
4902248520		1916087680
132435472		7047870336
626886000		5215920128
4502806048		
75764000		
142899552		
2037468800		
10963353856		

TABLE 3

$\|g\|$	1	3	3	9	9
g.	1	δ	α	9_1	9_2
χ_{10}	$2^{38}\cdot 3\cdot 5^4\cdot 23$	15360	-2555904	-12	24
χ_{11}	$2^{38}\cdot 3\cdot 5^4\cdot 23$	15360	-2555904	-12	24
χ_{12}	$2^{38}\cdot 23\cdot 31\cdot 47$	-35840	-1114112	-20	16
χ_{13}	$2^{38}\cdot 23\cdot 31\cdot 47$	-35840	-1114112	-20	16
χ_{14}	$2^{39}\cdot 11\cdot 19\cdot 23$	51200	-1441792	8	8

$\|g\|$	27	15	15	5	5	5	21	35	7
g		$\delta\tilde{a}$	αa	$\tilde{a}$	a		$s\alpha$	sa	s
χ_{10}	0	0	-4	0	3200	0	-1	1	-64
χ_{11}	0	0	-4	0	3200	0	-1	1	-64
χ_{12}	1	0	-12	384	384	-1	1	-1	64
χ_{13}	1	0	-12	384	384	-1	1	-1	64
χ_{14}	-1	0	8	-384	2816	1	-2	2	-128

$\|g\|$	33	55	11	39	13	17	19	23	23	31	31	47	47
g													
χ_{10}	1	-1	4	0	3	-1	-1	0	0	1	1	-1	-1
χ_{11}	1	-1	4	0	3	-1	-1	0	0	1	1	-1	-1
χ_{12}	1	-1	4	1	1	1	-1	0	0	0	0	0	0
χ_{13}	1	-1	4	1	1	1	-1	0	0	0	0	0	0
χ_{14}	0	0	0	-1	2	-2	0	0	0	1	1	-1	-1

REFERENCES

1. R. Brauer, *Investigations on group characters*, Ann. of Math. 42 (1941), 936-958.

2. ______________, *On groups whose order contains a prime number to the first power*, part I, Amer. J. Math. 64 (1942), 401-420.

3. P. Landrock, *The characters of a non-principal 2-block of Fischer's new simple group* $F_2(?)$, to appear.

4. L. L. Scott, *Modular permutation representations*, Trans. Amer. Math. Soc. 175 (1973), 101-121.

5. H. Wielandt, *Finite Permutation Groups*, Academic Press, (New York), 1964.

UNIVERSITY OF ILLINOIS AT CHICAGO CIRCLE
CHICAGO, ILLINOIS

A SETTING FOR THE LEECH LATTICE

BY

JOHN McKAY

The construction of the Leech lattice can be put in a setting which yields even unimodular lattices for all dimensions $d = 8k$ for k odd.

For $k = 1$ we derive the root lattice for E_8 and for $k = 3$, the Leech lattice. Some remarks concerning further lattices (in particular 40-dimensional) will be given.

CONCORDIA UNIVERSITY
MONTREAL, QUEBEC

THE SUBMODULE STRUCTURE OF WEYL MODULES FOR GROUPS OF TYPE A_1

BY

R. CARTER AND E. CLINE*

Let G^* be a simple simply-connected algebraic group over $\mathbb{C}$ and V_λ be the irreducible rational G^*-module with dominant weight λ. Let $V_{\lambda,Z}$ be a minimal admissible lattice in V_λ [3, p. 155] and $V_{\lambda,K} = V_{\lambda,Z} \otimes_Z K$ where K is an algebraically closed field of characteristic $p \neq 0$. The vector space $V_{\lambda,K}$ is a rational module for the group $G(K)$ of K-rational points of the simple simply-connected algebraic group of the same type as G^*. The modules $V_{\lambda,K}$ have been the subject of a number of recent investigations [1, 2, 4, 5]. They have been called Weyl modules since their dimensions are given by the classical formula of H. Weyl. The Weyl modules are not in general irreducible, but each one has a unique maximal submodule whose quotient is the irreducible rational

---

*Author partially supported by NSF grant GP-29411-3.

G(K)-module with dominant weight λ. The submodule structure of the Weyl modules is apparently quite complicated. In this note we discuss the submodule structure of $V_{\lambda,K}$ in the simplest case, *viz.*, when $G(K) = G = SL_2(K)$ has type A_1.

We view G concretely as the set of 2×2 matrices of determinant 1 over K, letting $X_\alpha(t) = \begin{pmatrix} 1 & t \\ 0 & 1 \end{pmatrix}$, $X_{-\alpha}(t) = \begin{pmatrix} 1 & 0 \\ t & 1 \end{pmatrix}$ for $t \in K$, and letting T denote the subgroup of diagonal matrices of the form $\begin{pmatrix} s & 0 \\ 0 & s^{-1} \end{pmatrix}$ for $s \in K^\times$. Let $X^*(T)$ denote the character module of the algebraic group T. Then $X^*(T)$ is a free Z-module of rank 1 with generator λ_1, where $\lambda_1: \begin{pmatrix} s & 0 \\ 0 & s^{-1} \end{pmatrix} \to s$ for $s \in K^\times$. Thus λ_1 is the usual fundamental dominant weight for T.

For each integer $m > 0$, the Weyl module V_m is the rational G-module of dimension m with high weight $(m-1)\lambda_1$, and a basis $v_0, v_1, \ldots, v_{m-1}$ on which the elements $x_{\pm\alpha}(t)$ act by the rules

$$x_\alpha(t)v_i = \Sigma_{j=0}^{i} \binom{m-1-i+j}{j} t^j v_{i-j}$$

and

$$x_{-\alpha}(t)v_i = \Sigma_{j=0}^{m-1-i} \binom{i+j}{j} t^j v_{i+j} .$$

Each vector v_i is a weight vector for the torus T with weight $(m-1-2i)\lambda_1$.

1. The Composition Factors of V_M

For each $j = 1, 2, \ldots,$ define $\rho_j: Z^+ \to Z$ by setting $\rho_j(m) = m - 2r$ where $m = kp^j + r$, $k \geqslant 0$ and $0 \leqslant r < p^j$.

(1.1) Definition: The function ρ_j is an *m-admissible reflection* provided $k \not\equiv 0 \bmod p$. The integer j is the *exponent* of ρ_j. A strictly decreasing sequence of integers $m, \rho_{e_0}(m), \ldots, \rho_{e_s} \circ \rho_{e_{s-1}} \circ \ldots \circ \rho_{e_0}(m)$ is an *m-admissible* sequence iff

(i) $0 < e_s < e_{s-1} < \ldots < e_0$;

(ii) for each j, ρ_{e_j} is $\rho_{e_{j-1}} \circ \ldots \circ \rho_{e_0}(m)$-admissible.

Let $V(m;p) = V$ denote the set of integers z which appear in some m-admissible sequence.

The composition factors of V_m are characterized by the

(1.2) Theorem: Let z be a positive integer. The weight $(z-1)\lambda_1$ is the high weight of a composition factor of V_m if and only if $z \in V(m;p)$.

2. The Structure Graph of a Weyl Module

We shall construct several graphs; the end product gives information about the submodule structure of V_m. If G is a graph, then $V(G)$ and $E(G)$ denote the vertices and edges of G respectively.

THE REFLECTION GRAPH $R(m;p) = R$: Here we set $V = V(R) = V(m;p)$, the set of integers appearing in the m-admissible sequences. An ordered pair (s,t) in $V \times V$ lies in $E(R)$ iff $s > t$ and s,t are adjacent in some m-admissible sequence. The *exponent* of the edge (s,t), written $e(s,t)$, is the exponent of the reflection mapping s to t.

For each vertex z in V, there is a unique path $< z,m >$ from m to z in R, hence R is a directed tree with maximal vertex m. The *parity* of z, $p(z)$, is the residue class modulo 2 of the length of the path $< z,m >$.

We regard the trivial subgraphs T of R, consisting of a vertex z and no edges, as subtrees of R. We set $v_T = z$ here. If T is a nontrivial subtree of R, it has a unique maximal vertex, v_T, and a unique edge $E_T = (v_T,t)$ whose exponent is maximal among the exponents of the edges of T. We call v_T the *leading vertex* of T, and E_T the *leading edge* of T. The *parity* of T is given by $p(T) = p(v_T)$, and the *exponent* of T, $e(T)$ is given by

$$e(T) = \begin{cases} e(E_T) & \text{if } T \text{ has an edge,} \\ \upsilon_p(m) & \text{otherwise,} \end{cases}$$

where for any integer q, $\upsilon_p(q)$ is the highest power of p dividing q.

THE CELL GRAPH $\Gamma(m;p) = \Gamma$:

(2.1) DEFINITION: A subtree T of R is a *cell* iff one of the following holds:

(i) T is trivial;

(ii) T is the maximal subtree of R with leading edge E_T .

The set of cells in R is partially ordered by inclusion, and is bijective with the disjoint union $E(R) \cup V(R)$. If C is a cell in R, then the following hold:

(2.2A) Either C is minimal, or C contains exactly two maximal subcells.

(2.2B) Either $C = R$, or C is maximal in a unique cell D.

(2.2C) If C, D are cells, $C \neq D$ and if C', C'', D', D'' are maximal in C, D respectively, then

$$\{C', C''\} \cap \{D', D''\} = \phi.$$

(2.2D) There is a unique chain

$$C = C_0 \subseteq C_1 \subseteq \dots \subseteq C_\ell = R$$

in the partially ordered set of cells in R with the property that C_i is maximal in C_{i+1} for $i < \ell$.

(2.3) DEFINITION: Let $C \neq R$ be a cell of exponent e, and let $\mathcal{D}$ be the unique cell in which C is maximal. Let $f = e(\mathcal{D})$. We say C is a j-*cell* for each j, $e \leqslant j < f$. If $r = e(R)$, then R is the unique r-cell. For each j, $\upsilon_p(m) \leqslant j \leqslant r$, we denote the set of j-cells by V^j.

We can now construct the cell graph $\Gamma(m;p) = \Gamma$. For each j, $\upsilon_p(m) \leqslant j \leqslant r$, let Γ^j denote a set bijective with V^j; assume the sets Γ^j are mutually disjoint, and for each j, let θ_j; $\Gamma^j \to V^j$ be a fixed bijection. Then

$$V(\Gamma) = \bigcup_{\upsilon_p(m) \leqslant j \leqslant r} \Gamma^j .$$

A pair $(v,w) \in \Gamma^j \times \Gamma^{j-1}$ for some $j > \upsilon_p(m)$ is an *edge* in Γ iff $\theta^j(v) \supseteq \theta^{j-1}(w)$. There are no other edges in Γ. It follows that Γ is a directed tree. If $v \in \Gamma^j$ for some j, the *parity* and *exponent* of v are given respectively by $p(v) = p(\theta^j(v))$, and $e(v) = e(\theta^j(v))$; the unique path from $\rho = (\theta^r)^{-1}(R)$ to v is denoted $< v,\rho >$.

THE STRUCTURE GRAPH $\mathcal{S}(m;p) = \mathcal{S}$: Fix j, $\upsilon_p(m) \leqslant j \leqslant r$. We construct a graph $\mathcal{S}^j$ on Γ^j as follows: if $v, w \in \Gamma^j$, we may assume the subgraph $< v,\rho > \cup < w,\rho >$ is given in figure 1, with $\ell = r-j+1$. The pair $(v,w) \in \Gamma^j \times \Gamma^j$ is an edge in $\mathcal{S}^j$ if and only if $p(v_t) = 0$, $p(w_t) = 1$, and for $k < t$, $p(v_k) = p(w_k)$. The *structure graph* of V_m is $\mathcal{S}(m;p) = \mathcal{S}^{\upsilon_p(m)} = \mathcal{S}$.

For each vertex $z \in \mathcal{V}$, let M_z denote the rational irreducible G-module with high weight $(z-1)\lambda_1$. Let $S(z)$ denote the set of vertices t of $\mathcal{S}$ for which there exists a path in $\mathcal{S}$ from z to t.

(2.8) THEOREM: Let $S \subseteq \mathcal{V}$. For $s \in S$, let $\mu_s = (s-1)\lambda_1$ be the corresponding weight of T on V_m. Let $v_s \in (v_m)_{\mu_s} - \{0\}$, and set $L_S = \sum_{s\in S} KGv_s$. If $z \in \mathcal{V}$, then M_z is a composition factor of L_S if and only if

$$z \in \bigcup_{s\in S} S(s) .$$

Further, every submodule of V_m is of the form L_S for some subset S of $\mathcal{V}(m;p)$.

FIGURE 1

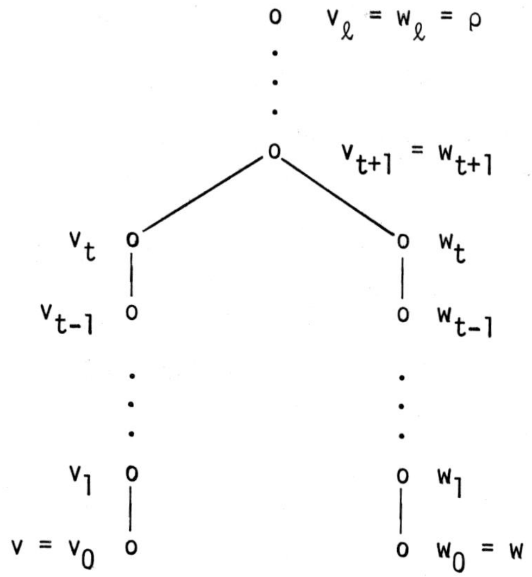

REFERENCES

1. R. Carter and G. Lusztig, *On the modular representations of the general linear and symmetric groups*, Math. Zeit. 136 (1974), 193-242.

2. J. Humphreys, *Modular representations of classical Lie algebras and semisimple groups*, J. Algebra 19 (1971), 51-79.

3. ______________, *Introduction to Lie Algebras and Representation Theory*. Springer-Verlag, 1972.

4. M. Payne. Work in progress.

5. B. Srinivasan, *On the modular characters of the special linear groups* $SL(2,p^n)$, Proc. Lond. Math. Soc. 14 (3) (1964), 101-114.

UNIVERSITY OF VIRGINIA
CHARLOTTESVILLE, VIRGINIA

ON THE 1-COHOMOLOGY OF FINITE GROUPS OF LIE TYPE

BY

WAYNE JONES AND BRIAN PARSHALL

1. INTRODUCTION: This paper contains a summary of recent work by the authors (together with E. Cline and L. Scott) on the problem of calculating $H^1(G_\sigma, V)$ where G_σ is a finite group of Lie type over $K = GF(q)$ and V is a minimal $\overline{K}G_\sigma$-module. For more details, proofs of the theorems, and references to the work of other authors in this area, the reader is referred to [1], [2], [3]. Our results are summarized in the table in §6.

2. DESCRIPTION OF THE GROUPS. Let $k = GF(p)$ be a prime field, $\overline{k}$ its algebraic closure. Let G be a simply connected simple algebraic group over $\overline{k}$. Let $\sigma : G \to G$ be a surjective rational endomorphism with finite fixed point set G_σ. It is known that σ stabilizes a Borel subgroup B and a maximal torus $T \leqslant B$. Let Σ be the set of roots of T in

G, and let $\Delta(\subseteq \Sigma^+)$ be the fundamental system of Σ defined by B. The comorphism σ^* of $\sigma|_T$ defines a permutation π of Σ stabilizing Δ and Σ^+ such that $\sigma^* \pi(\alpha) = q(\alpha)\alpha$ for powers $q(\alpha)$ of p. Extending σ^* to a linear map $\sigma^*: E = R \otimes X^*(T) \to E$ (where $X^*(T)$ denotes the group of rational characters on T) we can write

$$\sigma^* = r\tau \tag{1.1}$$

where $r \in R$ and τ is an isometry of the Euclidean space E. When π = identity, the G_σ are the finite universal (split) Chevalley groups over $K = GF(q)$, q = any $q(\alpha)$. The remaining types ($\pi \neq$ identity) constitute the "twisted groups" of Steinberg, Ree, and Suzuki.

The isometry τ leads to a decomposition of E as $E = E_0 \perp E_\tau$ where E_0 = subspace of τ-trace 0, E_τ = fixed-point subspace of τ. Let

$$j: E \to E_\tau \tag{1.2}$$

be the projection onto E_τ along E_0. Then $\Phi = j(\Sigma)$ is an "abstract root system" in E_τ (in the wide sense, allowing multiples other than ± 1 of a root to appear). For $\alpha \in \Sigma$, let $\bar{\alpha}$ be the unique element of $R^+ j(\alpha) \cap \Phi$ of minimal length. Let $\bar{\Sigma} = \{\bar{\alpha} | \alpha \in \Sigma\}$, $\bar{\Delta} = \{\bar{\alpha} | \alpha \in \Delta\}$ = fundamental system for Φ.

3. Modules and Lower Bounds.

Given $\alpha \in \Delta$, let $\lambda_\alpha \in X^*(T)$ denote the fundamental dominant weight on T corresponding to α. If $\rho_\lambda\colon G \to GL(V(\lambda))$ is the irreducible rational representation of dominant weight $\lambda = \Sigma n_\alpha \lambda_\alpha$, the restriction $\rho_\lambda|_{G_\sigma}$ remains irreducible for those λ for which $0 \leqslant n_\alpha \leqslant q(\alpha) - 1$.

Definition. A representation $\rho_\lambda|_{G_\sigma}$ is *minimal* if λ is minimal (among the non-zero dominant weights) relative to the partial order $\chi \geqslant \psi$ iff $\chi - \psi \in Z^+\Delta$.

These minimal λ are those appearing in the "Dominant weight" column of the table in §6.

For a dominant weight λ, let $\mathcal{U}(\lambda)$ denote the indecomposable Weyl module of G defined by the irreducible complex representation of dominant weight λ of the complex simple Lie algebra associated to G.

Theorem 1. Assume $\rho_\lambda|_{G_\sigma}$ is a minimal representation and let $X(\lambda)$ be a maximal G-submodule of $\mathcal{U}(\lambda)$. Assume (1) if π = identity and $r = 2$ then rank $(G) > 1$; (2) if $\pi \neq$ identity, then $r \geqslant 3$. Then $\dim_{\overline{K}} X(\lambda) \leqslant \dim_{\overline{K}} H^1(G_\sigma, V(\lambda))$.

4. Upper Bounds.

We describe two methods used for determining upper bounds on $\dim_{\overline{K}} H^1(G_\sigma, V(\lambda))$. The first, discussed in 4a, assumes $r > 3$, and is applicable to non-minimal modules (and can in fact be used to obtain the vanishing of $H^1(G_\sigma,V)$ in many instances). The second method, discussed in 4b, is an inductive procedure which does not depend on the size of the field (but depends heavily on the minimality of $V(\lambda)$). We also point out that the method of 4b to date has only been tried for the split Chevalley groups.

4a. Let $\beta \in \mathrm{Hom}(T_\sigma,\overline{k}^\times)$. Let $k(\beta)$ denote the finite extension of $k = GF(p)$ generated by $\beta(T_\sigma)$. The character β makes $k(\beta)$ into a kT_σ-module. If U is a kT_σ-module, T_σ acts with Z-*weight* β *on* U if there exists an isomorphism $x_\beta\colon k(\beta) \to U$ of kT_σ-modules. For $\gamma \in \mathrm{Hom}(T_\sigma,\overline{k}^\times)$, $k(\gamma)$ is isomorphic to $k(\beta)$ as a kT_σ-module iff $\gamma = \beta^\phi$ for some $\phi \in \mathrm{Gal}\,(\overline{k}/k)$. In this case we say γ and β are *Galois equivalent* and we write $\beta \sim \gamma$.

For $\overline{\alpha} \in \overline{\Sigma}$, we define $\underset{\sim}{\overline{\alpha}}$ to be the Z-weight $\beta|_{T_\sigma}$ where $\beta \in \Sigma$ is chosen so that $\overline{\alpha} = \overline{\beta} = j(\beta)$. The importance of this concept lies in the following considerations. Let V be a $\overline{k}G$-module. For $\chi \in X^*(T)$, let n_χ denote the multiplicity of the χ-weight space of T in V. Then

THEOREM 2. $\dim_{\overline{k}} H^1(G_\sigma,V) \leqslant \sum_{\overline{\alpha}\in\overline{\Delta}} \sum_{\chi|_{T_\sigma} \sim \underset{\sim}{\overline{\alpha}}} n_\chi - \dim_{\overline{k}} V^{T_\sigma}/V^{T_\sigma U_\sigma}$.

Here $U = R_u(B)$ is the unipotent radical of B.

For $\rho_\lambda\colon G \to GL(V(\lambda))$ irreducible of dominant weight

λ , let $P(\lambda)$ denote the set of weights of T in $V(\lambda)$. Given $\chi, \psi \in X^*(T)$ we write $\chi \sim \psi$ iff $\chi|_{T_\sigma} \sim \psi|_{T_\sigma}$. Now let λ be a minimal dominant weight. In order to apply Theorem 2 it is necessary to determine when $\alpha \sim \mu$ for $\alpha \in \Sigma$, $\mu \in P(\lambda)$. This is done in [1, §3] when G_σ is split. We wish to describe briefly how this is achieved in the non-split situation.

The permutation π decomposes Δ into orbits, say $\Delta = \Delta_1 \cup \ldots \cup \Delta_\ell$. Let $\alpha_i \in \Delta_i$ be a representative. We introduce an auxiliary k-split torus of dimension ℓ:

$$\widetilde{T} = GL_1\ (\overline{k}) \times \ldots \times GL_1\ (\overline{k}) .$$

Let $s_i: \widetilde{T} \to GL_1(\overline{k})$ be the <u>ith</u> coordinate projection morphism. Define a morphism $\eta: \widetilde{T} \to T$ by

$$\eta(t) = \prod_{i=1}^{\ell} \prod_{j=1}^{m-1} \sigma^j h_{\alpha_i}\ (s_i(t))$$

where $h_{\alpha_i} : GL_1(\overline{k}) \to T$ is the coroot corresponding to α_i . Here m = order of τ . If $K = GF(r^m)$, one checks that

$$\eta(T_K) = T_\sigma .$$

Let η^* be the comorphism of η . For $\chi \in X^*(T)$ write

$$\eta^*(\chi) = \sum_{i=1}^{\ell} n_i\ (\chi) s_i .$$

Then one verifies that for $\chi , \psi \in X^*(T)$, $\chi \sim \psi$ iff there exists a power p^d, $1 \leqslant p^d \leqslant r^m$, with

$$(3.1) \qquad n_i(\chi) \equiv n_i(\psi) \pmod{r^m - 1}$$

for all i. Since the images under η^* of the roots and fundamental dominant weights on $X^*(T)$ can easily be tabulated, (3.1) can be used to systematically determine the Galois equivalences. In the case of the Steinberg groups, it is also important to make use of the subtorus S of T annihilated by $X_0 = E_0 \cap X^*(T)$ (see (1.2)). S is a σ-stable torus (it is in fact the "maximal split torus" of Borel and Tits). Let $j\colon X^*(T) \to X^*(S)$ be the restriction map of character modules. For $\chi, \psi \in X^*(S)$ write $\chi \sim \psi$ if $\chi|_{S_\sigma}$ is Galois equivalent to $\psi|_{S_\sigma}$. Now for $\chi, \psi \in X^*(T)$, if $\chi \sim \psi$, necessarily $j(\chi) \sim j(\psi)$. This fact, together with the results of [1, §3] applied to S, considerably simplifies the calculations involving η^* [2; §3].

4b. Assume in this section that G_σ is split (π = identity), and enumerate Δ as in §6A. For each type we distinguish a fundamental root α as follows: for $A - D$, take $\alpha = \alpha_n$; for E_n, take $\alpha = \alpha_2$; and for F_4, take $\alpha = \alpha_4$ (no choice is made for B_2 or G_2). Let $\hat{\Delta} = \Delta - \{\alpha\}$, $\hat{\Sigma} = Z\hat{\Delta} \cap \Sigma$, and $\hat{W} = \langle w_\beta | \beta \in \hat{\Delta} \rangle$. Denote by $\hat{G}_\sigma$ ($< G_\sigma$) the split Chevalley group generated by the one-parameter root groups U_β for

$\beta \in \hat{\Delta}$. Fix a minimal $\overline{k}G_\sigma$-module $V = V(\lambda)$.

If L is the kernel of the restriction homomorphism

$$\text{res}: H^1(G_\sigma, V) \to H^1(\hat{G}_\sigma, V|_{\hat{G}_\sigma}) ,$$

then

$$\dim_{\overline{k}} H^1(G_\sigma, V) \leqslant \dim_{\overline{k}} L + \dim_{\overline{k}} H^1(\hat{G}_\sigma, V|_{\hat{G}_\sigma}) .$$

It is possible therefore to obtain upper bounds on

$$\dim_{\overline{k}} H^1(G_\sigma, V)$$

by determining L , determining $V|_{\hat{G}_\sigma}$, and then using induction on the rank of G .

The main result concerning the structure of L is given in

THEOREM 3. a) $\dim_{\overline{k}} L \leqslant 1$

b) $L = 0$ except possibly when $\lambda \in Z\Sigma$, α is short, and $V_0^{\hat{G}_\sigma} = 0$ (where V_0 = zero weight space of T in V).

Next let $\{\mathcal{O}_j \mid 1 \leqslant j \leqslant m\}$ be the set of $\hat{W}$-orbits of weights in $P(\lambda)$ such that $\mathcal{O}_j \cap \hat{\Sigma} = \emptyset$. Also, set

$$\mathcal{O}_0 = (P(\lambda) \cap \hat{\Sigma}) \cup \{0\} .$$

Then for $0 \leqslant j \leqslant m$, set

$$V(\mathcal{O}_j) = \sum_{\mu \in \mathcal{O}_j} V_\mu$$

where V_μ denotes the μ-weight space of T in V. Now we state the main result concerning the structure of $V|\hat{G}_\sigma$.

THEOREM 4. a) $V = \oplus \sum_{j=0}^{m} V(\mathcal{O}_j)$

b) For $j > 0$, $V(\mathcal{O}_j)$ is a trivial module or an irreducible minimal $\overline{k}\hat{G}_\sigma$-module (corresponding to some minimal dominant weight $\lambda_{\mathcal{O}_j} \notin Z\hat{\Sigma}$).

c) Assume $G_\sigma \neq SL(3,2)$. Let $\lambda_{\mathcal{O}_0}$ = maximal short root in $\hat{\Sigma}$. Then $V(\mathcal{O}_0)$ is a $\overline{k}\hat{G}_\sigma$-module isomorphic to $V(\lambda_{\mathcal{O}_0})$ except if $\dim_{\overline{k}} V(\mathcal{O}_0)^{\hat{G}_\sigma} = 1$ when one of the following two possibilities occurs:

1) $V(\mathcal{O}_0)/V(\mathcal{O}_0)^{\hat{G}_\sigma} \simeq V(\lambda_{\mathcal{O}_0})$;

2) $V(\mathcal{O}_0)/V(\mathcal{O}_0)^{\hat{G}_\sigma}$ contains a unique submodule of codimension 1 (which is isomorphic to $V(\lambda_{\mathcal{O}_0})$.

Combining Theorems 3 and 4 and using the long exact sequence of cohomology to determine $H^1(\hat{G}_\sigma, V(\mathcal{O}_0))$, we have

THEOREM 5. If $G_\sigma \neq SL(3,2)$ or $Sp(6,2)$, then

$$\dim_{\overline{k}} H^1(G_\sigma, V) \leqslant i + \sum_{j=0}^{m} \dim_{\overline{k}} H^1(\hat{G}_\sigma, V(\lambda_{\mathcal{O}_j}))$$

where

$$i = \begin{cases} 0 & \text{if } \lambda \notin Z\Sigma \\ 1 & \text{if } V(\mathcal{O}_0) \simeq V(\lambda_{\mathcal{O}_0}) \\ 0 & \text{if } V(\mathcal{O}_0)/V(\mathcal{O}_0)^{\hat{G}_\sigma} \simeq V(\lambda_{\mathcal{O}_0}) \text{ and } \dim_{\overline{k}} V(\mathcal{O}_0)^{\hat{G}_\sigma} = 1 \\ -1 & \text{if } V(\mathcal{O}_0)/V(\mathcal{O}_0)^{\hat{G}_\sigma} \not\simeq V(\lambda_{\mathcal{O}_0}) . \end{cases}$$

5. DETERMINATION OF $H^1(G_\sigma, V)$, V MINIMAL. With the exception of the low rank cases indicated by an asterisk in the table in §6 and the symplectic groups $C_n(2^m)$, $\dim_{\overline{k}} H^1(G_\sigma, V)$ is given by the lower bound determined in §3. In most cases, this is proved by comparing this lower bound with the upper bound in §4. In using the results of 4b, the low rank cases for the induction can be determined either by using 4a (when $r > 3$) or by direct computation. The exceptional cases indicated above are handled individually (in most cases by examining special isomorphisms with other groups).

As an illustration of the method of 4b, let $G_\sigma = E_7(p^m)$ and let V be a minimal $\overline{k}G_\sigma$-module. Assume it is known that for a minimal (or trivial) $\overline{k}A_6(p^m)$-module $\hat{V}$ we have

$$\dim_{\overline{k}} H^1(A_6(p^m),\hat{V}) = \begin{cases} 1 & \text{if } p = 7 \text{ and } \hat{V} = \text{adjoint module} \\ 0 & \text{if otherwise} . \end{cases}$$

Also assume that if V and $\hat{V}$ are both adjoint modules, then

$$V(\mathcal{O}_0) \simeq \hat{V} \qquad \text{if } p = 2$$

$$V(\mathcal{O}_0)/V(\mathcal{O}_0)^{\hat{G}_\sigma} \simeq \hat{V} \text{ and } \dim_{\overline{k}} V(\mathcal{O}_0)^{\hat{G}_\sigma} = 1 \qquad \text{if } p \nmid 14$$

$$V(\mathcal{O}_0)/V(\mathcal{O}_0)^{\hat{G}_\sigma} \neq \hat{V} \qquad \text{if } p = 7 .$$

Then from Theorem 5,

$$\dim_{\overline{k}} H^1(E_7(p^m),V) \leqslant \begin{cases} 1 & \text{if } p = 2 \text{ and } V = \text{adjoint module} \\ 0 & \text{otherwise.} \end{cases}$$

Finally, that equality holds here is proved using Theorem 1 together with a determination of $X(\lambda)$ (see [1; §1]).

6. TABLES

A. The fundamental dominant weights are enumerated as follows:

A_n

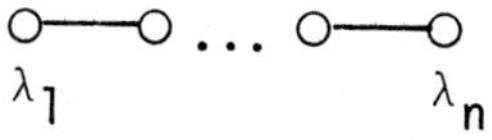

B_n

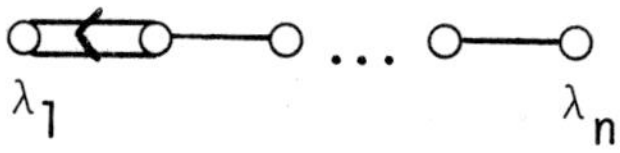

C_n

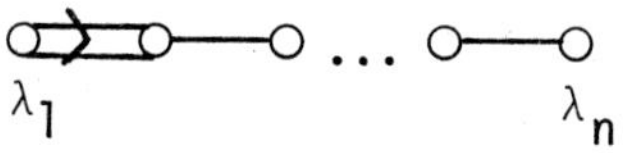

D_n

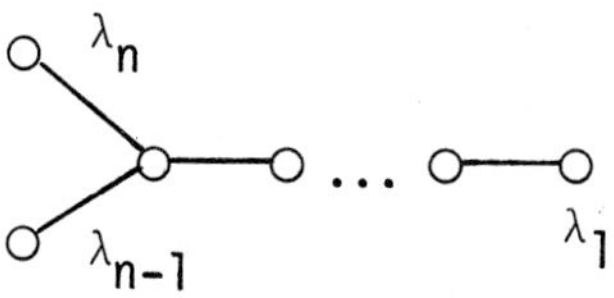

E_n

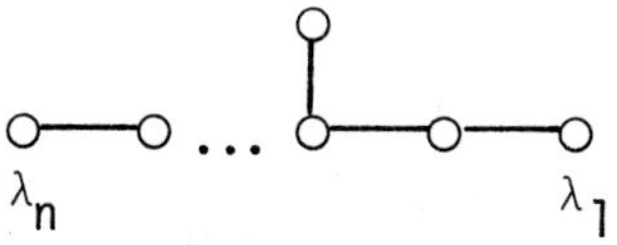

F_4

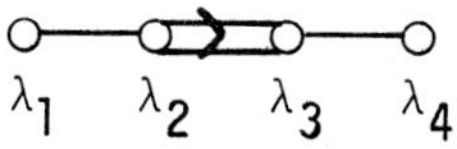

G_2

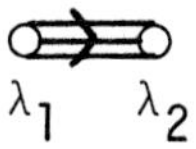

B. TABLE OF RESULTS. An asterisk (*) next to an entry under $\dim_{\bar{k}} H^1(G_\sigma, V)$ indicates that one or more exceptions to the stated dimension occurs. These are listed in C.

G_σ	char k = p	Dominant Weight	$\dim_{\bar{k}} V$	$\dim_{\bar{k}} H^1(G_\sigma, V)$
$A_n(q)$	arbitrary	$\lambda_i,\ 1\leqslant i\leqslant n$	$\binom{n+1}{i}$	0 (*)
	$p\nmid n+1$	$\lambda_1+\lambda_n$	$n(n+2)$	0 (*)
	$p\mid n+1$	$\lambda_1+\lambda_n$	$n(n+2)-1$	1 (*)
$^2A_n(q)$ $q>3$	arbitrary	$\lambda_i,\ 1\leqslant i\leqslant n$	$\binom{n+1}{i}$	0
	$p\nmid n+1$	$\lambda_1+\lambda_n$	$n(n+2)$	0
	$p\mid n+1$	$\lambda_1+\lambda_n$	$n(n+2)-1$	1
$B_n(q)$ $n\geqslant 3$	arbitrary	λ_1	2^n	0
	2	λ_n	$2n$	1
	$\neq 2$	λ_n	$2n+1$	0
$C_n(q)$ $n\geqslant 2$	2	λ_n	$2n$	1
	$\neq 2$	λ_n	$2n$	0
	$p\mid n$	λ_{n-1}	$(n-1)(2n+1)-1$	1
	$p\nmid n$	λ_{n-1}	$(n-1)(2n+1)$	0 (*)
$^2C_2(2^{2n+1})$	2	$\lambda_1,\ \lambda_2$	4	1 (*)

(Cont.)

$D_n(q)$ $n>3$	arbitrary	$\lambda_i, i=1,n-1,n$	$2n$ if $i=1$ 2^{n-1} if $i\neq 1$	0
	$\neq 2$	λ_2	$(2n-1)n$	0
$D_{2n}(q)$	2	λ_2	$2n(4n-1)-2$	2
$D_{2n+1}(q)$	2	λ_2	$(2n+1)(4n+1)-1$	0
${}^2D_n(q)$ $n>3$, $q>3$	arbitrary	$\lambda_i, i=1,n-1,n$	$2n$ if $i=1$ 2^{n-1} if $i\neq 1$	0
${}^2D_{2n}(q)$ $n>1$, $q>3$	2	λ_2	$2n(4n-1)-2$	2
	$\neq 2$	λ_2	$2n(4n-1)$	0
${}^2D_{2n+1}(q)$ $q>3$	2	λ_2	$(2n+1)(4n+1)-1$	1
	$\neq 2$	λ_2	$(2n+1)(4n+1)$	0
${}^3D_4(q)$ $q>3$	arbitrary	$\lambda_i, i=1,2,4$	8	0
	2	λ_2	26	2
	$\neq 2$	λ_2	28	0
$E_6(q)$	arbitrary	λ_1, λ_6	27	0
	3	λ_2	77	1
	$\neq 3$	λ_2	78	0

(Cont.)

$^2E_6(q)$ $q>3$	arbitrary	λ_1 , λ_6	27	0
	3	λ_2	77	1
	$\neq 3$	λ_2	78	0
$E_7(q)$	arbitrary	λ_7	56	0
	2	λ_1	132	1
	$\neq 2$	λ_1	133	0
$E_8(q)$	arbitrary	λ_8	248	0
$F_4(q)$	3	λ_4	25	1
	$\neq 3$	λ_4	26	0
$^2F_4(2^{2n+1})$ $n\neq 0, 1$	2	λ_4	26	0
$G_2(q)$	2	λ_2	6	1
	$\neq 2$	λ_2	7	0
$^2G_2(3^{2n+1})$	3	λ_2	7	0 (*)

C. EXCEPTIONS TO THE ABOVE TABLE.

(a) $\dim_{\bar{k}} H^1(A_1(2^n), V(\lambda_1)) = 1$ for $n > 1$

(b) $\dim_{\bar{k}} H^1(A_2(2), V(\lambda_i)) = 1$ for $i = 1,2$

(c) $\dim_{\bar{k}} H^1(A_3(2), V(\lambda_2)) = 1$

(d) $\dim_{\bar{k}} H^1(A_1(5), V(2\lambda_1)) = 1$

(e) $\dim_{\bar{k}} H^1(A_1(2), V(2\lambda_1)) = 0$

(f) $\dim_{\bar{k}} H^1(C_2(3), V(\lambda_1)) = 1$

(g) $\dim_{\bar{k}} H^1({}^2C_2(2), V(\lambda_i)) = 0$, $i = 1,2$

(h) $\dim_{\bar{k}} H^1({}^2G_2(3), V(\lambda_2)) = 1$

REFERENCES

1. E. Cline, B. Parshall, L. Scott, *Cohomology of finite groups of Lie type I*, Publ. Math. I.H.E.S. no. 45.

2. ______________, *Cohomology of finite groups of Lie type II*, Advances in Math., to appear.

3. Wayne Jones, *Cohomology of finite groups of Lie type*, thesis, Univ. Minnesota, to appear.

UNIVERSITY OF VIRGINIA
CHARLOTTESVILLE, VIRGINIA

FIELD AUTOMORPHISMS AND MAXIMAL SUBGROUPS OF FINITE CHEVALLEY GROUPS

BY

N. BURGOYNE, R. L. GRIESS, JR., AND R. LYONS

If G is a group, p is a prime, and $\lambda \in \mathrm{End}(G)$, write $O^{p'}(G)$ for the subgroup of G generated by all p-elements, and write G_λ for $C_G(\lambda)$.

THEOREM 1. Let G be a simple algebraic group over an algebraically closed field of characteristic $p \neq 0$. Let λ be a surjective endomorphism of G such that G_λ is finite. Exclude the exceptional cases $G_\lambda \cong A_1(2)$, $A_1(3)$, or ${}^2C_2(2)$. If M is a finite group such that $O^{p'}(G_\lambda) \leqslant M \leqslant G$, then there exists a positive integer n such that

$$O^{p'}(G_\mu) \leqslant M \leqslant G_\mu ,$$

where $\mu = \lambda^n$.

THEOREM 2. Let $G = {}^sL(q)$ be a finite simple Steinberg variation defined over the field $GF(q^s)$ of characteristic p. Let λ be a standard automorphism of G obtained from an automorphism of $GF(q^s)$. Exclude the exceptional cases $G = {}^2A_{2n}(2^m)$, ${}^2A_2(3)$, or ${}^2A_2(5)$. Let M be a group such that $O^{p'}(G_\lambda) \leqslant M \leqslant G$. If

$$p \nmid |M:O^{p'}(G_\lambda)| ,$$

then $M \leqslant G_\lambda$. If λ has order s, then either $M \leqslant G_\lambda$ or $M = G$.

COROLLARY. Let G be a finite simple Chevalley group, Steinberg variation, Suzuki group, or Ree group defined over a field k. Let λ be an automorphism of G obtained from an automorphism of k of prime order r. Exclude $G = A_1(2^r)$, $A_1(3^r)$, ${}^2C_2(2^r)$, ${}^2A_2(3)$, ${}^2A_2(5)$, and (if $r = 2$) ${}^2A_{2n}(2^m)$. Then G_λ is a maximal subgroup of G and is the only maximal subgroup containing $O^{p'}(G_\lambda)$.

THE UNIVERSITY OF CALIFORNIA AT SANTA CRUZ
SANTA CRUZ, CALIFORNIA

RUTGERS UNIVERSITY
NEW BRUNSWICK, NEW JERSEY

ON THE DEGREES OF CERTAIN CHARACTERS OF CHEVALLEY GROUPS

BY

C. T. Benson, L. C. Grove, and D. B. Surowski

J. A. Green, in [6], defined a semilinear automorphism J on the generic algebra A of a group G with a BN-pair, and applied it to the dimension functions of constituents of 1_B^G. We generalize his results for (untwisted) Chevalley groups by associating J with the linear character of A that corresponds to the Steinberg character of G and defining an analogue J_τ of J associated with each linear character τ of A. Each automorphism J_τ is then applied to dimension functions of arbitrary characters in the principal series of G.

Systematic use of the automorphisms J_τ can greatly facilitate the explicit computation of dimension functions, even for constituents of 1_B^G (in which case the results that follow also apply to Chevalley groups of twisted type). These dimension functions for $G = F_4(q)$, $E_6^1(q^2)$, $E_6(q)$, and $E_7(q)$ are listed in [2] or [9]. Most of the dimension functions for constituents of 1_B^G for G of classical type were computed by P. Hoefsmit in [7], the remainder by C. Benson and D. Gay in [1]. Since the rank 2 case was covered by W. Feit and G. Higman in [5] the only case that remains is $E_8(q)$.

Suppose, then, that G is an untwisted Chevalley group over a finite field F with q elements. Then G has a split BN-pair with $B \cap N = H$ a maximal split torus, and $B = HU$, a semidirect product. Assume that the Coxeter system (W,R) of G is simple, hence that R is the union of conjugacy classes R_1 and R_2 (we agree that R_2 is empty if the Dynkin diagram of W is simply laced). For each $r \in R$ let u_r be an indeterminate, with $u_r = u_s$ if r and s are conjugate in W, and set $k_0 = \mathcal{Q}(u_r : r \in R)$, $\mathcal{Q}$ the rational field.

We proceed essentially as in [10] to construct a generic algebra E_0 over k_0 corresponding to the Hecke algebra $H_{\mathcal{Q}}(G,U)$. The algebra E_0 has basis $\{e_n : n \in N\}$, with multiplication determined by

$$e_h e_n = e_{hn},$$

$$e_{n(r)}e_{n(w)} = \begin{cases} e_{n(r)n(w)} & \text{if } \ell(rw) > \ell(w), \\ u_r e_{n(r)n(w)} + \dfrac{u_r - 1}{q - 1} \sum\{e_{h(r,t)}e_{n(w)} : t \in F^*\} & \\ & \text{if } \ell(rw) < \ell(w), \end{cases}$$

for $h \in H$, $n \in N$, $r \in R$, $w \in W$, and $n(w)$, $h(r,t)$ determined by the Chevalley homomorphism (see [10], p. 344). Then E_0 is an associative algebra which specializes to an algebra isomorphic with $H_{\mathcal{Q}}(G,U)$ under the specialization $u_r \to q$, $r \in R$.

Suppose ψ is a linear character of H and that $\tilde{\psi}$ is its extension to B for which $\tilde{\psi}|U = 1_U$. Set $k = k_0(\psi)$, $E = k \otimes_{k_0} E_0$, and $\hat{e}_\psi = |H|^{-1}\sum\{\psi(h)^{-1}e_h : h \in H\}$.

THEOREM 1: The subalgebra $\hat{e}_\psi E \hat{e}_\psi$ of E specializes, under $u_r \to q$, $r \in R$, to an algebra isomorphic with the centralizer ring over $\mathcal{Q}(\psi)$ of $\tilde{\psi}^G$.

Set $W_\psi = \{w \in W : \psi^w = \psi\}$, the stabilizer of ψ in W. The subalgebra $\hat{e}_\psi E \hat{e}_\psi$ has basis $\{a_w = \hat{e}_\psi e_{n(w)} \hat{e}_\psi : w \in W_\psi\}$, and the structure constants relative to this basis are in k_0. Let $M(\psi)$ be the k_0-algebra having the same basis and structure constants.

Let A be the generic algebra over k_0 corresponding to $H_{\mathcal{Q}}(G,B)$, with basis $\{b_w : w \in W\}$, as discussed in [4]. Then A has linear characters ν and ε defined by $\nu(b_r) = u_r$, $\varepsilon(b_r) = -1$, $r \in R$. In the non-simply laced case A has two additional linear characters σ_1 and σ_2. If $r \in R_1$, $s \in R_2$ then $\sigma_1(b_r) = u_r$, $\sigma_1(b_s) = -1$; $\sigma_2(b_r) = -1$, $\sigma_2(b_s) = u_s$.

For each linear character τ of A define an automorphism J_τ of k_0 as follows:

$$J_\nu(u_r) = u_r \; ; \quad J_\varepsilon(u_r) = u_r^{-1} \; ;$$

$$J_{\sigma_1}(u_r) = u_r \text{ if } r \in R_1 \, , \; J_{\sigma_1}(u_r) = u_r^{-1} \text{ if } r \in R_2 \; ;$$

$$J_{\sigma_2}(u_r) = u_r^{-1} \text{ if } r \in R_1 \, , \; J_{\sigma_2}(u_r) = u_r \text{ if } r \in R_2 \; .$$

Each J_τ can be extended to an automorphism of a field K, finite over k, that is a splitting field for $M(\psi)$.

THEOREM 2. The automorphism J_τ on K can be extended to an involutory semilinear automorphism of $M^K(\psi) = K \otimes_{k_0} M(\psi)$ satisfying

$$J_\tau(a_w) = \frac{\tau(b_w)}{\nu(b_w)} a_w \; , \; w \in W_\psi \; .$$

REMARK: If $\psi = 1_H$ then J_ε is the semilinear automorphism of A^K defined by Green in [6].

If χ is an irreducible character of $M^K(\psi)$ define a *dimension function* (or *generic degree*)

$$d_\psi(\chi) = \frac{\sum\{\nu(b_w) \, : \, w \in W\}\deg(\chi)}{\sum\{\nu(b_w)^{-1}\chi(a_w)\chi(\hat{a}_w) \, : \, w \in W_\psi\}} \, ,$$

where $\hat{a}_w = \hat{e}_\psi e_{n(w)^{-1}} \hat{e}_\psi$. Then $d_\psi(\chi)$ specializes, under $u_r \to q$, $r \in R$, to the degree of the irreducible constituent of $\tilde{\psi}^G$ that corresponds to χ (see [3], p. 405).

THEOREM 3: If e is a primitive idempotent in $M^K(\psi)$, affording the irreducible character χ let χ' be the character afforded by $e' = J_\tau(e)$. If $w \in W_\psi$ then

$$J_\tau(\chi(a_w)) = \frac{\tau(b_w)}{\nu(b_w)} \chi'(a_w) .$$

The dimension functions $d(\tau)$ for the linear characters τ of A have been determined by R. Kilmoyer (see [8]).

THEOREM 4: If χ and χ' are related as in Theorem 3 then

$$d_\psi(\chi') = d(\tau) J_\tau(d_\psi(\chi)) .$$

The formula in Theorem 4 reduces to Green's formula (8f) in [6] when $\psi = 1_H$ and $\tau = \varepsilon$.

The proofs of the theorems will appear in [2].

REFERENCES

1. C. Benson and D. Gay, *On dimension functions of the generic algebra of type* D_n , to appear.

2. C. Benson, L. Grove, and D. Surowski, *Semilinear automorphisms and dimension functions for certain characters of finite Chevalley groups*, to appear in Math. Z.

3. C. Curtis and T. Fossum, *On centralizer rings and characters of representations of finite groups*, Math. Z. 107, 402-406 (1968).

4. C. Curtis, N. Iwahori, and R. Kilmoyer, *Hecke algebras and characters of parabolic type of finite groups with* BN-*pairs*, I.H.E.S., Math. Publ. No. 40, 81-116 (1972).

5. W. Feit and G. Higman, *The non-existence of generalized polygons*, J. of Algebra 1, 114-131 (1964).

6. J. Green, *On the Steinberg characters of finite Chevalley groups*, Math. Z. 117, 272-288 (1970).

7. P. Hoefsmit, *Representations of Hecke algebras of finite groups with* BN-*pairs of classical type*, Ph.D. dissertation, Univ. of British Columbia, Vancouver, B.C.(1974).

8. R. Kilmoyer, *Some irreducible complex representations of a finite group with a* (B,N) *pair,* Ph.D. dissertation, M.I.T., Cambridge, Mass. (1969).

9. D. Surowski, Ph.D. dissertation, Univ. of Arizona, Tucson, Ariz. (1975).

10. T. Yokonuma, *Sur la structure des anneaux de Hecke d'un groupe de Chevalley fini,* C. R. Acad. Sci. Paris 264, A344-A347 (1967).

UNIVERSITY OF ARIZONA
TUCSON, ARIZONA

PART III

REPRESENTATIONS

THE MAIN PROBLEM OF BLOCK THEORY

BY

J. L. ALPERIN*

1. INTRODUCTION

The primary aim of block theory is the discovery and verification of properties of the character table of a finite group. The method is that of congruences, of reduction modulo a prime p. Therefore, results about characters involving the prime p are often provable by these methods. There have been great successes and, of course, many open questions. It is easy, however, to summarize most of them in one main problem which has been the motivation for a great amount of research:

MAIN PROBLEM OF BLOCK THEORY. *If* G *is a finite group then give simple rules which determine, in terms of the* p-*local subgroups, the values of the characters on the* p-*singular elements.*

It is often useful in character theory to rephrase statements in terms of the character table. In this case we

---

*Partially supported by National Science Foundation Grant GP-37575X.

are considering the submatrix consisting of the columns indexed by the conjugacy classes of p-singular elements and we are asking to describe it readily in terms of the p-local subgroups.

In the case that the Sylow p-subgroup of the group G is cyclic the theory provides a beautiful and simple answer. In the general case there are not as yet even any comprehensive conjectures. Therefore, we shall devote ourselves to a much more modest question: To what extent is the *size* of this submatrix of the character table determined by the p-local subgroups?

As for the number of columns there is a completely satisfactory answer. The conjugacy classes of p-elements are determined completely by p-local subgroups by the fusion theorem. Thus, if $x_1, \ldots, x_s$ are representatives of the classes of non-identity p-elements and $x_{i1} = x_i, x_{i2}, \ldots$ are representatives of the classes of p-singular elements of $C(x_i)$ with p-part x_i then all the x_{ij} form a set of representatives of all the p-singular classes of elements of G.

Turning to the rows we are asking to what extent is the number of characters determined by the p-local subgroups. However, the alternating groups A_4 and A_5 have the same 2-local structure and different numbers of characters, so we must rephrase our question. Let p^n be the highest power of p dividing the order $|G|$ of G. The characters of degree divisible by p^n are known to vanish on the p-singular elements. Therefore it is important only to determine the submatrix of the character table given by the characters of degree not divisible by p^n and the columns of p-singular elements. Hence, in our question about size we should only ask if the number of characters of degrees not divisible by p^n is determined by the p-local subgroups. It seems likely that the answer is "yes."

These characters are partitioned into a collection of p-blocks $B_1, \ldots, B_r$, the p-blocks of positive defect. The number of such blocks is determined by the p-local subgroups by the First Main Theorem on Blocks. Therefore, we wish to know the number of characters in a given p-block B of positive defect.

2. Size of a Block

Let D be the defect group of such a block B so that D is a non-identity p-subgroup of G. There is a celebrated question of Brauer in this connection.

PROBLEM. *Is* $|B|$, *the number of characters in* B, *at most* $|D|$?

Let F be an algebraically closed field of characteristic p and let A be the two-sided ideal of the group algebra FG which corresponds to B. Thus A is an indecomposable direct summand of FG . The following result of Brauer's shows the relevance of this algebra to our inquiry:

THEOREM. $|B| = \dim_F Z(A)$.

Here, as usual, $Z(A)$ is the center of the algebra A . Now, we shall relate $|D|$ to the structure of A . Let S be the subalgebra of A which is the image of the group algebra FD under the projection of FG onto A . We have that A is relatively S-projective, that is, if

$$0 \to U \to V \to W \to 0$$

is an exact sequence of A-modules which splits when considered as S-modules then it splits. We now have the following

QUESTION. *If* S *is a subalgebra of an algebra* A *and* A *is relatively* S-*projective when is* $\dim(Z(A)) \leqslant \dim(S)$?

This is not true in general at all as is apparent by considering semi-simple algebras. However, in our case we have a number of additional properties which might be helpful:

1. A has a unit element and this is contained in S .
2. The codimension of the radical of $Z(A)$ in $Z(A)$ is one as is the codimension of the radical of S in S .
3. A is projective as an S-module.

3. McKay's Conjecture

It seems likely that not only is the number of characters in B determined locally but that the number of characters of B of degree divisible exactly by any power of p is also. We know that if $|D| = p^d$ then every character in B is of degree divisible by p^{n-d}, not all by p^{n-d+1} and none by p^n. In particular, putting all the blocks together there should be a way to determine locally the number of characters of G of degree not divisible by p. This number is called the McKay number $m_p(G)$ of G in view of the following startling conjecture of John McKay [4]:

Conjecture 1. *If* G *is a finite simple group and* S *is a Sylow* p-*subgroup of* G *then* $m_p(G) = m_p(N_G(S))$.

However, much more seems to be true, namely

Conjecture 2. *If* G *is any finite group and* S *is a Sylow* p-*subgroup of* G *then* $m_p(G) = m_p(N_G(S))$.

There seems to be no reason to limit oneself to simple groups. It is interesting that we have here a very easily stated conjecture about all finite groups which is not easily decided from a possible classification of all simple groups.

Since the prime p enters and one is dealing with characters it is natural to expect that there is a theorem on blocks that is even stronger. Here's a guess:

CONJECTURE 3. *If* B *is a* p-*block of the finite group* G *and* b *is the* p-*block of* N(D) *corresponding to* B, *then the number of characters of height zero in* B *equals the number of characters of height zero in* b .

We only mention this in passing for the experts. There is some evidence in the way of examples and results for certain types of defect groups D . We shall limit ourselves to only considering the second conjecture for the rest of the paper; when we speak of the conjecture we shall mean it.

To conclude this section let's survey the evidence for the conjecture. First, if the Sylow p-subgroup S is cyclic then it is correct;this is a consequence of very deep theorems and seems to be just as deep. Moreover, in this case, if $M_p(G)$ and $M_p(N(S))$ are the sets of characters of degree not divisible by p there is *no* natural one-to-one correspondence between them; their cardinalities, $m_p(G)$ and $m_p(N(S))$, are equal. In the case that p = 2 various special types of S have been dealt with by similar means.

If the group G has a normal p-complement then the conjecture is valid. This is a consequence of a theorem of Glauberman's and is in fact almost an equivalent statement in this case. We shall go into this in the next section. If the group G is solvable and if the index $|G : N(S)|$ is odd then again the conjecture is true by a result of Isaacs [3] . This is derived as a corollary at the end of a long study. The hypothesis on oddness may not be necessary. Moreover, in this case there is a choice-free one-to-one correspondence between $M_p(G)$ and $M_p(N(S))$.

Finally, there has been work done on just calculating $m_p(G)$ for various groups G . In particular, there are the papers of McKay [5] and W. C. Simpson [6]. In the last section we shall calculate $m_p(GL(n,q))$ for p dividing q and then verify that the conjecture holds for these groups by another calculation.

4. Glauberman's Theorem [2]

This deals with the situation of a group A acting on a group G where A and G are of relatively prime orders and where A is assumed to be solvable. The theorem states that there is a one-to-one correspondence between the A-invariant characters of G and the characters of $F = C_G(A)$. Now to prove this theorem it suffices to establish it in the case that A is a p-group for a prime p , provided that we produce a natural one-to-one correspondence between the characters. Indeed, if this holds and we have A and G in general we can choose a normal p-subgroup A_0 of A and apply induction and what we have to A_0 acting on G and A/A_0 acting on $C_G(A_0)$.

Therefore, let's concentrate our attention on the case of A a p-group. Hence, the semi-direct product $H = A \cdot G$ is a p-nilpotent group with Sylow p-subgroup A and normal p-complement G. There is now a nice description of the characters and the p-blocks of H which goes back to unpublished work of Brauer, but which follows from Fong's theory of characters of p-solvable groups. We start with an orbit $\mathcal{O}$ of characters of G under the action of A. Let ζ be an element of $\mathcal{O}$ and let D be the stabilizer of ζ in A. There is an extension of ζ to a character ζ^* of DG. If we multiply this extension by a character of D lifted to DG and induce to H the result is an irreducible character of H. Moreover, each character of H arises in this way once and only once. Finally, all the characters that arise this way starting with a single orbit $\mathcal{O}$ are exactly the set of characters in a single p-block which has defect group D.

Now let's apply this to calculate $m_p(H)$. Recall that the characters of H are determined by a representative ζ of an orbit of characters of G under A and a character δ of the stabilizer D of ζ. In fact, if ζ^* is the extension of ζ to DG and we denote the lift of δ to DG by δ also then the corresponding character of H is the induced character $(\zeta^*\delta)^H$. Its degree is $\zeta^*(1)\,\delta(1)\,|H:DG|$. This is just $\zeta(1)\delta(1)|A:D|$, which is not divisible by p if, and only if, $|A:D| = 1$ and $\delta(1) = 1$, since $\zeta(1)$ divides $|G|$. But $|A:D| = 1$ exactly when $\{\zeta\}$ is an orbit of characters of G under A . Hence, $m_p(H)$ is just the product of the number of A-invariant characters of G and $|A:A'|$, the latter being the number of linear characters of A . Now $N_H(A) = A F$ and $m_p(AF)$ is equal to the product of the number of characters of F and the number of linear characters of A . Thus $m_p(H) = m_p(N(A))$ if, and only if the number of A-invariant characters of G equals the number of characters of F .

Let's now give a new proof of that result together with a natural map between the two sets of characters. This, as we explained above, gives a new proof of Glauberman's theorem.

THEOREM. *If* $AG = H$ *is a* p*-nilpotent group with Sylow* p*-subgroup* A *and normal* p*-complement* G *then there is a one-to-one correspondence between the* A*-invariant characters of* G *and the characters of* $F = C_G(A)$. *Moreover, if* ζ *is an* A*-invariant character of* G *and* β *is the corresponding character of* F *then* β *is the unique character of* F *occurring in the restriction* $\zeta|F$ *of* ζ *to* F *with multiplicity not divisible by* p .

The first part of the theorem is immediate! Our description of the characters and p-blocks of H implies that there is a one-to-one correspondence between the p-blocks of

H with defect group A and the A-invariant characters of G. In fact, if ζ is an A-invariant character of G then $\{\zeta\}$ is an orbit and the set of characters constructed from this orbit is one of the p-blocks with defect group A. Any larger orbit gives a p-block with a defect group properly contained in A. Hence, by the first main theorem on blocks we have a one-to-one correspondence between the p-blocks of H with defect group A and the p-blocks of $N(A) = AF$ with defect group A. But AF is a direct product (or, it's also p-nilpotent) so there is a one-to-one correspondence between the p-blocks of AF with defect group A and the characters of F. Composition of all these correspondences gives the desired first part of the proof.

Next, let ζ be an A-invariant character of G and let β be a character of F which corresponds to it. Let ζ^* be an extension of ζ to H and let β^* be an extension of β to AF. Since ζ^* and β^* are in p-blocks which correspond it follows that for any element f of F we have the congruence

$$|f^H|\ \zeta^*(f)/\ \zeta^*(1) \equiv |f^{AF}|\ \beta^*(f)/\ \beta^*(1)$$

modulo a suitable prime divisor of p in a number field. This implies right away that

$$|f^G|\ \zeta(f)/\ \zeta(1) \equiv |f^F|\ \beta(f)/\ \beta(1)\ .$$

Thus,

$$\zeta(f) \equiv |f^F|/|f^G|\cdot\zeta(1)/\beta(1)\cdot\beta(f),$$

$$\zeta(f) \equiv |C_G(f):C_F(f)|\cdot|F|/|G|\cdot\zeta(1)/\beta(1)\cdot\beta(f) .$$

Now, let ε be the character of the permutation representation of G on the cosets of F. Since no elements of F not conjugate in F are conjugate in G (by a well known argument using Sylow's theorem) we have that $\varepsilon(f) = |C_G(f):C_F(f)|$. We deduce that

$$\zeta(f) \equiv \varepsilon(f)\cdot|F|/|G|\cdot\zeta(1)/\beta(1)\cdot\beta(f) .$$

There is an integer r_ζ strictly between 0 and p such that

$$r_\zeta \equiv |F|/|G|\cdot\zeta(1)/\beta(1) .$$

Thus, we have

$$\zeta(f) \equiv r_\zeta\ \varepsilon(f)\ \beta(f) .$$

Now the principal p-blocks of H and AF correspond so if ζ is the principal character then so is β. Applying the third congruence above, we deduce that

$$1 \equiv |F|/|G|\cdot\varepsilon(f) .$$

But $|G:F|$ is congruent to one modulo p, by Sylow's theorem, so the last two congruences yield

$$\zeta(f) \equiv r_\zeta\ \beta(f) .$$

Using this congruence in calculating the inner product of $\zeta|F$ and any character of F completes the proof.

5. The General Linear Group

Let q be a power of the prime p and set $G = GL(n,q)$. Let $B = B(n,q)$ be the Borel subgroup of G, namely the group of upper triangular matrices so that B is the normalizer of the Sylow p-subgroup U consisting of all strictly upper triangular matrices (triangular with 1's on the main diagonal).

THEOREM. $m_p(G) = m_p(B)$.

We shall do this by calculating both invariants. There does not seem to be any natural one-to-one correspondence between the two sets $M_p(G)$ and $M_p(B)$ of characters.

LEMMA 1. *There is a one-to-one correspondence between the conjugacy classes of* p'*-elements of* G *and the characters of* G *not divisible by* p .

Here's a sketch of a proof due to J. Olsson. The conjugacy classes of G , namely the similarity classes of nonsingular matrices, are parametrized in terms of partitions and polynomials; the characters are similarly parametrized by the theory of J. A. Green. The p'-elements are exactly the matrices diagonalizable over an algebraic closure of the field of q-elements. Hence, replacing each partition used to describe such elements by the dual partition we arrive at the parameters of the characters of degree prime to p .

LEMMA 2. *The number of conjugacy classes of* p'*-elements of* G *is* $q^n - q^{n-1}$.

Two diagonalizable matrices are similar if,and only if, they have the same characteristic polynomial. Moreover, a matrix is non-singular if, and only if, the constant term of its characteristic polynomial is non-zero. And, if s is a monic polynomial of degree n over $GF(q)$ then its companion matrix has characteristic polynomial s . Hence, the number of conjugacy classes of p'-elements of G is equal to the number of monic polynomials over $GF(q)$ with non-zero constant term, namely $q^{n-1}(q-1) = q^n - q^{n-1}$.

Hence, putting these two results together, we have that $m_p(G) = q^n - q^{n-1}$. Now we turn to B .

LEMMA 3. $m_p(B)$ *is equal to the number of characters of* B/U' .

If β is any character of B then by Clifford's theorem the restriction $\beta|U$ is a multiple of the sum of the characters of U in an orbit of the characters of U under B. That is,

$$\beta|U = m_\beta\ (\xi_1 + \cdots + \xi_t) \ .$$

Now U is a p-group so that if β has degree not divisible by p then all the ξ_i must be linear and so β has kernel containing U'. On the other hand suppose β has kernel containing U' . It follows that $\beta(1) = m_\beta \cdot t$; but m_β is a projective degree of B/U' and t divides $|B/U|$ so β has degree not divisible by p.

Now we introduce some very useful notation. We let $U(i)$, $1 \leqslant i \leqslant n-1$, be the subgroup of U of all elements whose j, $j+1$ entries are zero for $j=1, \cdots, n-1$ except perhaps $j=1$. Thus if $n=4$ then $U(2)$ consists of all the following matrices:

$$\begin{pmatrix} 1 & 0 & * & * \\ 0 & 1 & * & * \\ 0 & 0 & 1 & 0 \\ 0 & 0 & 0 & 1 \end{pmatrix}.$$

Now U' consists of all the elements of U whose j, $j+1$ entries are zero without exception. It follows that U/U' is the direct product of the groups $U(1)/U', U(2)/U', \cdots U(n-1)/U'$ each of which is isomorphic with the additive group of the field $GF(q)$. It is also true that if γ is any linear character of U then there exist unique linear characters γ_i, $i=1, \cdots, n-1$, of $U(i)$ containing U' in their kernel such that $\gamma=\gamma_1 \cdots \gamma_{n-1}$ in the obvious sense. We let the *support* of γ be the set of all i, between 1 and $n-1$, such that γ_i is *not* the principal character of $U(i)$.

LEMMA 4. *Two linear characters of* U *are conjugate in* B *if, and only if, they have the same support.*

Since each $U(i)$ is a normal subgroup of B it follows that if the linear character γ of U has support $I \subseteq \{1, \ldots, n-1\}$ then any B-conjugate of γ also has support I. An easy calculation also shows that any character with support I is conjugate to γ by a diagonal matrix. Or, this can be seen by counting. For, it easy to see that the stabilizer of γ in B has index $(q-1)^{|I|}$ while the set of characters with support I has cardinality $(q-1)^{|I|}$ also.

LEMMA 5. *The number of characters of* B/U' *is equal to the sum*

$$\sum_{l=0}^{n-1} \binom{n-1}{i} (q-1)^{n-1} .$$

Indeed, if I is a subset of $\{1, \ldots, n-1\}$ then the number of characters of U which have support I is $(q-1)^{i}$. Hence, the stabilizer in A, the set of diagonal matrices of G, of one of these linear characters has order $(q-1)^{n-i}$. The construction of characters of a p-nilpotent group works just as well for the group $B/U' = AU'/U' \cdot U/U'$. Since A is abelian we deduce that the orbit of characters with support I produces exactly $(q-1)^{n-i}$ characters of B/U'. The number of subsets of an $n-1$ element set of size i is the binomial coefficient $\binom{n-1}{i}$ so the lemma holds.

Looking at the sum it is now clear that it is just $(q-1)((q-1)+1)^{n-1}$ which is just $(q-1)q^{n-1} = q^{n} - q^{n-1}$. Putting the third and fifth lemmas together we have that $m_p(B) = q^n - q^{n-1}$ and the theorem is proved.

REFERENCES

1. R. Brauer and W. Feit, *On the number of irreducible characters of finite groups in a given block,* Proc. Nat. Acad. Sci. 45 (1959), 361-5.

2. G. Glauberman, *Correspondences of characters for relatively prime operator groups,* Can. J. Math. 20 (1968), 1465-88.

3. I. M. Isaacs, *Characters of solvable groups and symplectic groups,* Amer. J. 95 (1973) 594-635.

4. J. McKay, *A new invariant for simple groups,* Notices A.M.S. 18 (1971), 397.

5. J. McKay, *Irreducible representations of odd degree,* J. Alg. 20 (1972), 416-8.

6. W. C. Simpson, *Irreducible odd representations of* PSL(n,q), J. Alg. 28 (1974), 291-5.

UNIVERSITY OF CHICAGO
CHICAGO, ILLINOIS

ON PROJECTIVE REPRESENTATIONS OF FINITE WREATH PRODUCTS

BY

K. B. FARMER

1. INTRODUCTION.

In this paper, we outline the construction of the factor sets and projective representations of wreath products. These methods and Schur's representation groups are used to discuss the class of projectively monomial groups. We are concerned only with finite groups and representations over the complex field C .

We begin by defining projective representations and describing wreath products. In section 3, we discuss how to construct a complete set of inequivalent, irreducible projective representations of a wreath product. Projectively monomial groups are discussed in section 4.

Part of the work discussed in this paper was done jointly with Professor J. R. Durbin.

2. DEFINITIONS.

A projective representation with factor set ω of the group G of degree n over the field F is a map $D: G \to GL(n,F)$ such that for all x and y in G

$$D(x)\, D(y) = \omega(x,y)\, D(xy) ,$$

where $\omega : G \times G \to F^*$, the nonzero elements of F . The set of all factor sets of G forms an Abelian group. Under a natural equivalence relation, the set of equivalence classes is again an Abelian group, called the multiplier of G and denoted $M(G)$. We identify $M(G)$ with a complete set of inequivalent factor sets of G .

The wreath product $A \wr B$ of the group A with the group B is the semidirect product of the direct sum of $|B|$ copies of A by B . In $A \wr B$, B acts on the direct sum K called the base group by permuting summands.

3. CONSTRUCTION.

Our goal is to obtain the representations of $A \wr B$ in terms of those of A and B . A combination of work by Specht [6], Aisenberg [1], and Kerber [3] does exactly this for linear representations.

We begin the construction of the linear representations of $A \wr B$ by considering the base group K . Its irreducible linear representations are tensor products of those of its components, each identified with A. We choose a subgroup B_1 of B and associate with B_1 an irreducible linear representation of K. This is done by choosing a function $\phi: B \to \hat{A}$, a complete set of inequivalent, irreducible linear representations of A, such that ϕ is constant on the left cosets of B_1 and not constant on the left cosets of any subgroup strictly containing B_1. The representation T corresponding to B_1 and ϕ is defined by

$$T(k) = \phi_{b_1}(k(b_1)) \otimes \cdots \otimes \phi_{b_n}(k(b_n)) ,$$

for k in K and $\{b_i \mid 1 \leqslant i \leqslant n\} = B$. We associate two subgroups with T: the inertia group $H_T = \{x \in A \wr B \mid T^x \sim T\}$ and the inertia factor $H_T^* = \{b \in B \mid T^b \sim T\}$. Note that T^x is a representation of K defined as $T^x(k) = T(x^{-1}kx)$ for k in K. By our choice of ϕ, $B_1 = H_T^*$ and $KB_1 = H_T$. A representation of KB_1 is obtained in two steps. First we extend T to a representation $\tilde{T}$ of KB (see [3]). We next choose an irreducible linear representation V of B_1 and define T^* on KB_1 as

$$T^*(kb) = \tilde{T}(kb) \otimes V(b)$$

for all kb in KB_1. Then $T^* \uparrow A \wr B$ is an irreducible linear representation of $A \wr B$. All such representations are obtained by varying B_1 over all subgroups of B and varying ϕ and V over all possible choices with respect to B_1. To obtain a complete set of inequivalent representations we pick only one representation from each orbit of linear representations of K, where the orbit of T is $\{T' \in \hat{K} | T' \sim T^b$ for some $b \in B\}$.

To use this induction method for projective representations, we must keep in mind these complications. Just as linear representation theory varies from group to group, projective representation theory varies from factor set to factor set. Thus, we must apply the induction procedure to each ω in $M(G)$, each time obtaining a complete set of inequivalent irreducible ω-representations. Secondly, we must express each factor set in $M(A \wr B)$ in terms of three functions: a factor set of K, a factor set of B, and a third function depending on the factor set of K. Thirdly, a projective representation of a direct sum is not related in a simple way to the representations of the summands as in the linear case. In fact, we must apply the induction method each time we add a copy of A in order to obtain a projective representation of K.

Let us assume that we have the factor set ω of $A \wr B$ decomposed in terms of the factor set γ of K, α of B, and a third function β: $B \times K \to C(\text{mod } 1)$. Furthermore, let D be an irreducible γ-representation of K. The construction of a complete set of irreducible ω-representations of $A \wr B$

is based on the method given by Mackey in [4]. We obtain the inertia group H_D and inertia factor H_D^* of D by considering x-conjugates of D , where

$$D^x(k) = \frac{\omega(x^{-1},kx)\omega(k,x)}{\omega(x^{-1},x)} \cdot D(x^{-1}kx)$$

for k in K . Note that the x-conjugate of D is defined so that it has the same factor set γ as D . To extend D to a representation of $H_D = K\,H_D^*$, we use the fact that for each $b \in H_D^*$ there exists a nonsingular matrix, say S_b , such that

$$D^b(k) = S_b^{-1} D(k)\, S_b$$

for all k in K. The irreducibility of D implies that the map $b \to S_b$ is a λ-representation of H_D^* for some factor set λ of H_D^* . Let $\{V_i \mid 1 \leq i \leq n\}$ be a complete set of inequivalent, irreducible $\alpha\lambda^{-1}$-representations of H_D^* . The set $\{U_i \mid 1 \leq i \leq n\}$, where U_i is defined on kb in H_D by

$$U_i(kb) = D(k) \cdot S_b \otimes V_i(b)$$

is a set of ω-representations of H_D . By induction we lift this set to a set of irreducible ω-representations $\{U_i \uparrow A \sim B\}$ of $A \sim B$. Projective induction is slightly different from linear induction; again, care must be taken to preserve factor sets. To obtain a set of inequivalent ω-representations, we must consider orbits of the γ-representations of K .

We conclude this section with remarks concerning the

multiplier decomposition. For abelian-wreath-cyclic groups, we establish a one-to-one correspondence between the factor sets and certain lower triangular matrices (see [2]). This correspondence is useful in two ways. The matrices for the factor sets of K which appear in a decomposition of some ω in $M(A \wr B)$ exhibit symmetric properties. Also, these matrices yield information about the number and degrees of the corresponding irreducible projective representations.

4. Projectively Monomial Groups.

A group is projectively monomial if every projective representation is equivalent to a monomial representation. A useful tool in studying the class of projectively monomial groups is Schur's representation group. The group G^* is a representation group of G if $Z(G^*)$ contains a subgroup M of order $|M(G)|$, if there is an epimorphism $\phi: G^* \to G$ with kernel M, and if for every projective representation D of G there is a linear representation T of G^* such that $D(\phi(g^*)) = T(g^*)$ for all g^* in G^*.

We prove in [2] that G is projectively monomial if and only if G^* is monomial. Using this fact and Huppert's result on monomial groups (see [5]), we see that finite supersolvable groups are projectively monomial. Using this representation group approach, we also find projectively monomial groups which are not supersolvable.

On the other hand, the class of projectively monomial groups is strictly contained in the class of monomial groups. We obtain examples of monomial groups which are projectively monomial via the induction construction outlined in section 2.

REFERENCES

1. N. N. Aizenberg, *On the representations of the wreath product of finite groups*, Ukrain Mat. Z. 13, no. 4 (1961) 5-12 (Russian).

2. J. R. Durbin and K. B. Farmer, *On Induced Projective Representations of Finite Groups*, to appear.

3. A. Kerber, Representations of Permutation Groups, *Lecture Notes in Mathematics*, Vol. 240, Berlin-New York: Springer-Verlag 1971.

4. G. W. Mackey, *Unitary representations of group extensions*, I., Acta. Math. 99 (1958) 265-311.

5. W. R. Scott, *Group Theory*, Prentice-Hall, Englewood Cliffs, N.J., 1964.

6. W. Specht, *Eine Verallgemeinerung der Permutationsgruppen*, Math. Z. 37 (1933) 321-341.

UNIVERSITY OF FLORIDA
GAINESVILLE, FLORIDA

THEOREMS RELATING FINITE GROUPS AND DIVISION ALGEBRAS

BY

CHARLES FORD

The connection between finite groups and rational division algebras is apparent from Wedderburn's Theorems. This relationship has been explicated by R. Brauer [5] and E. Witt [17] and in the recent work of several authors. This paper will present some recent results, Ford [11], which extend the Brauer-Witt results and the work of Ford [8], and M. Benard and M. Schacher [3]. The results identify a very special class of groups which determine the structure of the division algebras. We begin this paper with two examples which illustrate the alternatives described by the main theorem. Proofs will be published in the Journal of Algebra.

We first present a set of finite groups whose faithful irreducible complex representations have rational enveloping algebras which are division algebras. The first examples of such groups appeared in a 1930 paper of Brauer [4]. In 1955 all such groups were classified by Amitsur [1].

We will use Q to denote the rational field. For a prime p and integer m denote by m_p the p-part of m, that is the highest power of p dividing m. We will use absolute value bars to denote order. ε_m will denote a primitive m-th root of unity.

EXAMPLE 1. Consider a group which generates a division algebra whose dimension over its center is a power of a prime p. If p is odd then Amitsur shows that the group is the semi-direct product of two cyclic groups.

We now describe the first example in a series of definitions and deductions. Let n be a positive integer, let p and q be primes with $(q-1)_p = p^n$, and let F be a group of order qp^{2n} with cyclic Sylow subgroups. Assume that conjugation by an element of order p^{2n} induces an automorphism of order p^n of the normal q-Sylow subgroup. Let Z be the center of F, (Z has order p^n), let H be the cyclic normal subgroup of F of order qp^n, let λ be a faithful linear complex character of H, let $\xi = \lambda^F$ be the irreducible induced character of F of degree p^n, let $Q(\xi)$ be the field extension of Q generated by the values of ξ on F, and let Δ be the rational enveloping algebra of a representation affording ξ.

The following has been proved about Δ. (See Ford [10]).

(i) Δ is a simple algebra of dimension p^{2n} over its center $Q(\xi)$.

(ii) $\varepsilon_{p^n} \in Q(\xi)$.

(iii) Δ is a division algebra.

The important result here is (iii), that Δ is a division algebra.

The smallest such group of odd order has $p = 3$, $q = 7$ and $|F| = 63$. The group and the division algebra are generated by the two matrices

$$\begin{vmatrix} \varepsilon_7 & & \\ & \varepsilon_7^2 & \\ & & \varepsilon_7^4 \end{vmatrix} \qquad \begin{vmatrix} 0 & 0 & \varepsilon_3 \\ 1 & 0 & 0 \\ 0 & 1 & 0 \end{vmatrix}$$

where ε_3 and ε_7 are primitive 3rd and 7th roots of unity respectively.

EXAMPLE 2. The second example is a family of groups each of which generates an algebra of dimension p^{4n} over its center. The algebra contains a division algebra of dimension p^{2n} over the center. None of these groups contains a subgroup or a factor group isomorphic to any of the groups in example 1, or to any of the groups, classified by Amitsur, which generate a division algebra.

Let n be a positive integer and let p, q, and r be primes such that

$$(q-1)_p = (r-1)_p = p^n.$$

Let $E = Q(\varepsilon_{p^n},\varepsilon_q,\varepsilon_r)$, let σ be an automorphism of E of order p^n fixing ε_{p^n} and ε_r, let τ be an automorphism of order p^n fixing ε_{p^n} and ε_q, and let k be the subfield of E fixed by σ and τ.

We will define a group and an algebra generated by the roots of unity and two elements u_σ and u_τ such that:

$$u_\sigma^{-1}\,\gamma\,u_\sigma = \gamma^\sigma \quad \text{for all } \gamma \in E .$$

$$u_\tau^{-1}\,\gamma\,u_\tau = \gamma^\tau \quad \text{for all } \gamma \in E .$$

$$|u_\sigma| = |u_\tau| = p^n .$$

$$u_\tau^{-1}\,u_\sigma^{-1}\,u_\tau\,u_\sigma = \varepsilon_{p^n} .$$

Let F be the group generated by ε_{p^n}, ε_q, ε_r, u_σ, and u_τ ($|F| = qrp^{3n}$), and let H be the subgroup generated by ε_{p^n}, ε_q, and ε_r. Z, the center of F, is $\langle\varepsilon_{p^n}\rangle$. Next, let λ be a faithful linear character of H, let $\xi = \lambda^F$ (ξ is an irreducible character of F of degree p^{2n}), and let $Q(\xi)$ be the field extension of Q generated by the values of ξ. Finally, let

$$\Delta = \sum_{i,j=1}^{p^n} E\,u_\sigma^i\,u_\tau^j .$$

Then (1) Δ is a simple algebra isomorphic to the rational enveloping algebra of a representation affording ξ and:

(i) $Q(\xi) \cong$ center of Δ ,

(ii) $[\Delta : Q(\xi)] = p^{4n}$,

(iii) $\varepsilon_{p^n} \in Q(\xi)$.

Also (2) Δ contains a division algebra component of dimension p^{2n} over the center provided one of the following congruences does not have integer solutions:

(i) $x^p \equiv q(\text{mod } r)$

(ii) $x^p \equiv r(\text{mod } q)$

These algebras are discussed in [13] for $n = 1$. The statement (2) above will be proved later in this paper from the main result.

Now suppose we are given a prime p and an integer n. According to the Dirichlet density theorem, the arithmetic progression $1 + xp^n$ contains primes whose density is $1/\varphi(p^n)$. The subset of these primes of the form $1 + xp^{n+1}$ has density $1/\varphi(p^{n+1})$. Thus there are infinitely many primes r with $(r-1)_p = p^n$. Since the integers modulo r form a cyclic group we can find an integer b whose order modulo r is p^n . According to the Chinese Remainder theorem we can find an integer t satisfying

$$t \equiv 1 \pmod{p^n}$$

$$t \equiv b \pmod{r} .$$

It is easy to see that t is relatively prime to pr. We now choose a prime q in the arithmetic progression $t + xp^n r$ which is not of the form $t + xp^{n+1} r$. Since the densities of primes in the two progressions are $1/\varphi(p^n r)$ and $1/\varphi(p^{n+1} r)$ respectively, there are an infinite number of choices for q. We have $(q - 1)_p = p^n$ and $q \equiv b (\mathrm{mod}\ r)$. From the definition of b we know that the congruence

$$x^p \equiv q (\mathrm{mod}\ r)$$

has no integral solution.

Thus given p and n we can find infinitely many q and r with which to construct a group of the type in example 2. The corresponding algebra contains a division algebra of dimension p^{2n}.

GENERAL CASE. We wish to discuss the group algebra of an arbitrary finite group. The division algebras which occur are closely related to certain sections of a group whose structure is a blend of the two types of groups just discussed. A section of a group is a factor group of a subgroup.

We now turn to the general case where we assume the following: G is a finite group, χ is an irreducible complex character of G, k is an algebraic number field containing the values of χ, and Λ is the k-enveloping algebra of a representation affording χ. Then Λ is a central simple k-algebra whose division algebra component has dimension m^2 over k for some integer $m = m_k(\chi)$ known as the Schur index of χ or of Λ over k.

The work of Brauer and Witt relates Δ to the structure of G. Assume $m_p = p^n$ for a prime divisor p of m.

BRAUER-WITT THEOREM. There exists a section F of G with the following special structure:

1. F contains a cyclic normal self-centralizing subgroup H of index a power of p.

Therefore F/H is isomorphic to an abelian p-group of automorphisms of H.

Let λ be any faithful linear character of H and $\xi = \lambda^F$. Then $F/H \cong \mathrm{Gal}(k(\lambda)/k(\xi))$.

2. The character ξ has the properties

(i) $m_k(\xi) = p^n$,

(ii) $p \nmid (\xi^G, \chi)$,

(iii) $p \nmid [k(\xi):k]$.

See Curtis and Reiner [7, Theorem 70.28, p. 477].

This theorem has been extended by the following two results:

THEOREM (Ford [8,9]). For all primes p, p^n divides $|Z(F)|$. Therefore $\varepsilon_{p^n} \in k(\xi)$.

THEOREM (Benard and Schacher [3]). For all primes $\varepsilon_{p^n} \in k = k(\chi)$. Therefore $\varepsilon_m \in k$.

The first theorem was proved for odd p in [8, Theorem 2, p. 637]. This proof also holds for $p = 2$ provided $\sqrt{-1} \in k$. When $\sqrt{-1} \notin k$ it is shown in [9, p. 4] and [3, proof of Lemma 1] that $n = 0$ or 1. If $n = 1$ then 2 must divide $|H|$ and since H is cyclic and normal in F we must have 2 dividing $|Z(F)|$. The first theorem is therefore valid for all primes p.

From the group F in the Brauer-Witt Theorem we can construct an algebra Γ. Let Γ be the enveloping algebra of a representation affording ξ and let

$$G = \mathrm{Gal}(k(\lambda)/k(\xi)) \cong F/H .$$

Then Γ is a central simple $k(\xi)$ algebra and

$$\Gamma = (k(\lambda)/k(\xi), G)$$

is the crossed product of the field $k(\lambda)$ with G.

LOCALIZATION. The Schur index of Γ is the least common multiple of the indices of the localizations of Γ at the primes of k. All finite primes containing the same rational prime q give the same index, which we call the q-local index, according to Benard [2, Theorem 1]. When $q = 2$ the local index is either 1 or 2 and can equal 2 only if $\sqrt{-1} \notin k$ and $p = 2$. See Witt [17] or Janusz [14, Prop. 3.2, p. 22]. The infinite primes give index 1 or 2 with index 2 only when $\sqrt{-1} \notin k$.

We shall ignore infinite primes and the case $p = q = 2$ by assuming either $\sqrt{-1} \in k$ or that the 2 part of the index is greater than 2, that is $n > 1$. When applying our results later we will treat the case $p = 2$ and $n = 1$ separately.

Since the index of Γ is a p-power, it must equal the q-local index for some prime q. The process of localization is to extend the base field k of Γ by the q-adic completion k_q. The resulting algebra Γ_q has the same index as an algebra Δ which is the crossed product of $k(\lambda)$ with a subgroup H of G. This algebra corresponds to a subgroup K of F. To describe K we proceed as follows: Let Q be the q-Sylow subgroup of H, let T be the complement to Q in H, and let α be the automorphism of T sending every element to its q-th power. We use Aut to denote the automorphism group.

Consider the sequence

$$H \to F \to \mathrm{Aut}(H) \xrightarrow{\text{res}} \mathrm{Aut}(T)$$

where the last map is restriction. Choose K with $H \subseteq K \subseteq F$ to be the largest subgroup of F satisfying

$$H \to K \to \mathrm{Aut}(H) \to \langle \alpha \rangle \ .$$

Let $\mathcal{H}$ be the subgroup of $\mathcal{G}$ such that

$$K/H \cong \mathcal{H} \subseteq \mathcal{G} \ .$$

$\mathcal{H}$ is the decomposition group for q .

REDUCTION THEOREM. For some prime q the crossed product algebra $\Delta = (k(\lambda),\mathcal{H})$ has index p^n where p^n divides $q - 1$.

This reduction has been used previously by many authors. See Witt [17], Lorenz [16] and Yamada [18]. The first important observation always made is that $\mathcal{H}$ is generated by two elements. To see this, observe that the kernel of the restriction to Aut(T) is isomorphic to a subgroup of Aut(Q) . Now Aut(Q) is cyclic for odd q and for q = 2 the automorphisms of Q fixing the element of order 4 form a cyclic group. Assuming $\sqrt{-1} \in k$ if q = 2, the intersection of the image $\mathcal{G}$ of F in Aut(H) with the subgroup of Aut(H) isomorphic to Aut(Q) is cyclic. This intersection is exactly the kernel of the restriction map into Aut(T). Thus H has this cyclic subgroup for which the factor group is also cyclic.

Let c be the smallest integer such that the image of K in $< \alpha >$ is generated by α^c. Choose u and v such that u is a preimage of α^c in G, v generates the preimage in G of the kernel of the restriction, and u and v lie in a p-Sylow subgroup of K. Then $K = < u,v,H >$.

Define an element z in K by the equation

(1) $$u^{-1}vu = v^{q^c}z .$$

If the elements u and v are replaced by other representatives of the same cosets of H, equation (1) above can be shown to hold for the same element z. We know for any y in H of order prime to q that $u^{-1}yu = y^{q^c}$. The element z defined above shows by how much this conjugation formula fails to apply for $y = v$, an element not in H. Our main theorem shows that z determines the index of Δ.

MAIN THEOREM. The element z is central in K. The order $|z|$ divides the order of v modulo H. If ψ is any faithful irreducible character of K, then $m_k(\psi) = p^n = |z|$.

A GENERAL THEOREM

An important consequence can be drawn from this theorem. We will use [,] to denote the commutator of group elements.

THEOREM A. Let G be a finite group with an irreducible complex character χ. Let k be an algebraic number field for which the index $m = m_k(\chi)$ has $m_p = p^n$. Then either

(i) p^{2n} divides the exponent of G, or

(ii) p^n divides the exponent of G'.

PROOF. We multiply the equation (1) by v^{-1} on the left to get $v^{-1}u^{-1}vu = v^{q^c-1}z$. Solving for z gives

$$z = [v,u]\, v^{-(q^c-1)} . \tag{2}$$

Since z is central, the two factors on the right commute. Therefore the product has order equal to the least common multiple of the orders of the factors. Hence either

(a) $|z|$ divides $|[v,u]|$, or

(b) $|z|$ divides $|v^{q^c-1}|$.

Now $|z| = p^n$. Thus if (a) holds, we have (ii) above. Suppose (b) holds. We know p^n divides $q - 1$ and $q^c - 1$. Therefore p^{2n} divides $|v|$ and (i) above holds. When $p = 2$ and $\sqrt{-1} \notin k$ then $n = 1$ or 0. Unless $n = 0$, some element in the factor set discussed in [8] is $\neq 1$ and the result follows.

EXAMPLE 1. The alternatives of this theorem are perfectly illustrated by examples 1 and 2. When $[v,u] = 1$, equation (2) becomes

$$z = v^{-(q^c-1)} .$$

This happens in example 1 where $u = 1$. The group T described before the Reduction Theorem is the center Z. Since Z and z both have order p^n, $Z = \langle z \rangle$. Since $p^n = (q-1)_p$ all elements y in T satisfy $y^{q-1} = 1$ and therefore $c=1$. Thus $z = v^{-(q-1)}$, and $|z| = (q-1)_p|v|$. It follows that $|v| = p^{2n}$ and v is a generator of the p-Sylow subgroup of F.

EXAMPLE 2. Let us identify for the groups in example 2 the elements u and v. For the q-localization the kernel of the restriction is generated by $v = u_\sigma$ which acts non-trivially only on ε_q. Next we must choose a preimage u of α^c, where α is the automorphism on $T = \langle \varepsilon_{p^n}, \varepsilon_r \rangle$ which sends every element to its q-th power. Since p^n divides $q-1$, ε_{p^n} is fixed by α. Regarding α and τ as automorphisms only on $\langle \varepsilon_r \rangle$, then $\langle \alpha^c \rangle = \langle \alpha \rangle \cap \langle \tau \rangle$. Let ℓ be the integer with $\langle \tau^{p^\ell} \rangle = \langle \alpha^c \rangle$. By possibly making a new choice for τ we can assume $\tau^{p^\ell} = \alpha^c$. Thus we choose $u = u_\tau^{p^\ell}$.

Continue to regard α and τ as automorphisms on

$<\varepsilon_r>$. Then ℓ is the largest integer such that α^c is a p^ℓ power in $<\tau>$. Now $(r-1)_p = p^n = |\tau|$ so $<\tau>$ is the p-Sylow subgroup of $Aut<\varepsilon_r>$. Therefore $<\alpha^c>$ is the p-Sylow subgroup of $<\alpha>$. Since $Aut<\varepsilon_r>$ is cyclic it follows that ℓ is the largest integer such that α is a p^ℓ power in $Aut<\varepsilon_r>$. Now α is the q-power map so ℓ is the largest integer for which

$$x^{p^\ell} \equiv q(\text{mod } r)$$

has an integral solution.

Since $v = u_\sigma$ has order p^n we have $v^{-(q^c-1)} = 1$. Thus equation (2) yields $[v,u] = z^{-1}$. From the equation

$$[u_\tau, u_\sigma] = \varepsilon_{p^n}$$

we substitute to get $[v,u]^{-1} = \varepsilon_{p^n}^{p^\ell}$. Therefore we have

$$z = \varepsilon_{p^n}^{p^\ell} \quad \text{and} \quad |z| = p^{n-\ell} .$$

Thus we have identified for the q-localization the subgroup K and the element z. Combining the results of the previous two paragraphs we see that a group in example 2 has q-local index $p^{n-\ell}$ where ℓ is the largest integer for which the congruence

$$x^{p^\ell} \equiv q(\text{mod } r)$$

has an integral solution. Similarly for the r-local index. Thus the index of the division algebra component is $p^{n-\ell}$ where ℓ is the minimum of ℓ_1 and ℓ_2 which are the largest integers for which the following respective congruences have integer solutions

$$\text{(i)} \quad x^{p^{\ell_1}} \equiv q(\text{mod } r)$$

$$\text{(ii)} \quad x^{p^{\ell_2}} \equiv r(\text{mod } q)\ .$$

RECIPROCITY

The congruences (i) and (ii) above are in some sense reciprocal. For $p = 2$ the law of quadratic reciprocity applies. If $n > 1$, then q and r are $\equiv 1$ (mod 4) and quadratic reciprocity says that $\ell_1 = 0$ iff $\ell_2 = 0$. This means the index is 2^n iff the q-local and r-local indices are both 2^n. Suppose $n = 1$. This means q and r are $\equiv 3$ (mod 4). Then according to quadratic reciprocity exactly one of ℓ_1, ℓ_2 is 1 and the other is 0. Thus the index is 2 for any choice of q and $r \equiv 3$(mod 4). Only in this case, when $Q(\xi)$ is real, can we have a non zero index at the infinite primes. The automorphism complex conjugation is $\sigma\tau$ and the corresponding algebra element is $u_\sigma u_\tau$. The index at infinite primes will be the order of $(u_\sigma u_\tau)^2$. We know $u_\tau^{-1} u_\sigma^{-1} u_\tau u_\sigma = \varepsilon_2 = -1$ and therefore $u_\tau u_\sigma = u_\sigma u_\tau \varepsilon_2$. Thus

$$(u_\sigma u_\tau)^2 = u_\sigma u_\tau u_\sigma u_\tau = u_\sigma u_\sigma u_\tau \varepsilon_2 u_\tau = \varepsilon_2 .$$

This shows the index at the infinite primes is always 2. Thus for any choice of q and $r \equiv 3 \pmod 4$ the index is always 2. The local index is two at primes extending exactly one of q or r.

The Artin Reciprocity Law gives conditions which theoretically determine when congruences (i) and (ii) have solutions for arbitrary p and n. This involves factoring the ideals generated by p, q and r in the ring of integers of $Q(\varepsilon_{p^n})$. The Gauss sum S for ε_q and σ over the field $Q(\varepsilon_{p^n})$ can be used to study the groups in example 2. This element S is well known from Galois theory as the generator of a radical extension of dimension p^n over $Q(\varepsilon_{p^n})$. The ideal in the ring of algebraic integers of $Q(\varepsilon_{p^n})$ generated by S^{p^n} factors into a product of powers of the $\varphi(p^n)$ different prime ideals containing q. The primes occur with different multiplicities, which range over all positive integers $\leqslant p^n$ which are relatively prime to p. See Lang [15, Theorem 10, p. 97]. This shows that in example 2 the q-local invariants are "uniformly distributed", a fact which has been proved in general by Benard [2].

Gauss sums have been very useful in studying division algebras as in Ford [12] and Janusz [14]. We believe that the interplay between division algebras, Gauss sums, and higher reciprocity can illuminate all three of these areas.

CONCLUDING REMARK

The Main Theorem attempts to describe the arithmetic structure of Δ in terms of the group theoretic structure of F. The biggest problem here is the choice of the integer c which can be chosen as the integer congruent modulo $|\alpha|$ to the residue class degree of $k(\xi)$ over Q. In special cases c can be described in terms of congruences among primes as we have done in examples 1 and 2 and as Amitsur does to achieve his explicit results. However if k is complicated, so is the problem of determining c. Even if we are only interested in the index over the rational field, the requirement that $k = k(\chi)$ in the reduction forces a dependence on the character χ of the original group in determining c.

REFERENCES

1. S. A. Amitsur, *Finite subgroups of division rings*, Trans. Amer. Math Soc. 18 (1955), 361-386.

2. M. Benard, *The Schur subgroup I*, J. of Algebra 22 (1972), 374-377.

3. ________ and M. Schacher, *The Schur subgroup II*, J. of Algebra 22 (1972), 378-385.

4. R. Brauer, *Untersuchungen über die arithmetischen Eigenschaften von Gruppen linearer Substutionen*, Math. Zeit. 31 (1930), 733-747.

5. ________, *On the algebraic structure of group rings*, J. Math. Soc. Japan 3 (1951), 237-251.

6. ________, *On the representations of groups of finite order*, Proc. Internat. Cong. Math., Cambridge, 1950 Vol. 2, 33-36.

7. C. Curtis and I. Reiner, *The Representation Theory of Finite Groups and Associative Algebras*, Interscience, New York, 1962.

8. C. Ford, *Some results on the Schur index of a representation of a finite group*, Can. J. Math 22 (1970), 626-640.

9. ________, *More on the Schur index of a representation of a finite group.* Unpublished.

10. ________, *Finite Groups and Division Algebras,* L'Enseignement Math. XIX (1973), 313-327.

11. ________, *Groups which determine the Schur index of a representation,* to appear in J. of Algebra.

12. ________, *Pure, normal, maximal subfields for division algebras in the Schur subgroup,* Bull. Amer. Math. Soc. 78 (1972), 810-812.

13. ________ and G. Janusz, *Examples in the theory of the Schur Group,* Bull. Amer. Math. Soc. 79 (1973), 1233-1235.

14. G. Janusz, *Generators for the Schur group of local and global number fields,* University of Illinois preprint, Urbana-Champaign, 1974.

15. S. Lang, *Algebraic Number Theory,* Addison Wesley, Reading, Massachusetts, 1970.

16. F. Lorenz, *Bestimmung der Schurschen Indizes von Charackteren endlicher Gruppen,* dissertation at Tübingen University, 1966.

17. E. Witt, *Die algebraische Struktur des Gruppenringes einer endlichen Gruppe über einem Zahlkörper,* J.Reine. Angew. Math. 190 (1952), 231-245.

18. T. Yamada, *Characterization of simple components of the group algebras over the p-adic number field,* J. Math. Soc. Japan 23 (1971), 295-310.

WASHINGTON UNIVERSITY
ST. LOUIS, MISSOURI

EXCEPTIONAL CHARACTERS OF FINITE GROUPS WITH A FROBENIUS SUBGROUP

BY

DAVID A. SIBLEY

We will discuss the answers to some problems involving exceptional characters of finite groups. In particular, the following situation is of interest.

HYPOTHESIS (*). G *is a finite group and* p *an odd prime. A Sylow* p-*group* P *of* G *is a trivial intersection set in* G *and its normalizer* $N_G(P) = N$ *is a Frobenius group with Frobenius kernel* P.

If in addition G has at least two classes of p-elements it is known that exceptional characters exist for both G and N . These may be described (for either G or N) as exactly those irreducible characters which are not constant on $P^{\#}$. There is a one-to-one correspondence between the exceptional characters λ of N and the exceptional characters Λ of G. If, in addition, P is abelian there is an integer c and a sign $\delta = \pm 1$ such that if λ corresponds to Λ , then

$$\lambda(g) = \delta\Lambda(g) + c$$

for all $g \in P^{\#}$. Our first theorem concerns the value of c.

THEOREM 1. *Assume hypothesis* (*) *holds, that* P *is abelian and* G *has at least three classes of* p-*elements. Then* $c = 0$.

This result is analogous to the situation when $|P| = p$ (see [1]). Unfortunately, our methods seem to yield no information when there are exactly two classes of p-elements.

The proof of Theorem 1 depends on a detailed analysis of class multiplication constants for G and N. A critical piece of information is that the constants for G and N involving the same three classes of p-elements are congruent modulo $|P|$. If $c \neq 0$ the usual character formula for these constants yields information about their values which is shown to be contradictory.

When P is non-abelian we can prove a similar theorem, although the proof is quite different. Here it is still known that the exceptional characters of G and N are in one-to-one correspondence, although at first sight this correspondence is less well behaved. Here

$$\lambda(g) = \delta\Lambda(g) + \theta(g)$$

for all $g \in P^{\#}$ where θ is a generalized character of N satisfying certain properties.

THEOREM 2. *Assume hypothesis* (*) *holds and that* P *is non-abelian. Then* $\delta = \pm 1$ *may be chosen so that* $\theta = 0$.

Thus, the only case of hypothesis (*) in which the exceptional character values for G on p-elements are not determined up to sign is when G has exactly two classes of p-elements.

Theorem 2 is proved by first showing that the exceptional characters of N are coherent, in the sense of Feit ([2], §31). This argument uses information about class multiplication constants similar to that used in the proof of Theorem 1.

The ideas used in the coherence proof generalize readily to handle more general isometries, and with some work one can say something about cases where N is not quite a Frobenius group. For example, one can solve the coherence problem for type V groups which arises in the proof of the solvability of groups of odd order [3] using these techniques.

Proofs of Theorems 1 and 2 will appear elsewhere.

REFERENCES

1. R. Brauer, *On Groups whose order contains a prime number to the first power,* Amer. J. Math. 64(1942), 401-420.

2. W. Feit, *Characters of Finite Groups,* Benjamin, New York-Amsterdam, 1967.

3. W. Feit and J. G. Thompson, *Solvability of groups of odd order,* Pacific J. Math. 13(1963), 775-1029.

PENNSYLVANIA STATE UNIVERSITY
UNIVERSITY PARK, PENNSYLVANIA

SIMPLE GROUPS WITH A CYCLIC SYLOW SUBGROUP

BY

LEO J. ALEX

Let G be a finite simple group satisfying the following hypotheses.

I. $|G| = p^k m$, p a prime, $(p,m) = 1$, k a positive integer.

II. If P is a Sylow p-subgroup of G, then $|N(P) : P| = 2$.

III. If $x \in P^{\#}$, then $C_G(x) = P$.

It is an immediate consequence of a recent result of S. Smith and A. Tyrer [5] that the Abelian Sylow p-subgroup P is cyclic.

In recent work [1], [2] the author has studied groups satisfying these hypotheses in the case $k = 1$. First of all in [1] the following result was obtained by restricting the possible prime divisors of m to be 2, 3, 5 or 7.

THEOREM 1. Let G be a finite simple group of order $2^a 3^b 5^c 7^d p$, satisfying hypothesis II. Then G is isomorphic to PSL(3,4) or PSL(2,q) with q = 5, 7, 9, 8, 16, 25, 27 or 81.

In [2] these groups were considered with no explicit restriction on the divisors of m. However, the degree of a non-identity ordinary irreducible character in the principal p-block, $B_0(p)$, was restricted. The main result here is the following theorem.

THEOREM 2. Let G be a finite simple group of order pm, p a prime, $(p,m) = 1$, such that G satisfies hypothesis II. If G has a non-identity ordinary irreducible character χ, in $B_0(p)$ with $\chi(1) \leqslant 25$, then G is isomorphic to $PSL(2,q)$ with $q = 5, 7, 9, 8, 11, 13, 16, 23$ or 25.

In this paper, Theorems 1 and 2 are extended to the case $k > 1$. Here we add hypothesis III that the Sylow p-subgroup is strongly self-centralizing.

In Chapter IV of his 1974 Ph.D. thesis [4], D.C. Morrow used exceptional character theory to derive information about character values for characters in $B_0(p)$, for groups satisfying hypotheses I, II, III. This information is essentially identical to the corresponding character information provided by Brauer [3] in the case $k = 1$.

Using this character information and the techniques developed in [1], [2], the following extensions of Theorems 1 and 2 are obtained.

THEOREM 3. Let G be a finite simple group of order $2^a\, 3^b\, 5^c\, 7^d\, p^k$, such that G satisfies hypotheses II and III, then G is isomorphic to PSL(3,4) or PSL(2,q) with q = 5, 7, 9, 8, 16, 25, 27, 49 or 81.

Note that PSL(2,49) appears in Theorem 3 but does not appear in Theorem 1.

The following is the extension we obtain of Theorem 2.

THEOREM 4. Let G be a finite simple group satisfying hypotheses I, II and III such that G has a non-identity character, χ, in $B_0(p)$ with $\chi(1) \leqslant 25$. Then G is isomorphic to PSL(2,q) with q = 5, 7, 9, 8, 11, 13, 16, 17, 19, 23 or 25.

Notice that Theorem 4 gives two groups, PSL(2,17) and PSL(2,19), not given by Theorem 2.

The first step in proving Theorems 3 and 4 is consideration of the equation relating the degrees of the ordinary irreducible characters in the principal p-block. This degree equation for $B_0(p)$ has the form

$$1 + x = y \tag{1}$$

where $(xy,p) = 1$. After the solutions to equation (1) are found, they are considered in turn to determine what groups, if any, are involved.

The class multiplication constants, a_{uvw}, where u and v are p-regular elements and w is a p-singular element are extremely productive here. It turns out that

$$(2)\qquad a_{uvw} = \frac{|G|\,[\chi(u) - x][\chi(v) - x]}{|C(u)|\,|C(v)|\,x(x+1)}$$

where χ is a non-identity character of $B_0(p)$ having degree x. Since these coefficients must be positive integers, they give quite precise information about centralizers of elements of G. In particular a very good bound on $|G|$ in terms of the character degree x can be obtained. The following Lemma gives this bound.

LEMMA 5. Suppose G satisfies hypotheses I, II, III with degree equation (1) for $B_0(p)$. Let q, r be primes such that $(q,pxy) = 1$, $(q^{a+1},|G|) = q^a$, $(r^{c+1},xy) = r^c$, $(r,p) = 1$, $(r^{b+1},|G|) = r^b$. Then

1) $q \leqslant (x + 3)/2$, $r \leqslant x + 1$, $p^k \leqslant x + 2$,

2) a is even, $b \equiv c \pmod 2$, and

3) $q^{a/2} - q^{a/2-1} < x$, $r^{(b+c)/2} - r^{(b+c)/2-1} \leqslant x$.

This Lemma follows immediately from Theorem 2.8 of [2] and the character information developed in [4].

We next sketch the proofs of Theorems 3 and 4. If G satisfies the hypotheses of Theorem 3 with $k > 1$, then the character relations for $B_0(p)$ imply that the only possible solutions to the degree equation (1) are $(x,y) = (7,8)$ with $|P| = 9$ and $(x,y) = (48,49)$ with $|P| = 25$. In the former case it is well known that G must be isomorphic to PSL(2,8), and in the latter case a count of Sylow 5-subgroups implies that $|G| = 2^4\ 3 \cdot 5^2\ 7^2$. It is then an easy matter using subgroup information provided by the class multiplication constants (2) to verify that G is isomorphic to PSL(2,49). Of course, if $k = 1$, Theorem 3 reduces to Theorem 1. Thus if G satisfies the hypotheses of Theorem 3, then G is isomorphic to PSL(3,4) or PSL 2,q) with $q = 5$, 7, 9, 8, 16, 25, 27, 49 or 81.

Next if G satisfies the hypothesis of Theorem 4 with $k > 1$, then the character relations for $B_0(p)$ imply that the only possible solutions to (1) are $(x,y) = (7,8)$ with $|P| = 9$, $(x,y) = (10,11)$ with $|P| = 9$, $(x,y) = (16,17)$ with $|P| = 9$, $(x,y) = (19,20)$ with $|P| = 9$, $(x,y) = (23,24)$ with $|P| = 25$, and $(x,y) = (25,26)$ with $|P| = 9$ or 27.

When $(x,y) = (7,8)$, we proceed as in the proof of Theorem 3 to conclude G is isomorphic to PSL(2,8). When $(x,y) = (10,11)$, Lemma 5 implies then 5 divides $|G|$ to the first power only. Then consideration of $B_0(5)$ leads to a contradiction.

When $(x,y) = (16,17)$, a count of Sylow 3-subgroups and of Sylow 17-subgroups yields $|G| = 2^4\ 3^2\ 5^2\ 7^2\ 17$ or $2^4\ 3^2\ 17$. In the first case the class multiplication constant (2) where $u = v$ is an involution yields that $|C_G(u)| = 2^4\ 5 \cdot 7$. Then if v is an element of order 7 in $C_G(u)$, the class multiplication constant a_{vvw} gives a contradiction. In the second case, it is easy to verify that G is isomorphic to PSL(2,17).

When $(x,y) = (19,20)$, a count of Sylow 3-subgroups and of Sylow 19-subgroups together with Lemma 3.5 of [2] yields $|G| = 2^2\,3^2\,5 \cdot 19$. It follows easily that G is isomorphic to PSL(2,19).

When $(x,y) = (23,24)$, a count of Sylow 5-subgroups and of Sylow 23-subgroups together with Lemma 3.6 of [2] yields a contradiction.

Next when $(x,y) = (25,26)$ if $|P| = 27$, block separation applied to $B_0(13) \cap B_0(3)$ implies that $B_0(13)$ contains at least 14 characters, a contradiction. If $|P| = 9$, a count of Sylow 3-subgroups and of Sylow 13-subgroups yields $|G| = 2^5\,3^2\,5^2\,7^2\,13$ or $2^3\,3^2\,5^2\,7^2\,11^2$. In the first case a close consideration of the class multiplication constant a_{uuw}, where u is an involution and w is a 3-singular element leads to a contradiction. In the latter case G must clearly be a known group, but this is not compatible with $|G|$.

When $k = 1$, Theorem 4 reduces to Theorem 2 so that we have verified that if G satisfies the hypotheses of Theorem 4, then G is isomorphic to PSL(2,q) with q = 5, 7, 9, 8, 11, 13, 16, 17, 19, 23 or 25.

This completes the sketches of the proofs of Theorems 3 and 4.

REFERENCES

1. L. J. Alex, *Simple groups of order* $2^a 3^b 5^c 7^d p$, Trans. Amer. Math. Soc. 173 (1972) 389-399.

2. __________, *Index two simple groups,* J. Algebra 31 (1974) 262-275.

3. R. Brauer, *On groups whose order contains a prime number to the first power,* I, II, Amer. J. Math. 64 (1942) 401-440.

4. D. C. Morrow, *A characterization of* SL (3,8), Doctoral dissertation, University of Virginia, 1974.

5. S. D. Smith and A. P. Tyrer, *Finite groups with a certain Sylow normalizer,* I, II, J. Algebra 26 (1973) 343-367.

STATE UNIVERSITY OF NEW YORK
ONEONTA, NEW YORK

ON FINITE LINEAR GROUPS IN DIMENSION AT MOST 10

BY

WALTER FEIT*

1. INTRODUCTION

Let G be a finite group which has a faithful irreducible quasi-primitive unimodular complex representation of degree n . If $n \leqslant 7$ the structure of G is known by the work of Blichfeldt [1], Brauer [2], Lindsey [9] and Wales [11]. In this note we announce some results which cover the cases that n = 8, 9 or 10.

In case n = 8 the proof depends to a large extent on computer work. I wish to thank Sidnie M. Feit who did all the necessary programming and ran the programs on the Yale Computer.

The work in case n = 8 was just about completed when I became aware of a paper by C. W. Huffman [7]. This paper

---

*The work on this paper was partially supported by NSF Contract GP-33591.

made it possible to simplify much of the work. Recently announced results of Huffman and Wales [8] produced further simplifications.

The work of Huffman and Wales [8] is absolutely essential for the proof of Theorem B and Theorem C(I) in the next section. Once their results are available the main difficulty in proving these Theorems is to show that $7^2 \nmid |G|$. This was done by S. Doro in his thesis [3]. Given these results Theorem B becomes an almost trivial application of Block Theory and Theorem C(I) is proved in a similar way though is technically more complicated. I had proved Theorem C(II) some years ago. The above mentioned results can be used to simplify the argument considerably.

In the proofs of Theorem A and Theorem C(I) a recent result of M. Hall [6] is also very helpful.

2. Statement of Results

If X is a group with $Z(X) \subseteq X'$ let $\tilde{X}$ denote a covering group of X . That is to say $Z(\tilde{X}) \subseteq \tilde{X}'$ and $\tilde{X}/Z \approx X$ for some subgroup Z of $Z(X)$.

Theorem A. *Let* G *be a finite group which has a faithful irreducible quasi-primitive unimodular complex representation of degree* 8. *Then either* $|G| = 2^a \cdot 3^b \cdot 5^c$ *or* $G = HZ(G)$ *with* $|Z(G)| \mid 8$ *and one of the following holds.*

(i) $H \approx SL_2(17)$. *There are two algebraically conjugate characters in* $Q(\sqrt{-17})$ *and the corresponding representations can be written in any extension field of* $Q(\sqrt{-17})$ *which splits the classical quaternions.*

(ii) H *contains a normal* 2-*group* T *which is extra special of order* 2^7. *Furthermore* H/T *is isomorphic to a subgroup of* $O_6^+(2) \approx S_8$ *of order divisible by* 7 .

(iii) H *contains a normal* 2-*group* T *with* $T = T_0 \circ Z$, *where* T_0 *is extra special of order* 2^7 *and* Z *is cyclic of order* 4 . *Furthermore* H/T *is isomorphic to a subgroup of* $Sp_6(2)$ *of order divisible by* 7 . [o *denotes central product.*]

(iv) $H \approx SL_2(7) \circ A$ *or* $\tilde{A}_7 \circ A$ *with* $|Z(\tilde{A}_7)| = 2$, *where* A *has a faithful irreducible quasi-primitive unimodular complex representation of degree* 2.

(v) $H \approx SL_2(7)$, $SL_2^{\pm}(7)$, $PSL_2(7)$ or $PGL_2(7)$. *There is one rational character of* $SL_2(7)$ *which has two extensions to a rational character of* $SL_2^{\pm}(7)$, *the corresponding representations can be written in any quadratic imaginary extension of* Q *which splits the classical quaternions.* $PSL_2(7)$ *has one rational representation of degree* 8 *which has two extensions to a rational representation of* $PGL_2(7)$.

(vi) $H \approx SL_2(8)$ *or the extension of* $SL_2(8)$ *by a field automorphism of order* 3. *There is a unique representation of* $SL_2(8)$, *this is rational. There are* 3 *extensions of this to the larger group, one rational and the other two conjugate in* $Q(\sqrt{-3})$.

(vii) $H \approx \tilde{A}_8, \tilde{A}_9,$ *or* $\tilde{S}_8$ *with* $|Z(H)| = 2$. *Both* $\tilde{A}_8$ *and* $\tilde{A}_9$ *have a unique irreducible faithful character of degree* 8 . *The corresponding representation can be written in any field which splits the classical quaternions. There are two nonisomorphic groups* $\tilde{S}_8$ *and the unique character of* $\tilde{A}_8$ *of degree* 8 *extends to both of these in two ways.*

(viii) $H \approx A_9$ *or* S_9 . *There is a unique irreducible character of* A_9 *of degree* 8 . *This extends in two ways to a character of* S_9 . *The corresponding representations can all be written in* Q .

(ix) $H \approx W(E_8)'$ *or* $W(E_8)$. [$W(E_8)$ *is the Weyl group of type* E_8] . *There is a unique irreducible character of* $W(E_8)'$ *of degree* 8. *This extends to* $W(E_8)$ *in two ways. The corresponding representations can all be written in* Q .

(x) $H \approx W(\tilde{E}_7)' \approx \widetilde{Sp}_6(2)$ *with* $|Z(H)| = 2$. *There is a unique irreducible character of degree* 8 *which can be written in* Q .

It should be remarked that the groups and characters described in (ii) and (iii) exist. See Griess [5]. The subgroups of $Sp_6(2)$ whose orders are divisible by 7 are either contained in S_8 or are isomorphic to $Sp_6(2)$, $G_2(2)$ or $G_2(2)' \approx SU_3(3)$.

COROLLARY A. *If* G *satisfies the hypotheses of Theorem A then either* $|G| = 2^a \cdot 3^b \cdot 5^c$ *or every composition factor of* G *is a known simple group.*

C. W. Huffman and D. B. Wales have recently given a complete classification of groups of order $2^a \cdot 3^b \cdot 5^c$ which have a faithful irreducible quasi-primitive unimodular complex representation of degree 8. Thus Theorem A together with their result completes the classification of 8 dimensional complex groups.

THEOREM B. *Let* G *be a finite group which has a faithful irreducible quasi-primitive unimodular complex representation of degree* 9. *Then either* $|G| = 2^a \cdot 3^b \cdot 5^c$ *or* $G = HZ(G)$ *with* $|Z(G)| \mid 9$ *and one of the following holds.*

(i) $H \approx PSL_2(17)$. *There are two algebraically conjugate irreducible characters in* $Q(\sqrt{17})$ *and the corresponding representations can be written in* $Q(\sqrt{17})$.

(ii) $H \approx PSL_2(19)$. *There are two algebraically conjugate irreducible characters in* $Q(\sqrt{-19})$ *and the corresponding representations can be written in* $Q(\sqrt{-19})$.

(iii) $H \approx A_{10}$ *or* S_{10} . *There is a unique irreducible character of* A_{10} *of degree* 9 . *This has two extensions to* S_{10} . *The corresponding representations can all be written in* Q .

(iv) $H \approx PSL_2(7) \times A$, *where* A *has a faithful irreducible quasi-primitive unimodular complex representation of degree* 3 .

(v) H *has a subgroup* H_0 *of index* 2 *with* $H_0 \cong PSL_2(7) \times PSL_2(7)$.

THEOREM C. *Let* G *be a finite group which has a faithful irreducible quasi-primitive unimodular complex representation of degree* 10 .

(I) *Either* $|G| = 2^a \cdot 3^b \cdot 5^c \cdot 11^d$ *with* $d \leqslant 1$ *or every composition factor of* G *is a known simple group.*

(II) *If* $11 \mid |G|$ *and* $|Z(G)| \mid 2$ *then every composition factor of* G *is a known simple group.*

3. Outline of Proofs

Suppose that G is a counterexample to one of Theorems A, B or C(I); by standard arguments it may be assumed that G = G' and $G/Z(G) = \overline{G}$ is simple. In case n = 8 or 9 $|G| = 2^a \cdot 3^b \cdot 5^c \cdot 7$. In case n = 10, $|G| = 2^a \cdot 3^b \cdot 5^c \cdot 7 \cdot 11^d$ with $d \leq 1$ by results of Doro [3] and Lindsey [10]. The known simple groups of such orders can be listed. An inspection of their character tables and those of their covering groups shows that G is not a counterexample. Thus it may be assumed that $\overline{G}$ is an unknown simple group.

Let P be a S_7-group of G. Let $C = C_G(P) = P \times A$, let $N = N_G(P)$ and $e = |N:C|$. In case n = 8, [4, Theorem 8.6] implies that A = Z(G). By using the results of Huffman and Wales [8] the same argument shows that A = Z(G) for n = 9. A technically more complicated but similar argument also shows that A = Z(G) for n = 10. Thus $n \equiv \pm 1$ or $\pm e$ (mod 7). Hence for n = 8, e = 2, 3 or 6: for n = 9, e = 2: for n = 10, e = 3.

If e = 2 then the degrees in the principal 7-block are 1, x, 1+x with x and 1+x both of the forms $2^a \cdot 3^b \cdot 5^c$ and $x \equiv 1$ or -2 (mod 7). A simple arithmetical Lemma shows that either x = 8 or 15 and so $\overline{G}$ has an irreducible character of degree 8 or 16. By block separation this implies that a S_2-group of $\overline{G}$ has order at most 16 and so $\overline{G}$ is a known simple group.

If $e = 3$ the possible degrees in the principal block of $\overline{G}$ are limited. Various arguments of a standard type eliminate all possible sets of degrees except 1, 24, 27, 50. A result of M. Hall [6] shows that this cannot occur.

This leaves the case that $e = 6$, $n = 8$. At this point the machine comes into play. Initially there are more than a thousand sets of possible degrees. If one then builds in various tests based on a large variety of known theorems the machine produces about 20 sets of degrees which are handled on a case-by-case basis. The argument has to be broken up into cases depending on how many algebraically conjugate characters of degree 8 there are in a block. If there are many, then the degrees in the principal block are all small ($\leqslant 64$). If there is only one, then the group is isomorphic to a subgroup $GL_8(2)$. Most of the final cases are eliminated in fairly short order by using properties of involutions. A couple of them require fairly complicated additional arguments. Recently announced results that imply that $2^{11} \mid |\overline{G}|$ should simplify much of this work considerably.

The proof of Theorem C(II) is mostly by methods similar to those mentioned above. We will sketch the proof of a special case which is slightly different. Suppose that χ is rational valued where χ is the character afforded by the representation of order 10. If the Schur index of χ is 1, the result follows from [4, Theorem C]. Suppose that the Schur index of χ is 2. Let m_q denote the Schur index of χ at the q-adic completion of Q. If $m_q \neq 1$ then χ must be reducible modulo q. It can be shown directly that χ is irreducible modulo q for $q \neq 2$. Hence by the theory of rational division algebras $m_2 = 2$ and χ is a sum of 2 conjugate 5 dimensional Brauer characters over the field of 4 elements. Thus $G/Z(G)$ is isomorphic to a subgroup of $U_5(2)$. It is then not difficult to show that $G \approx U_5(2)$. The group $U_5(2)$ is known to have an irreducible character of degree 10. Thus the group arises in a natural manner.

The details used in the arguments sketched above will appear elsewhere.

REFERENCES

1. H. F. Blichfeldt, *Finite Collineation Groups*, University of Chicago Press, Chicago, 1917.

2. R. Brauer, *Über endliche lineare Gruppen von Primzahlgrad*, Math. Ann. 169 (1967), 73-96.

3. S. Doro, *On finite linear groups in nine and ten variables*, Thesis, Yale University, 1975.

4. W. Feit, *On integral representations of finite groups*, Proc. L.M.S. (3) 29 (1974), 633-683.

5. R. Griess Jr., *Automorphisms of extra special groups and nonvanishing degree 2 cohomology*, Proceedings of the Gainesville Conference, North Holland Amsterdam-London, 1972, 68-73.

6. M. Hall Jr., *Nonexistence of a finite group with a specified 7-block*, these Proceedings.

7. C. W. Huffman, *Linear groups containing an element with an Eigenspace of codimension two*, (to appear).

8. ____________ and D. Wales, *Linear groups containing an involution with two eigenvalues -1*, (to appear).

9. J. H. Lindsey II, *Finite linear groups of degree six*, Can. J. Math. 23 (1971), 771-790.

10. ____________, *Projective groups of degree less than 4p/3 where centralizers have normal Sylow p-subgroups*, Transactions A.M.S. 175 (1973), 233-247.

11. D. B. Wales, *Finite linear groups in seven variables*, Bull. A.M.S. 74 (1968), 197-198.

YALE UNIVERSITY
NEW HAVEN, CONNECTICUT

NONEXISTENCE OF A FINITE GROUP WITH A SPECIFIED 7-BLOCK

BY

MARSHALL HALL JR.

1. INTRODUCTION. In his investigations of finite groups with a faithful matrix representation of degree 8 over the complex field, Walter Feit [4] was obliged to consider a possible group in which a Sylow 7-subgroup was self-centralizing of order 7 and had a normalizer of order 21 and the degrees in the principal 7-block were 1, 27, 50, and two exceptional 24's. Under the restrictions implied by his investigation, the writer was able to show that no such group exists.

At the group theory conference in Park City, Utah, January 1975, the question was raised as to whether or not a group with this block existed, without assuming any further restrictions. In this note it is shown that no such group exists. The writer wishes to thank Stephen Smith for pointing

---

This research was supported in part by NSF Grant GP 36230X.

out a generalization of a formula due to Burnside whose application has greatly simplified this note.

2. General Methods Used

Most of the methods used here are described in the writer's article, "A search for simple groups of order less than one million" [5].

The principle of "block separation" appears in a paper of Brauer-Tuan [2] and Stanton [7]. Let p and q be different primes and suppose that the group G contains no element of order pq. Let $|G| = p^a q^b g'$, $(g', pq) = 1$ and suppose that $\zeta_1, \ldots, \zeta_k$ are characters of G such that

2.1) $$\sum_{i=1}^{k} a_i \zeta_i(x) = 0$$

for all p-regular elements x. Then

2.2) $$\sum a_i \zeta_i(x) = 0 , \qquad \zeta_i \in B(q) \quad \text{a } q\text{-block} ,$$

for all q-singular elements x. Furthermore

2.3) $$\sum a_i \zeta_i(1) \equiv 0 \ (\mathrm{mod}\ q^b), \ \zeta_i \in B(q) .$$

In the same paper [2] Brauer-Tuan showed that if p divides $|G|$ to the first power only then for a character χ in the principal p-block $B_0(p)$ and for an element v centralizing the Sylow p-subgroup $S(p)$, we have

2.4) $$\chi(v) = m + p\,\theta(v)\ ,$$

where $m \geqslant 0$ is a rational integer and θ is some character of the cyclic group $\langle v \rangle$ of order prime to p. Of course $m + p\theta(1) = \chi(1)$.

In the group ring $G(Q)$ of G over the rational field Q let C_i be the sum of the elements in the ith conjugacy class. It is well known that the C_i are a basis for the center of $G(Q)$. Then

2.5) $$C_i C_j = \sum_k a_{ijk} C_k\ ,$$

where the a_{ijk} are non-negative integers. These coefficients can be evaluated by a formula due to Burnside [3, p. 316].

2.6) $$c(x_i, x_j, x_k) = a_{ijk} = \frac{|G|}{c(x_i)c(x_j)} \sum_\chi \frac{\chi(x_i)\chi(x_j)\overline{\chi(x_k)}}{\chi(1)}\ ,$$

where x_s is an element of the sth class, $\overline{\chi(x_k)}$ is the complex conjugate of $\chi(x_k)$, the summation is over all irreducible characters χ of G, and $c(x_s)$ is the order of the centralizer of x_s. This formula is particularly useful when there is a class most of whose characters vanish. It has been pointed out to me by Stephen Smith that the same proof can be used to derive further formulae of the same general kind.

If

2.7) $$C_iC_jC_k = \sum_t b_{ijkt}\, C_t \ ,$$

where the b_{ijkt} are non-negative integers, then

2.8) $$c(x_i,x_j,x_kx_t) = b_{ijkt} = \frac{|G|^2}{c(x_i)c(x_j)c(x_k)} \sum_\chi \frac{\chi(x_i)\chi(x_j)\chi(x_k)\overline{\chi(x_t)}}{\chi(1)^2} \ .$$

Schur [6] showed that if G is faithfully represented by a character χ of degree n, χ is rational on p elements, and p^a is the highest power of p dividing $|G|$, then

2.9) $$a \leqslant \left[\frac{n}{p-1}\right] + \left[\frac{n}{p(p-1)}\right] + \cdots + \left[\frac{n}{p^i(p-1)}\right] \cdots \ .$$

3. The Specific Block

We assume that G is a finite group with a self centralizing Sylow 7 subgroup $S(7)$ and that the degrees in the principal 7 block $B_0(7)$ are 1, 27, 50, and two exceptional 24's. Necessarily a 7 normalizer $N(7)$ will have order 21. $N(7)$ will be a group $< a,b >$ where

3.1) $$a^7 = 1,\ b^3 = 1,\ b^{-1}ab = a^2 \ .$$

It follows readily that G has no subgroup of index 2. The given 7 block characters have the form

3.2)

	1	a	a^{-1}
ρ_0	1	1	1
ρ_1	27	-1	-1
ρ_2	50	1	1
ρ_3	24	$\frac{-1+\sqrt{-7}}{2}$	$\frac{-1-\sqrt{-7}}{2}$
ρ_4	24	$\frac{-1-\sqrt{-7}}{2}$	$\frac{-1+\sqrt{-7}}{2}$

Here all further irreducible characters χ have degrees which are multiples of 7, and for such characters $\chi(a) = \chi(a^{-1}) = 0$. Furthermore $\rho_0, \ldots, \rho_4$ are rational on all further elements of order prime to 7.

In Brauer-Tuan [2] it is shown that a character of degree p^s, $s \geqslant 1$ is not in the principal p-block $B_0(p)$. Hence ρ_1 is not in the principal 3 block. By application of block separation, as there is no element of order 21, relation 2.3) has as its only solution

$$1\rho_0(1) + 1\rho_2(1) + \frac{-1+\sqrt{-7}}{2}\rho_3(1) + \frac{-1-\sqrt{-7}}{2}\rho_4(1) =$$

$$1 + 50 - 24 = 27 \equiv 0 \pmod{q^b} .$$

Hence the exact power of 3 dividing $|G|$ is 3^3 .

If we write $h(x) = |G|/c(x)$, then Brauer's condition [1] that two characters χ and ρ be in the same p-block is

3.3) $$\frac{h(x)\chi(x)}{\chi(1)} \equiv \frac{h(x)\rho(x)}{\rho(1)} \pmod{P}$$

for all x and P a prime ideal dividing p. Hence if ρ_2 is in the principal 5 block $B_0(5)$ we would have for x in the center of a Sylow 5 subgroup

3.4) $$\frac{h(x)\rho_2(x)}{50} \equiv \frac{h(x)\rho_0(x)}{\rho_0(1)} \equiv h(x) \pmod{P} .$$

Here $h(x)$ is prime to 5 so that this would require, as $\rho_2(x)$ is rational,

3.5) $$\frac{\rho_2(x)}{50} \equiv 1 \pmod{5}.$$

Clearly $\rho_2(x)$ would have to be a non-zero multiple of 25 and so $\rho_2(x) = \pm 25$, but this does not satisfy 3.5). Hence ρ_2 is not in the principal 5-block and, with $q = 5$, 2.3) yields

$$1 - 27 + \left(\frac{-1+\sqrt{-7}}{2}\right) 24 + \left(\frac{-1-\sqrt{-7}}{2}\right) 24 = -50 \equiv 0 \pmod{q^b} .$$

Hence 5^2 is the exact power of 5 dividing $|G|$.

Thus $|G|$ is divisible by exactly 3^3, 5^2, and 7. Applying Schur's condition 2.9) to ρ_3, $|G|$ may be divisible by at most 2^{46}, 11^2, 13^2, 17, 19, 23, but by no further primes.

Possible characters for p elements x_p with $p = 3$, 5, 11, 13, 17, 19, 23 excluding those for which $c(x_p, x_p, a) < 0$ are given here:

	1	a	$x_{3,1}$	$x_{3,2}$	$x_{3,3}$	$x_{3,4}$	$x_{3,5}$	$x_{5,1}$	$x_{5,2}$
ρ_0	1	1	1	1	1	1	1	1	1
ρ_1	27	-1	0	0	0	0	0	-3	2
ρ_2	50	1	5	2	-1	-4	-7	0	0
ρ_3	24	$\frac{-1+\sqrt{-7}}{2}$	6	3	0	-3	-6	4	-1
ρ_4	24	$\frac{-1-\sqrt{-7}}{2}$	6	3	0	-3	-6	4	-1

3.6)

	1	$x_{11,1}$	$x_{11,2}$	$x_{13.1}$	$x_{13,2}$	x_{17}	x_{19}	x_{23}
ρ_0	1	1	1	1	1	1	1	1
ρ_1	27	16	5	14	1	10	8	4
ρ_2	50	28	6	24	-2	16	12	4
ρ_3	24	13	2	11	-2	7	5	1
ρ_4	24	13	2	11	-2	7	5	1

As $\rho_1, \rho_2, \rho_3, \rho_4$ are rational on 7' elements no one of these can be exceptional in the principal p block $B_0(p)$ for a prime p dividing $|G|$ to the first power, so that for such a character ρ_i in $B_0(p)$, $\rho_i(x_p) = \pm 1$. Hence from the principle of block separation it follows that we cannot have 17 or 19 dividing $|G| = g$ from the values appearing in 3.6).

If τ is an involution, we cannot have $\tau'\tau'' = a$ with

τ', τ'' conjugates of τ since then we would have $\tau' \, a \, \tau' = \tau'' \, \tau' = a^{-1}$. This is a conflict, as $N(7)$ is of order 21 and contains no involution. Hence for an involution we must have $c(\tau,\tau,a) = 0$. Since G does not have a subgroup of index 2, in the diagonal representation of an involution we will have an even number of -1's which leads to $\chi(\tau) \equiv \chi(1) \bmod 4$. Using these conditions there are exactly four possible characters for involutions τ_i.

3.7)

		τ_1	τ_2	τ_3	τ_4
ρ_0	1	1	1	1	1
ρ_1	27	-9	3	-21	15
ρ_2	50	-10	10	-30	30
ρ_3	24	0	8	-8	16
ρ_4	24	0	8	-8	16

Now if 23 divides g, then by block separation ρ_3 and ρ_4 are in the principal 23 block $B_0(23)$. As $c(x_{23}) \geqslant 1^2 + 4^2 + 4^2 + 1^2 + 1^2 = 35$ there must be a p element x_p, $p \neq 23$ centralizing x_{23} and from 2.4) we would have

$$\rho_4(x_p) = m + 23\,\theta(x_p) .$$

Here $m \equiv 1 \pmod{23}$, and so $\rho_4(x_p) \equiv 1 \pmod{23}$. Inspection of 3.6) and 3.7) shows that no such element exists. Hence we cannot have 23 dividing g.

We have now eliminated 17, 19, 23 as possible divisors of g. Let us now suppose that 13^2 divides g. Suppose that in an $S(13)$ there are r elements of type $x_{13,1}$ and s of type $x_{13,2}$. We now consider the restriction of ρ_1 to $S(13)$ and let m be the multiplicity of the identity in this restriction. This gives

3.8)
$$r + s = 168$$
$$27 + 14r + s = 169m \quad .$$

In addition, $r \equiv s \equiv 0 \pmod{12}$ since in a cyclic group of order 13 all 12 elements different from the identity have the same character. In particular $m \equiv 169m \equiv 27 \equiv 3 \pmod{12}$, and as $169m \leqslant 14(168) + 27 < 169 \cdot 15$ we have $m < 15$ so that $m = 3$. Thus $r = 24$, $s = 144$. Let $< u >< v >$ be cyclic subgroups of order 13 giving elements of type $x_{13,1}$. For ρ_4 we will have for the diagonal form

3.9) $$\rho_4(u) = 1^{12}, \zeta, \ldots, \zeta^{12}, \zeta = e^{2\pi i/13} ,$$

and $\rho_4(v)$ is of the same form. Since $S(13) = < u,v >$, if u and v both had a 1 in the same diagonal position every element of $S(13)$ would have a 1 in this position. But elements of type $x_{13,2}$ have $\zeta, \ldots, \zeta^{12}, \zeta, \ldots, \zeta^{12}$ as their diagonal form in $S(13)$. Hence with appropriate numbering of diagonal positions we have

$$\text{3.10)}\quad \begin{aligned} \rho_4(u) &= 1^{12}, \zeta, \zeta^2, \zeta^3, \zeta^4, \zeta^5, \zeta^6, \zeta^7, \zeta^8, \zeta^9, \zeta^{10}, \zeta^{11}, \zeta^{12} \\ \rho_4(v) &= \zeta, \zeta^2, \zeta^3, \zeta^4, \zeta^5, \zeta^6, \zeta^7, \zeta^8, \zeta^9, \zeta^{10}, \zeta^{11}, \zeta^{12}, 1^{12} . \end{aligned}$$

Now from ρ_1 we have $3 \nmid c(x_{13,1})$ and from ρ_2 we have $5 \nmid c(x_{13,1})$. Also with $c(x_{13,1}, \tau_i, a) = \frac{g}{c(x_{13,1})c(\tau_i)} W_i$ we have for $i = 1, 2, 3, 4$

$$\text{3.11)}\quad W_1 = 13/15,\ W_2 = 26/45,\ W_3 = 52/45,\ W_4 = 13/45 .$$

Hence if τ_i is in the center of an $S(2)$ then g and $c(\tau_i)$ are divisible by the same power of 2 so that the power of 2 dividing $c(x_{13,1})$ divides the numerator of W_i. Hence at most 2^2 divides $c(x_{13,1})$. On the other hand $c(x_{13,1}) \geqslant 1^2 + 14^2 + 24^2 + 11^2 + 11^2 = 1015$. Since this is greater than $2^2 \cdot 13^2 = 676$ some odd prime p different from 13 divides $c(x_{13,1})$. As 3, 5, 7 are not possible, the only possibility is $p = 11$. Hence $c(x_{13,1})$ is a multiple of 11 and a divisor of $2^2 \cdot 11^2 \cdot 13^2$. Hence $S(13)$ is normal in $C_G(x_{13,1})$ and an element of order 11 normalizing $S(13)$ necessarily centralizes $S(13)$ since $11 \nmid (13^2-1)(13^2-13)$. But then from 3.10), as the 24 linear representations of $S(13)$ are distinct, an element in the centralizer is necessarily of diagonal form. Therefore $\rho_4(uvw)$ contains a primitive 143rd root of unity on the diagonal which is not possible if the character is to be rational. Thus the assumption that $13^2 | g$ has led us to a contradiction.

Now suppose $13|g$ but $13^2 \nmid g$. Then from block separation, $\rho_1 \in B_0(13)$, and $\rho_1(x_{13}) = 1$ so that the 13 element is of type $x_{13,2}$. $c(x_{13,2}) \geq 1^2 + 1^2 + (-2)^2 + (-2)^2 + (-2)^2 = 14$ so that $x_{13,2} = x_{13}$ is not self centralizing and there is some p element x_p with $p \neq 13$ centralizing x_{13}. From 2.4) $\rho_1(x_p) = 1 + 13\theta(x_p)$. But inspection of 3.6) and 3.7) shows that there is no such element. We have now shown that 13 does not divide g.

If 11 divides g then 11^2 divides g since no degree 27, 50, 24 is of the form $11k \pm 1$ which would be necessary if 11 divided g to the first power exactly. Now suppose there are in an $S(11)$ r elements of type $x_{11,1}$, and s of type $x_{11,2}$. Then restricting ρ_1 to $S(11)$ we have

$$\text{3.12)} \qquad \begin{aligned} r + s &= 120 \\ 27 + 16r + 5s &= 121\,m \end{aligned}$$

where m is the multiplicity of the identity in this restriction. Also $r \equiv s \equiv 0 \pmod{10}$. This gives $m \equiv 7 \pmod{10}$. Since also $m < 17$ is easily shown, we have $m = 7$, and find $r = 20$, $s = 100$.

Now $c(x_{11,1}, \tau_i, a) = \dfrac{g}{c(x_{11,1})c(\tau_i)} W_i$ where

$$\text{3.13)} \qquad W_1 = 11/15,\ W_2 = 22/45,\ W_3 = 44/45,\ W_4 = 11/45 .$$

Hence 2^2 is the highest power of 2 possibly dividing $c(x_{11,1})$, and from ρ_1 and ρ_2 neither 3 nor 5 divides $c(x_{11,1})$. But $c(x_{11,1}) \geq 1^2 + 16^2 + 28^2 + 13^2 + 13^2 = 1379$. On the other hand we have shown that $c(x_{11,1})$ divides $2^2 \cdot 11^2 = 484$ as there are no further primes besides 2 and 11 to divide this order. This is a conflict and we conclude that 11 does not divide g.

At this stage we have eliminated all primes greater than 7 as possible divisors of g so that $g = 2^t \cdot 3^3 \cdot 5^2 \cdot 7$ with $t \leqslant 46$. We now apply 2.8) to restrict the power of 2. With $c(\tau_i, \tau_i, \tau_i, a) = \dfrac{g^2}{c(\tau_i)^3} W_i$ we find

3.14) $\quad W_1 = 8/5,\ W_2 = 64/135,\ W_3 = 512/135,\ W_4 = 8/135$.

If τ_i is in the center of an $S(2)$ then the numerator of W_1 must be divisible by the order of $S(2)$. Thus $t \leqslant 9$.

We now turn our attention to 5 elements. Let there be r of type $x_{5,1}$ and s of type $x_{5,2}$ in an $S(5)$. Then restricting ρ_1 and ρ_3 to $S(5)$ we have

$$r + s = 24$$

3.15) $$27 - 3r + 2s = 25\, m_1$$

$$24 + 4r - s = 25\, m_2$$

We immediately find $51 + r + s = 75 = 25(m_1 + m_2)$ giving

$m_1 + m_2 = 3$. Here r and s are both multiples of 4 (indeed a multiple of 20 for an element of order 25). We find that $m_1 \equiv 27 \equiv 3 \pmod 4$ so that $m_1 = 3$, $m_2 = 0$. This leads to $r = 0$, $s = 24$ so that all 5-elements are of type $x_{5,2} = x_5$. We now find with $c(x_5,\tau_i,a) = \frac{g}{c(x_5)c(\tau_i)} W_i$ such that

3.16) $\qquad W_1 = 5/3, W_2 = 10/9, W_3 = 20/9, W_4 = 5/9$.

Hence the highest power of 2 dividing $c(x_5)$ is at most 2^2, and as neither 3 nor 7 divides $c(x_5)$ we conclude that $c(x_5)$ divides $2^2 \cdot 5^2$. If an x_5 were in the intersection of two different $S(5)$'s then both of these would be in $C_G(x_5)$ which would contain $1 + 5k$ $S(5)$'s, and $1 + 5k | 2^2 \cdot 5^2$ which is possible only with $k = 0$. Hence the $S(5)$'s form a trivial intersection set and the number of $S(5)$'s is of the form $1 + 25k$. Since at most 2^2 divides $c(x_5)$ and from 3.14) no involution centralizes an entire $S(5)$ we conclude that an $S(5)$ is its own centralizer. Thus the normalizer $N(5)$ of an $S(5)$ has order $2^\alpha \cdot 3 \cdot 5^2$ or $2^\alpha \cdot 5^2$ with $\alpha \leqslant 5$. The number of $S(5)$'s is then $2^n \cdot 3^2 \cdot 7 \equiv 1 \pmod{25}$ or $2^n \cdot 3^3 \cdot 7 \equiv 1 \pmod{25}$. In the first case $n \equiv 1 \pmod{20}$ and in the second $n \equiv 14 \pmod{20}$. The second case is impossible as at most 2^9 divides g. In the first case $n = 1$ and

$$g = 2^{n+\alpha} \cdot 3^3 \cdot 5^2 \cdot 7 = 2^{1+\alpha} \cdot 3^3 \cdot 5^2 \cdot 7 .$$

Since $|N(7)| = 21$ the number of $S(7)$'s is

$$2^{1+\alpha}\cdot 3^2\cdot 5^2 \equiv 2^{1+\alpha} \equiv 1 \pmod 7.$$

Hence $1+\alpha \equiv 0 \pmod 3$ and as $\alpha \leqslant 5$ we have $g = 2^t\cdot 3^3\cdot 5^2\cdot 7$ with $2^t = 2^3$ or 2^6.

For τ_3 and τ_4 we cannot have $3^2|c(\tau_i)$ or $5^2|c(\tau_i)$. Thus $c(\tau_3)$ and $c(\tau_4)$ divide $2^t\cdot 3\cdot 5$ and so $2^6\cdot 3\cdot 5 = 960$. But from 3.7) we see that $c(\tau_3) > 960$ and $c(\tau_4) > 960$. Hence there cannot be involutions of type τ_3 or τ_4. If there is an involution of type τ_2 then as $3^2 \nmid c(\tau_2)$ and $5^2 \nmid c(\tau_2)$, it follows that $c(\tau_2)|2^t\cdot 3\cdot 5$. From 3.7), $c(\tau_2) \geqslant 238 > 120$ so that necessarily $2^t = 2^6$. Note that from 3.14) if an involution of type τ_1 is in the center of an $S(2)$ then necessarily $2^t = 8$. Hence an involution τ_2 is in the center of an $S(2)$ of order 2^6. As $2^6 = 2^{\alpha+1}$, $2^\alpha = 2^5$ and so a group of order 32 acts on the 24 non-identity elements of an $S(5)$. Hence for some x_5 we have $2^2|c(x_5)$, and so $c(x_5) = 2^2\cdot 5^2$. But now

$$3.17)\qquad c(x_5,\tau_2,a) = \frac{g}{c(x_5)c(\tau_2)}\cdot\frac{10}{9}.$$

Here $g = 2^6\cdot 3^3\cdot 5^2\cdot 7$, $c(x_5) = 2^2\cdot 5^2$ and $2^6|c(\tau_2)$. This is a conflict as the numerator in 3.17) on the right is divisible only by 2^7 while the denominator is divisible by 2^8. Hence there can be no involution of type τ_2.

We are now reduced to the case in which there are only involutions of type τ_1. From 3.14) it follows that $2^t | 2^3$ so that $2^t = 2^3$ and $g = 2^3.3^3.5^2.7 = 37,800$. Hence $c(\tau_1) | 2^3 \cdot 3^2 \cdot 5 = 360$ from 3.7) and as $c(\tau_1) \geqslant 1^2 + (-9)^2 + (-10)^2 = 182$ it follows that $c(\tau_1) = 360$. Write $H = C_G(\tau_1)$ so that $|H| = c(\tau_1) = 360$. We shall obtain our final conflict by observing that $c(x_5) = |C_G(x_5)|$ divides $2^2 \cdot 5^2$, but we shall show that for $x_5 \in H$, $|C_H(x_5)|$ is a multiple of 3. With $x_5 \in H$ we have $\tau_1 \in C_H(x_5)$. The number of S(5)'s in H is a divisor of $2^2 \cdot 3^2 = 36$ and so is 36, 6, or 1. If the number is 36 then $|N_H(x_5)| = |C_H(x_5)| = 10$ and so H has a normal 5 complement K of order 72. Here x_5 must normalize and so centralize some S(3) of K, and so H contains an element of order 15, whence $|C_H(x_5)|$ is a multiple of 3, giving the desired conflict. If the number is 6 then $N_H(5)$ is of order 60 and again H contains an element of order 15. If the number is 1 the $N_H(5)$ is of order 360 and so contains an element of order 15. Thus in all cases $|C_H(x_5)|$ is a multiple of 3 contrary to our earlier proof that $|C_G(x_5)| = c(x_5)$ is a divisor of $2^2 \cdot 5^2$.

REFERENCES

1. R. Brauer, *Zur Darstellungstheorie der Gruppen endlicher Ordnung,* Part I, Math Zeit. 63 (1956), 406-444; Part II, Math. Zeit. 72 (1959), 25-46.

2. R. Brauer and H. F. Tuan, *On simple groups of finite order,* Bull. Amer. Math. Soc., 51 (1943), 756-766.

3. W. Burnside, *The Theory of Groups,* 2nd ed., Cambridge U. Press, 1911.

4. Walter Feit, *On finite linear groups of degree* 8, to appear.

5. Marshall Hall Jr., *A search for simple groups of order less than one million,* Computational problems in Abstract Algebra, ed. John Leech, Pergamon Press, Oxford, 1969.

6. Issai Schur, *Über eine Klasse von endlichen Gruppen linearer Substitutionen,* Sitz. der Preussischen Akad. Berlin (1905), 77-91.

7. R. G. Stanton, *The Mathieu groups,* Canad. J. Math., 3 (1951), 164-174.

California Institute of Technology
Pasadena, California

LINEAR GROUPS CONTAINING AN ELEMENT WITH AN EIGENSPACE OF CODIMENSION TWO

BY

W. CARY HUFFMAN AND DAVID B. WALES

Let G be a finite group and $X: G \to GL_n(V)$ an irreducible representation of G over the complex vector space V of dimension n. We say that X is *quasiprimitive* if for all $N \lhd G$, $X|N$ splits into equivalent irreducible representations. H. H. Mitchell (*Amer. J. Math.* 36 (1914), 1-12.) classified all G having a faithful quasiprimitive representation X such that for some $g \in G$, $X(g)$ has exactly $n-1$ equal eigenvalues. These groups are the finite reflection groups. We will classify G such that for some $g \in G$, $X(g)$ has exactly $n-2$ equal eigenvalues.

Important tools in handling this problem are as follows. Let $g, h \in G$ such that $X(g)$ and $X(h)$ have eigenvalues $\beta_1, \beta_2, \beta, \ldots, \beta$. Then the subspaces of $X(g)$ and $X(h)$ corresponding to the eigenvalue β are of dimension $n-2$ and intersect in a subspace of dimension at least $n-4$. Hence $X|\langle g,h\rangle = X_1 \oplus (n-4)\xi$ where X_1 has degree 4 and ξ has degree 1 such that $\xi(g) = \xi(h) = \beta$. If we adjoin another such element to $\langle g,h \rangle$ we obtain a group with a six dimensional representation which is scalar on the complementary space. Continuing in this manner we can construct subgroups of small dimension. We attempt to determine $\langle g,h \rangle$ precisely in certain cases, and in other cases we try to obtain generators and relations to determine G . Often contradictions are obtained by applying a result of Blichfeldt which states that if X is faithful and quasiprimitive, then $G - Z(G)$ does not contain an element with one eigenvalue ξ such that all others are $\leq 60°$ away from ξ .

The results will now be described. Let $X:G \to GL_n(V)$ be faithful and quasiprimitive such that there exists $g \in G$ with $X(g)$ possessing eigenvalues $\beta_1, \beta_2, \beta, \ldots, \beta$ where $\beta_i \neq \beta$. As we are interested in $G/Z(G)$, we may assume $\beta = 1$. By Mitchell's result, β_i may be assumed to both be primitive $|g|^{th}$ roots of 1 . We say that X is *primitive* if there does not exist $m \geq 2$ proper nontrivial subspaces V_i such that $V = V_1 \oplus \ldots \oplus V_m$ and $X(h)$ permutes $\{V_i\}$ for all $h \in G$. Primitive groups of degree $n \leq 7$ are known.

We have four cases:

i) $|g| \geq 4$: Then X is primitive and $n \leq 4$.

ii) $X(g)$ has eigenvalues $\omega, \omega, 1, 1, \ldots, 1$ where $\omega = e^{2\pi i/3}$. Then X is primitive. Furthermore if $n \geq 8$, two such elements generate $SL_2(3)$ or commute. Using a result of Aschbacher and M. Hall, we obtain that if $n = 8$, $G/Z(G)$ is a direct product of a 4 dimensional group and a 2 dimensional group. If $n \geq 9$, G doesn't exist.

iii) $X(g)$ has eigenvalues $\omega, \bar{\omega}, 1, 1, \ldots, 1$. Again X is primitive. If $n \geq 8$, two such elements commute or generate A_4, $SL_2(3)$, or A_5. Using generators and relations for alternating groups and the results of Aschbacher-Hall and Stellmacher we obtain: if $n \geq 9$, then $G/Z(G) \simeq A_{n+1}$ or S_{n+1}; if $n = 8$, $G/Z(G) \simeq A_9, S_9, O_8^+(2)$, or an extension of $O_8^+(2)$.

iv) $X(g)$ has eigenvalues $-1,-1,1,1,\ldots,1$. Call such elements special involutions. If $n \geq 8$ one of the following holds:

a. There exists $g \in G$ where $X(g)$ is as in case iii).

b. If τ_1, τ_2 are different special involutions, then $|\tau_1\tau_2| = 2,3,4,$ or 5. If $|\tau_1\tau_2| = 4$, $(\tau_1\tau_2)^2$ is special or $(\tau_1\tau_2)^2 \in O_2(G)$. By Timmesfeld, $G/O_\infty(G)$ is known where $O_\infty(G)$ is the largest normal solvable subgroup of G.

We outline the proof of iv). Let τ_1, τ_2 be special.

Then $X|\langle\tau_1,\tau_2\rangle = X_1 \oplus X_2 \oplus (n-4)1_{\langle\tau_1,\tau_2\rangle}$ where X_i are of degree 2 acting on subspaces U_i. We first prove that $|\tau_1\tau_2| = 2,4$, or odd. To obtain a contradiction we may assume $|\tau_1\tau_2| = 2k$ where $k = 4$ or k is an odd prime. After obtaining information about certain subgroups H containing $\langle\tau_1,\tau_2\rangle$ we are able to prove that if τ is any special involution, either $X(\tau)$ fixes $U_1 \oplus U_2$ or $|\tau_1\tau| = 3$ or 4. We are then able to define $\gamma(\tau_1) = \langle(\tau_1\tau_2)^4\rangle$ to be a well-defined function independent of the choice of τ_2 whenever $|\tau_1\tau_2| = 2k$. We then show that if $\tau_3 = \tau_1{}^g$, $\tau_4 = \tau_2{}^g$ then $[\gamma(\tau_1), \gamma(\tau_3)] = 1$ and so $\langle\gamma(\tau_1)^G\rangle$ is a normal noncentral abelian subgroup of G contradicting quasiprimitivity. Using similar techniques we show $|\tau_1\tau_2| = 2, 3, 4$, or 5. When $|\tau_1\tau_2| = 4$ and $(\tau_1\tau_2)^2$ is not special, it suffices to show that if $\tau_3 = \tau_1{}^g$ and $\tau_4 = \tau_2{}^g$, then $\langle(\tau_1\tau_2)^2, (\tau_3\tau_4)^2\rangle$ is a 2-group. A variety of methods are involved in showing this; mainly generators and relations, coset enumeration, and construction of the actual matrices are used.

An application of these results is to the case of quasi-primitive linear groups of degree 8. Lindsey, Feit, and others have settled all cases except $|G| = 2^a\, 3^b\, 5^c$. Using these results we are able to determine G in the remaining case. At present $n = 9$ is being considered.

REFERENCES

1. W. C. Huffman, *Linear groups containing an element with an eigenspace of codimension two*, to appear in J. Algebra.

2. W. C. Huffman and D. B. Wales, *Linear groups of degree* n *containing an element with exactly* $n-2$ *equal eigenvalues*, to appear.

3. W. C. Huffman and D. B. Wales, *Linear groups containing an involution with two eigenvalues* -1, to appear.

DARTMOUTH COLLEGE
HANOVER, NEW HAMPSHIRE

AND

CALIF. INST. OF TECH.
PASADENA, CALIFORNIA

ON FINITE COMPLEX LINEAR GROUPS OF DEGREE $(q-1)/2$

BY

HENRY S. LEONARD, JR.

INTRODUCTION. Let G denote a finite group, let p be a fixed rational prime, let P denote a fixed Sylow p-subgroup of G, and let q denote $|P|$. We denote the normalizer of P by N, the centralizer of P by C, and the center of G by Z. Throughout this paper we shall assume:

HYPOTHESIS 1. *The centralizer of every non-identity element of* P *is the centralizer* C *of* P.

This condition implies that P is an abelian trivial intersection (t.i.) set. Furthermore, if the group of p'-elements of C is denoted by V then $C = P \times V$ and N/V is a Frobenius group with kernel P.

Under Hypothesis 1, D. A. Sibley [7] recently proved the following

THEOREM. *If* G *has a faithful complex representation*

of degree $d < (q-1)/2$ *then* $P \lhd G$.

This theorem was proved in 1942 by Brauer [1] in the important case that P has prime order. More recently it was proved in other special cases by Brauer and Leonard [2], Leonard [4 and 5], and Sibley [6].

Because of the groups $SL(2,q)$ the bound in Sibley's theorem is sharp. But the following conjecture is strongly suggested.

CONJECTURE. *Suppose* G *satisfies Hypothesis 1 and has a faithful complex representation of degree* $d = (q-1)/2$. *Then either* $P \lhd G$ *or* $G/Z \cong PSL(2,q)$.

This conjecture was proved in 1942 by Brauer [1] in the case that P has prime order. In the remainder of this paper we describe an effort to prove the conjecture, and we confirm the conjecture under certain quite restrictive conditions.

AN ATTEMPT TO PROVE THE CONJECTURE. In order to describe our attempt to prove the conjecture, we formulate

HYPOTHESIS 2. *The group* G *satisfies Hypothesis 1 and has a faithful complex representation of degree* $d=(q-1)/2$. *Furthermore* $C = P \times Z$ (that is, $V = Z$), N/P *is abelian, and subject to these conditions,* G *is a counterexample of minimal order to the conjecture. In particular* $P \ntriangleleft G$ *and* $G/Z \ncong PSL(2,q)$.

In the remainder of the paper we assume the above conditions. The assumptions on V and N/P are made primarily to facilitate the presentation. In this section we prove a series of propositions, most of which concern the properties of the characters of G. By using recently developed methods we shall prove nearly all the properties of the characters one would expect to need. Our proofs depend on methods of ordinary character theory as developed in [4] and [5]. Brauer's proofs of these results when P has prime order [1] depend on aspects of the modular representation theory which are not available under our more general assumptions.

The final phase of Brauer's proof of the conjecture when P has prime order [1] consists of the use of the information about the ordinary characters of G to determine the degrees of all the modular characters in the first p-block of G. From this information the proof is relatively easily completed. Of course under our more general assumptions Brauer's method is completely unavailable. While it seems probable that much of the character theoretic information described below will be needed in proving the conjecture, an entirely new approach is needed for the completion of the proof.

Let Λ denote the character afforded by the faithful representation given in Hypothesis 2. In what follows, familiarity with much of [4] will be assumed.

(1) *The character* Λ *is irreducible.*

PROOF. If Λ were reducible we could apply Sibley's theorem to its constituents, concluding that $P \lhd G$.

(2) *The character* Λ *is exceptional and* $\Lambda|_N$ *is irreducible.*

PROOF. We know that Λ is exceptional since otherwise $\Lambda(1) \geqslant q - 1$. It follows from [5, Theorem 2] that $\Lambda|_N$ is irreducible.

(3) *We have* $(N: C) = (q-1)/2$ *so* P *has exactly two classes of* G-*conjugate elements.*

PROOF. Proposition (2) implies that $\Lambda|_N$ is induced from C . Hence the result.

(4) *If* X_m *is an irreducible character of* G *with* $X_m \neq 1_G$ *then* P *is not contained in the kernel of* X_m .

PROOF. Suppose G does have a non-principal character X_m whose kernel H contains P . Since $P \ntriangleleft G$ and $H \neq G$, we must have $H/Z(H) \cong PSL(2,q)$. All conjugates of P are in H , and $P^{\#}$ has exactly two classes of conjugate elements relative both to G and to H . Therefore if $u \in G$ and $y \in P^{\#}$, then there exists $v \in H$ such that $y^u = y^v$. So $u \in vC$, and $G \subseteq HC = HZ$. Hence $G/Z \cong PSL(2,q)$, a contradiction. This completes the proof of (4).

NOTATION. For the members of the first p-block B_1 of G we use the notation introduced in [4, §1], and we use the notation introduced there for certain numbers associated with these characters. In particular B_1 contains two exceptional characters Λ_1 and Λ_2, and a certain number t of non-exceptional characters X_m, $m = 1, 2, \cdots, t$. Since $C = PZ$, the kernels of all these characters contain Z. As in [4,§2] we distinguish two cases:

$$\text{CASE 1.} \qquad 2c_1 - \varepsilon_1 > 0 .$$

$$\text{CASE 2.} \qquad 2c_1 - \varepsilon_1 < 0 .$$

Since $\varepsilon_1 = \pm 1$, one of these cases must hold.

(5) *If* $d_{1m} < 0$ *then* $f_{m1} = 0$, *no constituent of* $X_m|_N$ *has* P *in its kernel, and* $X_m(1) = -d_{1m}(q-1)$.

In case 1 *either*

$$c_1 = 0 \text{ and } \varepsilon_1 = -1$$

or

$$c_1 = 1 \text{ and } \varepsilon_1 = 1 .$$

In case 2, $b_1 = 0$ *and* $c_1 = -a_1$.

PROOF. We apply the results of [4, §2] to Λ. In [4,(2.7)] the left hand side is zero, so if $d_{1m} < 0$ then $f_{m1\ell} = 0$ for every ℓ and $f_{m1} = 0$. The first assertion now follows.

In case 1 we know $c_1 \geqslant 0$. If $c_1 > 0$ then, according to [4(1.2)], $b_1 > 0$ and then [4,(2.6)] implies that $2c_1 - \varepsilon_1 \leqslant \chi_{1\ell}(1) = 1$, since N/C is abelian. Hence the second assertion.

In case 2, [4,(2.7)] implies that $b_{1\ell} = 0$ for every ℓ , so that the third assertion follows from [4,(1.2)].

(6) *We have*

$$\sum_{d_{1m}>0} (d_{1m}^{2}+d_{1m}\, e_{m1}) + \begin{Bmatrix} b_1 \\ 0 \end{Bmatrix} = \frac{q-1}{2} \geqslant 2 \sum_{d_{1m}>0} d_{1m}\, e_{m1} + 1 + \begin{Bmatrix} 2a_1+\varepsilon_1 \\ 0 \end{Bmatrix}$$

where the first alternative occurs in Case 1 and the second in Case 2.

PROOF. Since the left hand side of [4,(2.7)] is zero, the above equation follows from [4,(1.6)]. The above inequality is proved in the same way that the corresponding inequality in [4,(4E)] is proved.

(7) *If* $d_{1m} > 0$ *then* $d_{1m} = 1$.

PROOF. Since N/C is abelian and V = Z , the numbers $x_{1\ell}$ in [4,§1] satisfy $x_{1\ell} = \chi_{1\ell}(1) = 1$. Applying [4,(1.4)] for each ℓ and applying (5) we obtain (7).

(8) *Let* u *denote the number of non-exceptional characters in* B_1 *for which* $d_{1m} > 0$. *Then*

$$u + \left\{ \begin{matrix} b_1 \\ \\ 0 \end{matrix} \right\} \geqslant 1 + \sum_{d_{1m} > 0} e_{m1} + \left\{ \begin{matrix} 2a_1 + \varepsilon_1 \\ \\ 0 \ . \end{matrix} \right.$$

PROOF. This is a reformulation of part of (6) which results from the use of (7).

(9) *In case 1 we may assume* $c_1 = 0$, $a_1 = b_1 = 1$, *and* $\varepsilon_1 = -1$, *so* $\Lambda_1(1) = \Lambda_2(1) = (q+1)/2$.

PROOF. According to (5) if $c_1 \neq 0$ then $c_1 = 1$ and $\varepsilon_1 = 1$. Hence we may assume $c_1 = 0$, since B_1 contains only two exceptional characters. Then in any case (5) implies $\varepsilon_1 = -1$, and according to [4,(1.2)] $a_1 = b_1$. If $X_m \neq 1_G$ then according to (4) $e_{m1} > 0$. Hence (8) implies $2a_1 + \varepsilon_1 \leqslant a_1$, so $a_1 = 1$ since $\varepsilon_1 = -1$.

(10) *If* $X_m \neq 1_G$ *and* $d_{1m} > 0$ *then* $e_{m1} = 1$ *and* $X_m(1) = q+1$.

PROOF. According to (4) $e_{m1} > 0$. Combining (8) and (9) we have

$$u \geqslant 1 + \sum_{d_{1m} > 0} e_{m1} ,$$

so that $e_{m1} = 1$. According to (7) $d_{1m} = 1$, so [3,(1.3)] implies $f_{m1} = 2$. Therefore $X_m(1) = q + 1$.

(11) *We have*

$$u = \begin{cases} (q-1)/4 & \textit{in case } 1, \\ (q+1)/4 & \textit{in case } 2. \end{cases}$$

In particular $q \equiv 1 \pmod 4$ *in case* 1 *and* $q \equiv -1 \pmod 4$ *in case* 2.

PROOF. Combining (6), (7), (9) and (10) we have

$$u + (u-1) + \begin{Bmatrix} 1 \\ 0 \end{Bmatrix} = \frac{q-1}{2},$$

where the first alternative occurs in case 1 and the second in case 2. Hence

$$2u = \frac{q-1}{2} + \begin{Bmatrix} 0 \\ 1 \end{Bmatrix},$$

and the assertion follows.

(12) *We have*

$$\Lambda \bar{\Lambda} = \sum_{d_{1m} > 0} X_m + \begin{cases} \Lambda_1 \\ 0 \end{cases}$$

where, in case 1, Λ_1 *denotes one of the two exceptional characters in* B_1.

PROOF. It follows from [4,(2.1) and (2.2)] that $\Lambda \bar{\Lambda}$ includes among its constituents the characters in the above sum. Upon applying (9), (10), and (11) to compare degrees, we find we have equality.

(13) *Let* Λ' *denote the* p-*conjugate of* Λ *different from* Λ . *Then*

$$\Lambda \overline{\Lambda}' = \sum_{d_{1m} < 0} -d_{1m}X_m + \begin{cases} 0 \\ \ell_1\Lambda_1 + \ell_2\Lambda_2 , \end{cases}$$

where $\ell_1 + \ell_2 = \varepsilon_1 - 2c_1$ *in case* 2.

PROOF. According to [4,(1.7)]

$$(c_1 - \varepsilon_1)^2 + c_1^2 + \sum d_{1m}^2 = \frac{q-1}{2} + 1.$$

By (7) and (11)

$$u = \sum_{d_{1m} > 0} d_{1m}^2 = \begin{cases} (q-1)/4 , \\ (q+1)/4 . \end{cases}$$

Hence

$$\sum_{d_{1m} < 0} d_{1m}^2 = \frac{q-1}{2} - 2c_1^2 + 2c_1\varepsilon_1 - \begin{cases} (q-1)/4 , \\ (q+1)/4 . \end{cases}$$

In case 1, (9) implies this is (q-1)/4. Furthermore (5) implies

$$\sum_{d_{1m} < 0} - d_{1m}X_m(1) = (q-1) \sum_{d_{1m}} d_{1m}^2 .$$

According to [4,(2.1)], $-d_{1m}X_m \subseteq \Lambda \overline{\Lambda}'$ if $d_{1m} < 0$. Equation (13) now follows in case 1 upon comparison of degrees.

In case 2, [4,(2.2)] and (12) imply $(\Lambda \overline{\Lambda}' , \Lambda_1 + \Lambda_2) = \varepsilon_1 - 2c_1$. According to [4] and (5),

$$\Lambda_1(1) = \Lambda_2(1) = \varepsilon_1(q-1)/2 - c_1(q-1) .$$

Although ε_1 and c_1 are not known in case 2, the proof can be completed in case 2 by the same method used in case 1.

(14) *We have* $|Z| \leqslant 2$, *and in case* 1, $|Z| = 2$.

PROOF. In case 1, $(q-1)/2$ is even, by (11), so each of the classes of elements of P is its own inverse since N/Z is a Frobenius group. Therefore Λ is real: $\Lambda = \overline{\Lambda}$. Since the kernel of each character in B_1 contains Z, equation (12) implies $z^2 = 1$ for every $z \in Z$. But Λ is faithful so Z is cyclic and $|Z| \leqslant 2$. Since Λ does not have Z in its kernel, $|Z| = 2$.

In case 2, $(q-1)/2$ is odd, $\Lambda \neq \overline{\Lambda}$, so $\overline{\Lambda} = \Lambda'$. Hence $\Lambda \overline{\Lambda}' = \Lambda \Lambda$. Now equation (13) implies $|Z| \leqslant 2$, completing the proof.

In case 2, I conjecture that $Z = 1$, but I don't have a proof. To simplify the discussion we shall assume from now on

HYPOTHESIS 3. *Hypothesis* 2 *holds and, in case* 2, $Z = 1$.

Under this assumption $\varepsilon_1 = 1$ and $c_1 = a_1 = b_1 = 0$ in case 2. Also (13) can be refined as follows.

(13') *Let* Λ' *denote the* p-*conjugate of* Λ *different from* Λ . *Then*

$$\Lambda\,\overline{\Lambda}' = \sum_{d_{1m} < 0} - d_{1m}X_m + \begin{cases} 0\,, \\ \overline{\Lambda}\,. \end{cases}$$

PROOF. In case 2, $\Lambda\,\overline{\Lambda}' = \Lambda\,\Lambda$. If $(\Lambda\,\Lambda,\Lambda) \geqslant 1$ then $(\Lambda\,\overline{\Lambda},\Lambda) \geqslant 1$, contrary to (12). Therefore (13) implies

$$(\Lambda\,\Lambda,\overline{\Lambda}) \geqslant 1\,,$$

and (13') now follows from (13).

(15) *If* χ_ℓ *is a non-principal linear character of* N *whose kernel contains* Z *, then in case* 1 *either*

$$\chi_\ell^G = \Lambda_1 + \Lambda_2 + \textit{characters of defect } 0$$

or

$$\chi_\ell^G = X_m + \textit{characters of defect } 0,$$

where χ_ℓ^G *denotes* χ_ℓ *induced to* G *and where* X_m *is a non-exceptional character in* B_1 *for which* $d_{1m} > 0$. *In case* 2 *only the second equation can occur.*

In both cases,

$$(1_N)^G = 1_G + \textit{characters of defect } 0.$$

PROOF. Suppose first that we have case 1. If $d_{1m} < 0$ then each constituent of $X_m|_N$ is induced from P , so

$$X_m|_{N - PZ} = 0\,.$$

According to (9), $c_1 = 0$ and $\varepsilon_1 = -1$. Therefore [4,(1D)(ii)] implies that

$$\Lambda_1 + \sum_{d_{1m} > 0} X_m$$

vanishes on $N - PZ$. If $d_{1m} > 0$ and $X_m \neq 1_G$ then $X_m|_N$ has exactly two linear constituents. And $\Lambda_1|_N$ has exactly one linear constituent. Since $u = (q-1)/4$, the non-exceptional characters X_m for which $d_{1m} > 0$, including 1_G, account for $1 + 2(u-1) = (q-3)/2$ linear characters of N. Hence the sum of the linear constituents of

$$(\Lambda_1 + \sum_{d_{1m} > 0} X_{1m})|_N \quad \text{is} \quad \sum \chi_\ell .$$

Since Λ_1 and Λ_2 agree on p'-elements, we have the result in case 1.

The proof in case 2 is entirely similar. In this case, Λ_1 and Λ_2 are not involved in the reasoning.

To complete the proof of the conjecture presumably requires other sorts of methods, but it is not at all clear to me what they should be. If E denotes a complement of P in N then one can prove that $C_G(E)$ is abelian, has the same exponent as E, and E is a direct factor of $C_G(E)$. If $\overline{E} = E/Z$ then $(N_G(\overline{E}) : C_G(\overline{E})) \leqslant 2$. But this is weak information.

Some Special Cases

1. Since the characters of N are known, the class multiplication constants in N/Z for the two classes of non-identity p-elements can be computed. Using the character theory developed in the preceding section, the same can be done in G/Z in terms of $((G : N) - 1)/q$, which we denote by e. If $p = 3$ then we may apply a theorem of Herzog [3, Theorem 4.1] which states that in certain cases these numbers are the same for G/Z as for N/Z. The conclusion in case 2 is that $e = 1$. Doubtless entirely analogous calculations imply in case 1 too that $e = 1$. By representing G/Z as a permutation group on the $q+1$ cosets of N/Z it is easily concluded that the conjecture is true when $p = 3$.

2. When $q = 5^2$ we can show that G/Z has exactly 6 characters of degree 24. By studying the principal 13-block of G/Z we can show that there is only one class of involutions in G/Z. Consequently the power of 2 in $(G : Z)$ is 8, and the conjecture holds.

3. If the Sylow 2-subgroup of G/Z has sectional 2-rank $\leqslant 4$, then G/Z is known and the conjecture is true. This enables us to show, for example, that if $q = 7^2$ then $(G : Z) > 135{,}000{,}000$.

REFERENCES

1. R. Brauer, *On groups whose order contains a prime number to the first power, I, II,* Amer. J. Math. 64 (1942) 401-420, 421-440.

2. R. Brauer and H. S. Leonard Jr., *On finite groups with an Abelian Sylow group,* Canad. J. Math. 14 (1962), 436-450.

3. M. Herzog, *On finite groups which contain a Frobenius subgroup,* J. Alg. 6 (1967), 192-221.

4. H. S. Leonard Jr., *Finite linear groups having an Abelian Sylow subgroup,* J. Alg. 20 (1972), 57-69.

5. ____________, *Idem II,* J. Alg. 26 (1973), 368-382.

6. D. A. Sibley, *Finite linear groups with a strongly self-centralizing Sylow subgroup,* J. Alg. 36 (1975), 158-166.

7. ____________, *Idem II,* to appear.

NORTHERN ILLINOIS UNIVERSITY
DEKALB, ILLINOIS

SYLOW AUTOMIZERS OF ODD ORDER

OR

AN APPLICATION OF COHERENCE

BY

STEPHEN D. SMITH

Let $P \in Syl_p(G)$, p an odd prime. The *automizer* of P in G is $N_G(P)/C_G(P)$. We investigate what may happen if this automorphism group of P has *odd* order. Coherence plays a crucial role in these investigations. Our results extend earlier work by G. Higman and D. Garland (to be published).

In a classical coherence proof, we try to show that the irreducible characters of G are closely related to the irreducible characters of a suitable p-local subgroup (usually N(P)) by their values *on* p-*elements*.

A typical situation to study is:

1. N(P) is a Frobenius group, with kernel P and complement E .
2. Centralizers of (non-1) elements of P lie in

P.

3. $N(P)$ controls fusion of elements of P.

If now the coherence problem can be completed, a typical application is:

If G has irreducible character χ, with $\chi(1)$ sufficiently small in terms of $|P|$, then $G = N(P)$.

Another familiar application, involving an easy class-multiplication constant calculation is:

Suppose $|E| = 2$. Then involutions of $G - O_{p'}(G)$ are conjugate to the involution of E.

One way of generalizing the analysis is to enlarge $C(P)$:

(1*) $N(P)/O_{p'}(C(P))$ is a Frobenius group as in (1).

(2*) Centralizers of elements of $P^{\#}$ lie in $P \cdot C(P)$.

The resulting calculations are not much more complicated than the first case, and similar applications emerge.

If we take into account the theory of p-blocks, we can relax even these assumptions about centralizers of p-elements. In particular, if we restrict attention to *principal blocks*, then we can effectively ignore p-regular cores. The situation:

(1**) $N(P)/O_{p'}(C(P))$ is a Frobenius group with kernel P and complement E.

(2**) For $x \in P^{\#}$, $C(x) = O_{p'}(C(x)) \cdot C_{N(P)}(x)$.

And now by the "coherence problem" we mean the attempt to relate the characters of the principal p-block of G to the characters of the principal block of N(P), via values *on p-singular elements*. It is to be emphasized that, while the techniques differ somewhat, the arithmetic involved is the same as in classical coherence problems. As an example we have the following

RESULT. If $|E| = 2$, the coherence problem is easy and involutions of $G-O_{p'}(G)$ are conjugate to those of $N-O_{p'}(N)$, where $N = N(P)$.

Suppose, in addition:

(3**) the coherence problem can be done.

(4**) the p-part of the Schur multiplier of G is not trivial.

Then calculations can be made in a suitable central extension H of G--these could not be done for G alone.

CONCLUSION. If $|E|$ is odd, then $G = O_{p'}(G) \cdot N(P)$.

The same conclusion for $|E| = 2$ was known previously. We see that, given the existence of a multiplier, the p-singular character values in these cases actually control the character values *on p-regular elements*.

Walter Feit points out that the calculation establishing the above can be rephrased to demonstrate:

Under a mild numerical assumption, described later, if

$|E|$ is even, then involutions of $G\text{-}O_{p'}(G)$ are conjugate to those of $N\text{-}O_{p'}(N)$.

This motivates asking: when can the coherence problem be done? We are led to the conjecture:

Assume (1**), (2**), and (4**). If $|E|$ is odd, then $G = O_{p'}(G)\cdot N(P)$.

By means of grisly calculations, we may establish the conjecture in the following cases:

a) P abelian

b) $p \geqslant 7$, $c\ell(P) \leqslant 2$, $|E| = 3$ or 5 .

This extends results of Higman and Garland. It now seems that remarkable new methods of D. Sibley may be adapted for the present situation, to solve the coherence problem for G . One can then try to solve the problem for an extension H of G— and establish the conjecture in general.

One ought to be able to prove the result with (1**) and (4**) replaced by:

(1***) $N(P)/O_{p'}(C(P))$ is a central extension of a Frobenius group.

For the calculation is really only concerned with the existence of weakly closed p-elements and the characters of $N(P)$. If, however, $P \cap Z(N(P)) \nleqslant Z(G)$, the numbers are more difficult, and may require additional hypotheses (e.g., E cyclic?)

In the case where $|E|$ is even, P must be abelian. Here the coherence problem is easy unless $|P| - 1 = |E|$.

THEOREM. Assume (1**), (2**), and (4**). If $|E|$ is even, then involutions of $G\text{-}O_{p'}(G)$ are conjugate to those of $N\text{-}O_{p'}(N)$, *except* possibly if $|E| = |P| - 1$ or if $\zeta(1) \equiv \pm 1 \pmod{|E|}$ where ζ is the unique E-invariant character of a suitable central extension of P , with a given (nontrivial) restriction to the center.

CALIFORNIA INSTITUTE OF TECHNOLOGY
PASADENA, CALIFORNIA

ON GROUPS OF CENTRAL TYPE

BY

JAY YELLEN

Let G be a finite group with center Z having an irreducible character χ such that $\chi(1)^2 = \lfloor G:Z \rfloor$. Then G is called a group of central type. K. Iwahori and S. Matsumoto [J. Fac. Sci. Univ. Tokyo, 10 (1964)] noted that G is a group of central type if and only if there is a central simple projective group algebra KH associated with G where H is finite. They conjectured that groups of central type are solvable.

Suppose G is a counterexample to the conjecture. We show that G/Z has a homomorphic image, say G/N, such that

$$S_1 \times \ldots \times S_m \leqslant G/N \leqslant \mathrm{Aut}(S_1 \times \ldots \times S_m)$$

where $S_i \cong S$ a non-abelian simple group and G/N acts transitively on the m components. If $[\mathrm{Aut}\ S:S] = 1$ then $G/N = H \sim T$ where $H \leqslant \mathrm{Aut}\ S$ and T is transitive. The following result provides information for the case $N = Z$ and is a possible first step in an induction proof of the conjecture.

THEOREM: Let G be a group of central type with

$$S_1 \times \ldots \times S_m \leqslant G/Z \leqslant \mathrm{Aut}(S_1 \times \ldots \times S_m)$$

where $S_i \cong S$ a non-abelian simple group containing a non-trivial cyclic Sylow p-subgroup. Suppose G/Z acts transitively on the m components. Then m is even.

COROLLARY: With the above hypothesis G/Z cannot have an abelian Sylow 2-subgroup.

COLORADO STATE UNIVERSITY
FORT COLLINS, COLORADO

PART IV

PERMUTATION GROUPS

TWO-TRANSITIVE EXTENSIONS OF SOME GROUPS

BY

PETER KORNYA

The two-transitive permutation group (G,Ω) is called a two-transitive extension of the group H if $G_\alpha \cong H$. The following theorem is proved:

THEOREM: Let (G,Ω) be a two-transitive extension of even degree of S_n, the symmetric group on n letters. Then $|\Omega| = n + 1$ and G is the symmetric group on Ω, except if $n = 6$ or 8.

The general problem is to investigate possible two-transitive extensions of H for $G(q) \leqslant H \leqslant \text{Aut } G(q)$, where $G(q)$ is a Chevalley group. The above theorem is useful in the case $|\Omega|$ is even and q is odd.

UNIVERSITY OF OREGON
EUGENE, OREGON

THE NON-EXISTENCE OF RANK-3 TRANSITIVE EXTENSIONS OF THE HIGMAN-SIMS SIMPLE GROUP

BY

SPYROS S. MAGLIVERAS

A finite group G is said to be a *transitive extension* of a group H if G has a representation as a transitive permutation group on a set Ω with point stabilizer G_α isomorphic to H, $\alpha \in \Omega$. The G_α-orbits in Ω are called the *suborbits*, and the suborbit lengths, the *subdegrees* of G. The *rank* of G is the number of suborbits. By $\underline{Ext_r(H)}$ we mean the collection of all isomorphism types of groups G which are rank-r transitive extensions of H.

The following result extends and completes [6]:

THEOREM: $Ext_3(HS) = \emptyset$.

In the sequel, HS denotes a group isomorphic to the Higman-Sims simple group, and the notation used follows [3], [4], [7]. Throughout, G is a rank-3 transitive extension of HS with subdegrees $1 \leq k \leq \ell$. Φ_k, Φ_ℓ denote the characters of the transitive representations of HS on the suborbits of lengths k , ℓ, respectively, and θ denotes the induced character $1_{HS}\uparrow^{G}$. A conjugacy class of HS is denoted by a symbol $\underline{n_i}$ where n is the order of an element in $\underline{n_i}$ and i an index to differentiate between classes of elements of the same order. If there is a unique conjugacy class of elements of order n , we denote the class by $\underline{n}$.

2. Preliminary

LEMMA 1. If G is a rank-3, primitive extension of a simple group H , and if N is a regular normal subgroup of G on which $G_x \cong H$ acts non-trivially, then N is elementary abelian.

PROOF. If G is a permutation group on Ω , then $|N| = |\Omega|$ implies that N is a minimal normal subgroup of G, and consequently characteristically simple.

The action of H on $N^* = N - \{1\}$ has two orbits and consequently, the non-identity elements of N can have at most two distinct orders k_1 , k_2 . It follows that at most two primes can divide $|N|$. Hence, N is solvable.

∎

LEMMA 2. (i) There are no imprimitive, rank-3 extensions of HS .

(ii) Every rank-3, transitive extension G of HS is simple.

(i) If k, ℓ, $k < \ell$, are the subdegrees for an imprimitive rank-3 extension, the blocks of imprimitivity would be of size $1 + k$ with $1 + k$ dividing ℓ. Consideration of all possible pairs $k < \ell$ shows that this is not possible.

(ii) Suppose that there is a *composite* group G which is transitive, rank-3 on Ω, with $G_x \cong HS$. Then, G is primitive, and possesses a regular normal subgroup N which by Lemma 1 is elementary abelian. Table I exhibits the possible suborbit lengths $k \leqslant \ell$ for which $n = 1 + k + \ell$ is a power of a prime, p^m, $m > 1$. In each of these cases the order of HS does not divide the order of $GL_m(p)$, hence such an extension is impossible.

TABLE I

All possible $n = 1 + k + \ell = p^m$, $m > 1$.

k	ℓ	n	p	m
176	352	529	23	2
1100	38500	39601	199	2
8800	30800	39601	199	2
36960	138600	175561	419	2
36960	177408	214369	463	2
36960	554400	591361	769	2
36960	739200	776161	881	2
36960	1478400	1515361	1231	2
46200	14784000	14830201	3851	2
73920	138600	212521	461	2
132000	147840	279841	23	4
138600	295680	434281	659	2
147840	443520	591361	769	2
184800	591360	776161	881	2
211200	1774080	1985281	1409	2
211200	5544000	5755201	2399	2
221760	369600	591361	769	2
221760	554400	776161	881	2
295680	295680	591361	769	2
396000	22176000	22572001	4751	2
462000	2956800	3418801	43	4
591360	924000	1515361	1231	2

3. PROOF OF THEOREM

Since a rank-3 transitive extension G of HS is simple, the Gorenstein-Harris characterization of HS by its Sylow-2 subgroup yields that [G: HS] must be even, hence exactly one of k , ℓ would be odd. There are precisely five inequivalent transitive permutation representations of HS of odd degree corresponding to the five conjugacy classes of subgroups containing the Sylow-2 subgroup of HS. The characters of these representations, except for the two cases where $|HS_\alpha| = 2^9 \cdot 3$, are tabulated in Table III. Consideration of the possible pairs $\{k,\ell\}$, with one of k , ℓ odd, shows that the only possible parameters are the ones indicated in Table II.

TABLE II

Case	n	k	ℓ	λ	μ
1	6876	1100	5775	154	180
2	11376	5600	5775	2728	2784
3	11826	4125	7700	1548	1380
4	23376	5775	17600	1422	1428
5	34476	5600	28875	1144	864
6	36576	5775	30800	654	960
7	75076	5775	69300	230	462
8	75076	28875	46200	11066	11130
9	98176	5775	92400	398	336
10	105876	28875	77000	7818	7896
11	797776	5775	792000	14	42

TABLE II (Cont.)

Case	s	t	f_2	f_3
1	20	-46	4775	2100
2	32	-88	8295	3080
3	183	-15	875	10950
4	63	-69	12175	11200
5	296	-16	1750	32725
6	15	-321	34925	1650
7	21	-253	69300	5775
8	105	-169	46200	28875
9	111	-49	30030	68145
10	111	-189	66605	39270
11	63	-91	471375	326400

TABLE III

Certain Permutation Characters of HS

1	2_1	2_2	4_1	4_2	4_3	8_1	8_2	8_3	3	6_1
100	20	0	4	8	0	2	0	0	10	2
1100	60	32	4	16	40	2	0	0	11	3
1100	60	20	8	8	0	2	2	2	20	0
5775	111	75	3	15	31	3	1	1	15	3
5600	160	0	8	0	0	0	4	0	20	4
5600	160	0	8	0	0	0	0	4	20	4
4125	125	45	5	21	5	1	1	1	30	2
7700	260	0	4	40	0	2	0	0	50	2
86625	705	225	9	57	105	5	1	1	0	0

TABLE III (Cont.)
Certain Permutation Characters of HS

6_2	12	5_1	5_2	5_3	10_1	10_2	20_+	20_-	15	7	11_+	11_-
0	0	5	0	0	0	0	0	0	0	2	1	1
5	1	0	10	0	2	0	0	0	1	1	0	0
2	0	5	0	0	0	0	0	0	0	1	0	0
3	1	0	0	25	0	1	1	1	0	0	0	0
0	0	5	0	0	0	0	0	0	0	0	1	1
0	0	5	0	0	0	0	0	0	0	0	1	1
0	2	0	0	0	0	0	0	0	0	2	0	0
0	0	10	0	0	0	0	0	0	0	0	0	0
0	0	0	0	0	0	0	0	0	0	0	0	0

We proceed to show that cases 1 through 11 in Table II are not possible.

CASE 1. Suppose that there is a rank-3 transitive extension G of HS with subdegrees 1100, 5775. Then, $|\Omega| = 1 + 1100 + 5775$ and $|G| = 2^{11}\cdot 3^4\cdot 5^3\cdot 7\cdot 11\cdot 191$. The characters of the two inequivalent representations of HS on 1100 points and the unique representation on 5775 points are listed in Table III. We have:

$$\theta(2_1) = 1 + \Phi_{1100}(2_1) + \Phi_{5775}(2_1) = 1 + 60 + 111 = 172 = 4\cdot 43$$

and,

$$\theta(2_2) = 1+\Phi_{1100}(2_2)+\Phi_{5775}(2_2) = 1+\{32 \text{ \underline{or} } 20\}+75 = 108 \text{ \underline{or} } 96 .$$

It follows that $\underline{2_1}$ and $\underline{2_2}$ do not fuse in G, and consequently 43 divides $|C_G(2_1)|$, a contradiction.

CASE 2. Suppose that there is a rank-3 transitive extension G of HS with subdegrees 5600 and 5775. Then $[G: HS] = 1 + 5600 + 5775 = 11376 = 2^4 \cdot 3^2 \cdot 79$. Again from the characters appearing in Table III we have:

$$\theta(2_1) = 1 + \Phi_{5775}(2_1) + \Phi_{5600}(2_1) = 272 = 2^4 \cdot 17$$

and,

$$\theta(2_2) = 1 + \Phi_{5775}(2_2) + \Phi_{5600}(2_2) = 76 = 2^2 \cdot 19 .$$

Thus, $\underline{2_1}$ and $\underline{2_2}$ do not fuse in G, and consequently 17 divides $|C_G(2_1)|$, a contradiction.

CASE 3. Let us assume now that there is a rank-3 transitive extension G of HS of degree $11826 = 1 + 4125 + 7700 = 2 \cdot 3^4 \cdot 73$. There is a unique representation of HS on 4125 points namely the one with point stabilizer the maximal subgroup in HS isomorphic to $Z_4 \times Z_4 \times Z_4 \cdot L_3(2)$. Also, there is a unique transitive representation of HS on 7700 points, with point stabilizer the group $2^4 \cdot A_6$ in $2^4 \setminus S_6 = G_{(\overset{\shortparallel}{2})}$. The characters of these representations are listed in Table III. We have:

$$\theta(2_1) = 1 + 125 + 260$$

and,

$$\theta(2_2) = 1 + 45 + 0 = 46 .$$

Hence, $\underline{2_1}$ and $\underline{2_2}$ do not fuse in G. It follows that 23 divides $|C_G(2_2)|$, a contradiction.

CASE 4. Suppose that HS has a rank-3 transitive extension G with subdegrees 5775 and 17600. Then $[G: HS] = 1 + 5775 + 17600 = 2^4 \cdot 3 \cdot 487$, and Sylow's equation for the prime 487 shows that 77 must divide the order of the centralizer of a Sylow-487 subgroup of G. Now, an element of order 11 in G must fix precisely one point of Ω because 11 divides $|G|$ squarefree, and divides both subdegrees, and $|HS|$. Hence, an element of order 11 cannot centralize an element of order 487.

CASE 5. If $H < HS$ with $|H| = 2^9 \cdot 3$ it can be shown that H must be contained in a group $C \cong (Z_4 \times Z_4 \times Z_4) \cdot L_3(2)$. Computation inside C shows that

$$|\underline{2_1} \cap H| = 79 \quad \text{and} \quad |\underline{2_2} \cap H| = 72 .$$

Hence,

$$\Phi_{28875}(2_1) = 395 \quad \text{and} \quad \Phi_{28875}(2_2) = 135 .$$

If G is a rank-3 transitive extension of HS with subdegrees 5600 and 28875, then $[G: HS] = 34476 = 2^2 \cdot 3 \cdot 13^2 \cdot 17$. We have:

$$\theta(2_1) = 1 + 395 + 160 = 2^2 \cdot 139$$

and,

$$\theta(2_2) = 1 + 135 + 0 = 8 \cdot 17 .$$

Again, $\underline{2_1}$ and $\underline{2_2}$ do not fuse in G and we reach the contradiction that 139 must divide $|G|$.

CASE 6. Here, $n = [G: HS] = 1 + 5775 + 30800 = 2^5 \cdot 3^2 \cdot 127$. Since an element of order 11 in G fixes exactly one point of Ω , 11 does not divide the order of the centralizer of a Sylow-127 subgroup in G . Furthermore, it can be shown that an element of order 5 of HS can fix at most 30 points in the transitive representation of HS on 30800 points. It follows that an element of order 5 of HS can fix at most $1 + 25 + 30$ points of Ω , and must fix at least one point of Ω since the Sylow-5 subgroup of G is in HS . It follows that no element of order 5 can centralize an element of order 127 in G . The fact that neither 5 nor 11 divides the order of the centralizer of a Sylow-127 subgroup of G contradicts Sylow's equation for the prime 127.

CASE 7. Suppose that there is a rank-3 transitive extension G of HS with subdegrees 5775 and 69300. Then, $[G: HS] = 75076 = 2^2 \cdot 137^2$. Since the primes 7 and 11 divide $|G|$, $|HS|$ squarefree, and also divide the subdegrees, an element of order 7 or 11 in G must fix exactly one point of Ω, and consequently neither 7 nor 11 divides the order of the centralizer in G of a Sylow-137 subgroup of G. Sylow's equation for the prime 137 shows that this is not possible.

CASE 8. Case 8 is handled exactly like Case 7.

CASE 9. If $H < HS$, with $|H| = 480$, then H is a subgroup of a conjugate of $C_{HS}(z)$, $z \in \underline{2_1}$, and has the form $(Z_4 \cdot A_5)\backslash Z_2$. On the basis of this information, $\Phi_{92400}(2_1) = 336$ and $\Phi_{92400}(2_2) = 60$ *or* 300. If a rank-3 transitive extension G of HS exists with subdegrees 5775, 92400, then we would have:

$$\theta(2_1) = 1 + \Phi_{5775}(2_1) + \Phi_{92400}(2_1) = 1 + 111 + 336 = 448$$

and,

$$\theta(2_2) = 1 + \Phi_{5775}(2_2) + \Phi_{92400}(2_2) = 1 + 75 + \begin{cases} 60 \\ 300 \end{cases} = \begin{cases} 8 \cdot 17 \\ 8 \cdot 47 \end{cases}.$$

Hence, $\underline{2_1}$ and $\underline{2_2}$ do not fuse in G and 47 or 17 would divide $|G|$, which is impossible since $[G:HS] = 2^7 \cdot 13 \cdot 59$.

CASE 10. Suppose that there is a rank-3 transitive extension G of HS of degree 105876 with subdegrees 28875 and 77000. Then $|G| = 2^{11} \cdot 3^4 \cdot 5^3 \cdot 7 \cdot 11 \cdot 17 \cdot 173$. Since the primes 7 and 11 occur in $|G|$ squarefree, and both divide $|HS|$, and the subdegrees, an element in G of order 7 or 11 must fix exactly one point of Ω . It follows that the order of the centralizer of $P \in Syl_{173}(G)$ is not divisible by 7 or by 11. Under this condition on $C_G(P)$, the solutions to Sylow's equation:

$$|G| = (1 + r \cdot 173) \cdot ce \cdot 173 , \quad e \mid 172 = 2^2 \cdot 43$$

are $(1 + r \cdot 173) = 3^2 \cdot 7 \cdot 11$, $2 \cdot 3^4 \cdot 5^3 \cdot 7 \cdot 11$, $2^6 \cdot 3^4 \cdot 5 \cdot 7 \cdot 11 \cdot 17$, $2^9 \cdot 3 \cdot 5^3 \cdot 7 \cdot 11 \cdot 17$. These possibilities are ruled out by the fact that in each case there are no solutions to the degree equation for the principal 173 block. [1].

CASE 11. Suppose that there exists a rank-3 transitive extension G of HS with subdegrees 5775 and 792000 . Then, $|G| = 2^{13} \cdot 3^2 \cdot 5^3 \cdot 7^2 \cdot 11 \cdot 17 \cdot 419$. The stabilizer of a point in the transitive representation of HS on 792000 points is a group of order 56 normalizing an elementary abelian subgroup of order 8 with all involutions from $\underline{2_1}$. It follows that

$$\Phi_{792000}(2_1) = 960 , \quad \text{and} \quad \Phi_{792000}(2_2) = 0 .$$

Whence,

$$\theta(2_1) = 1+111+960 = 1072 , \quad \theta(2_2) = 1+75+0 = 76 = 4 \cdot 19 ,$$

and $\underline{2_1}$ and $\underline{2_2}$ do not fuse in G . But this would imply that 19 divides $|C_G(2_2)|$, a contradiction.

REFERENCES

1. R. Brauer, *On groups whose order contains a prime number to the first power,* Part I, Am. J. Math. 64 (1942), 401-420. Part II, Am. J. Math. 64 (1942), 421-440.

2. D. Gorenstein and M. E. Harris, *A Characterization of the Higman-Sims simple group,* J. Algebra 24 (1973), 565-590.

3. D. G. Higman, *Finite permutation groups of rank 3,* Math. Z. 86 (1964), 145-156.

4. ______________, *Intersection matrices for finite permutation groups,* J. Algebra 6 (1967), 22-42.

5. S. S. Magliveras, *The subgroup structure of the Higman-Sims simple group,* Bul. Amer. Math. Soc. 77 (1971), 535-539.

6. ______________, *On transitive extensions of the Higman-Sims simple group,* J. Algebra 30 (1974), 317-319.

7. H. Wielandt, *Finite permutation groups,* "Lectures", University of Tubingen, 1954/55; English Transl. Academic Press, New York, 1964.

STATE UNIVERSITY OF NEW YORK
OSWEGO, NEW YORK

ON THE $n,2n$ PROBLEM OF MICHAEL FRIED

BY

L. L. Scott*

The following situation arises in a number-theoretic investigation of Michael Fried, ["On Hilbert's irreducibility theorem," J. Number Theory, 1974]:

G is a primitive permutation group on a set Ω of cardinality $2n$, and G also acts doubly transitively on a set Γ of cardinality n. In addition G contains an element x which acts as a product of two n-cycles on Ω and as a single n-cycle on Γ, and G is not transitive on $\Omega \times \Gamma$.

We prove in this situation that G has rank 3 on Ω, and this has the consequence that $2n = m^2 + 1$ for some integer m. (In particular, n must be odd.) The proof uses heavily the author's theory of orbital characters (cf. "Modular permutation representations," Trans. AMS, 1973) and the

---

*The results announced in this abstract are part of joint work in progress with Peter M. Neumann.

theory of centralizer rings. In addition, the following new result on characters of cyclic groups plays a key role:

Let X be a finite cyclic group, and χ a multiplicity-free rational-valued character of X. Assume $\chi(x) \neq 0$ for some $x \neq 1$ of X (that is, χ is not the regular character), and set $a = \text{G.C.D.}\{\chi(x) \mid x \in X, x \neq 1\}$. Then $a \mid |X|$ and χ is induced from a character of the subgroup of index a in X; in particular, if $a > 1$ then $\chi(x) = 0$ for x any generator of X.

The proof is obtained by considering modular representations of X for a prime p dividing a.

An immediate (and nontrivial†) corollary is that if χ above has $|\chi(x)|$ constant for $x \neq 1$ in X then $a = 1$, and $\chi(1) = 1$ or $|X| - 1$.

---

†Several of the participants at the conference attempted to find an alternate approach to this corollary, in response to the author's offer of a free dinner for the best proof. None succeeded during the conference, although I have now received an ingenious argument from Walter Feit, using difference sets and the 2-structure of X. (He gets a dinner.)

UNIVERSITY OF MICHIGAN
ANN ARBOR, MICHIGAN

AND

UNIVERSITY OF VIRGINIA
CHARLOTTESVILLE, VIRGINIA

BLOCK DESIGNS FROM FROBENIUS GROUPS AND PLANAR NEAR-RINGS

BY

GERHARD BETSCH AND JAMES R. CLAY*

The concept of a planar near-ring was introduced in [1], and further geometric interpretations were presented in [3, 7, 8, 11, 12, 13, 14]. In [11], Ferrero showed, for the first time, that balanced incomplete block designs can be obtained from certain planar near-rings, that each finite planar near-ring is derived from a Frobenius group, and that each finite Frobenius group gives many planar near-rings, all of which yield the same tactical configuration. Later [14] Ferrero uses a powerful result of Hall to show that each of these finite planar near-rings, or finite Frobenius groups, give several different partially balanced incomplete block designs. Some of Ferrero's results were extended in [7,8], but most of the work in this direction had been done or directed by Fer-

---

*The second author was supported by a Special Program Award administered by the Humboldt Foundation of West Germany.

rero [3, 11, 12, 13, 14]. The research has developed to a point where the important ideas have surfaced. Hence one can give a unified, fairly self contained, and streamlined development of these ideas. This we do here. Most of the results, and underlying ideas of the proofs, were taken from papers, written in Italian, by, or directed by, Ferrero [3, 11, 12,13, 14].

The reader that is not interested in the connection between planar near-rings and Frobenius groups should skip this next section.

1. BASIC PROPERTIES OF PLANAR NEAR-RINGS AND THEIR CONNECTION WITH FROBENIUS GROUPS.

DEFINITIONS AND NOTATION. Throughout this paper $N = (N,+,\cdot)$ will denote a (left) near-ring; i.e., $a(b + c) = ab + ac$. The symbol $\equiv_m$ denotes the equivalence relation defined in N by $a \equiv_m b$ if and only if $ax = bx$ for every $x \in N$. Let $A := \{a \in N | a \equiv_m 0\}$, and $N^* := N - A$. If the equivalence classes $N/\equiv_m$ defined by $\equiv_m$ are at least three in number, and if $a \not\equiv_m b$ implies the existence of a unique solution to any equation in N of the form

$$ax = bx + c, \quad a, b, c \in N,$$

then N is said to be *planar*. Examples and geometric interpretations are given in [1, 7].

From this point on, N will always denote a planar near-ring. The identity $0 \in N$ is a two sided zero; i.e., $0X = X0 = 0$ for all X. Hence the equation $aX = 0X + c$ is more simply written $aX = c$. For $a \not\equiv_m 0$, the unique solution to $aX = a$ is denoted by 1_a, and B_a: $\{x \in N^* | x1_a = x\}$. The fundamental ideas of planar near-rings are contained in the following theorem, the proof of which may be found in [1] or [7].

THEOREM 1. For a planar near-ring $(n,+,\cdot)$, we have

1) $N = A \cup [\bigcup_{a \in N^*} B_a]$;

2) $N^*B_a = B_a$ for $a \in N^*$;

3) for all $a \in N^*$, $(B_a,\cdot)$ is a group with identity 1_a;

4) the family $\{A\} \cup \{B_a | a \in N^*\}$ is pairwise disjoint;

5) if $a, c \in N^*$, then $\phi: B_a \to B_c$ defined by $\phi(x) = x1_c$ is a group isomorphism;

6) each 1_a is a left identity for $(N,+,\cdot)$.

We now turn our attention to the connection between planar near-rings and Frobenius groups.

For $a \in N$, let $f_a(x) = ax$. So $f_a = \zeta$, the zero endomorphism, if and only if $a \in A$, and f_a is an automorphism if and only if $a \in N^*$. If $f_a \notin \{1_N, \zeta\}$, then f_a is a fixed point free automorphism.

Let $\Phi = \{f_a | a \in N^*\}$. We shall see that Φ is isomorphic to any group B_a. Define $\psi: B_a \to \Phi$ by $\psi(c) = f_c$. Now $f_{b \cdot c}(x) = (b \cdot c) \cdot x = b \cdot (c \cdot x) = f_b \circ f_c(x)$ proves that ψ is a homomorphism. If $f_b = f_c$, then $bx = cx$ for all $x \in N$. Hence $b^{-1}bx = b^{-1}cx$, and so $b^{-1}c$ is a left identity for N. Hence ψ is injective. From 5) of theorem 1 we can conclude that ψ is surjective.

The existence of unique solutions to $ax = bx + c$ if $a \not\equiv_m b$ is equivalent to the surjectivity of $f_a - f_b$ if $f_a \neq f_b$, which in turn is equivalent to the surjectivity of $1_N - f_c$ if $f_c \in \Phi \backslash \{1_N\}$. (Injectivity is not needed since each $f_c \in \Phi \backslash \{1_N\}$ is fixed point free.)

So, Φ is a group of fixed point free automorphisms of $(N,+)$ such that if $f_a \in \Phi \backslash \{1_N\}$, then $1_N - f_a$ is surjective.

The orbits of Φ on $(N,+)$ are of two types: (a) those in A, and (b) those in N^*. From

$$f_{f_a(b)} = f_{a \cdot b} = f_a \circ f_b$$

we deduce: a) if $b \in A$, then $f_b = \zeta$, and so $f_a \circ f_b = \zeta$. Hence $f_a(b) \in A$; b) if $b \notin A$, $f_b \in \Phi$ and so $f_a \circ f_b \in \Phi$, ensuring $f_a(b) \notin A$.

Select pairwise disjoint sets of indices $\{0\}$, J, K such that i) $\{0\} = \{x_0\} = \Phi(0)$; ii) $j \in J$ implies x_j is a representative of the orbit $\Phi(x_j) \subset A$; iii) $k \in K$ implies x_k is a representative of the orbit $\Phi(x_k) \subset N^*$; iv) if i_1, $i_2 \in I_0 = \{0\} \cup J \cup K$, and $i_1 \neq i_2$; then $\Phi(x_{i_1}) \neq \Phi(x_{i_2})$. Then

$$A = [\cup\{\Phi(x_j) | j \in J\}] \cup \Phi(x_0)$$

and

$$N^* = \cup\{\Phi(x_k) | k \in K\}.$$

One now gets an "orbit selecting function"

$$\chi\colon I_0 \to \{0,1\}$$

that is surjective and such that $\chi(i) = 1$ if and only if $i \in K$; i.e., if and only if $\Phi(x_i) \subset N^*$.

For $a \in \Phi(x_k) \subseteq N^*$, we will show that $B_a = \Phi(x_k)$, and so there is a unique left identity $e_k \in \Phi(x_k)$ for each $k \in K$. Now $B_a = \{ca | c \in B_a\} = \{f_c(a) | c \in B_a\} \subseteq \{f_c(a) | f_c \in \Phi\} = \Phi(a) = \Phi(x_k)$. Hence $B_a \subseteq \Phi(x_k)$. By 2) of theorem 1, $\Phi(x_k) = \Phi(a) = N^*a \subseteq B_a$.

The planar near-ring N thus gives a pair of groups (N^+, Φ) where the elements of Φ act fixed point free on $(N,+) = N^+$. Theorem V.8.5 of Huppert [18] tells us that N^+ is the kernel of a Frobenius group G with complement Φ.

Theorem V.8.7 of [18] forces $(N,+)$ to be nilpotent.

Summarizing: from a planar near-ring $(N,+,\cdot)$ we get

1. disjoint subsets

$$A = \{x \in N | x \equiv_m 0\} \quad \text{and} \quad N^* = N \setminus A;$$

2. a group Φ of fixed point free automorphisms of N^+ such that $f_a \in \Phi \setminus \{1_N\}$ implies $1_N - f_a$ is surjective;

3. a bijection between an index set $I_0 = \{0\} \cup J \cup K$, where $\{0\}$, J, K are pairwise disjoint, and a set of representatives $\{x_i\}$, $i \in I_0$, of the orbits of Φ on N^+, in such a way that $\Phi(x_0) = \{x_0\} = \{0\}$, $j \in J$ implies $\Phi(x_j) \subseteq A$, $k \in K$ implies $\Phi(x_k) \subseteq N^*$;

$$A = \cup\{\Phi(x_j) | j \in J \cup \{0\}\}$$

and

$$N^* = \cup\{\Phi(x_k) | k \in K\};$$

4. a surjective map $\chi\colon I_0 \to \{0,1\}$ such that $\chi(i) = 1$ if and only if $\Phi(x_i) \subseteq N^*$;

5. a left identity $e_k \in \Phi(x_k)$ for each $k \in K$;

6. a Frobenius group G with kernel N^+ and complement Φ.

The converse is also true. I.e., if we have a Frobenius group with kernel N^+ (operation written additively) and complement Φ such that $1_N - f$ is surjective on N for each $f \in \Phi \setminus \{1_N\}$, we can construct numerous planar near-rings $(N,+,\cdot)$ depending on our choice of $\chi: I_0 \to \{0,1\}$ and selection $e_k \in \Phi(x_k)$ of left identities.

To be more precise, let G be a Frobenius group with kernel N^+, operation written additively, and complement Φ. We can, and do, think of Φ as a group of fixed point free automorphisms acting on N.. Also assume that $1_N - f$ is surjective for $f \in \Phi \setminus \{1_N\}$. Suppose there is a bijection $i \to x_i$ from an index set I_0 with $0 \in I_0$ and a set of representatives $\{x_i | i \in I_0\}$ of all the orbits of Φ on N with $\{x_0\} = \{0\} = \Phi(0)$. Let $\chi: I_0 \to \{0,1\}$ be a surjective map with $\chi(0) = 0$, and define $K = \chi^{-1}(1)$, $J = I_0 \setminus K \cup \{0\}$. Then $\{0\}$, J, K are pairwise disjoint and $\{0\} \cup J \cup K = I_0$. Define $A = \cup\{\Phi(x_j) | j \in J \cup \{0\}\}$ and $N^* = \cup\{\Phi(x_k) | k \in K\}$. Then $N^* = N - A$. For each $k \in K$, select $e_k \in \Phi(x_k)$ arbitrarily. There remains to define $\cdot$ so that $(N,+,\cdot)$ is a near-ring.

For $a \in A$, define $a \cdot x = 0$ for all $x \in N$. If $a \in N - A = N^*$, then there is exactly one $k \in K$ such that $a \in \Phi(x_k)$ and there is exactly one $f \in \Phi$ such that $f(e_k) = a$, since Φ is fixed point free. Define $a \cdot x = f(x)$ for all $x \in N$. One can show directly or by using results of [6] that $(N,+,\cdot)$ is a planar near-ring, that $A = \{x \in N | x \equiv_m 0\}$, and that each e_k is a left identity.

Fundamental to the geometric interpretations of a planar near-ring are the following

DEFINITIONS. For $0 \neq a, b \in N$, the sets $Na + b = \{xa + b \mid x \in N\}$ are called *blocks*. If $b = 0$, then the Na are called *basic blocks*.

LEMMA 2. For a planar near-ring N, the basic blocks are the $\Phi(x_i) \cup \Phi(x_0) = \{0\} \cup \Phi(x_i)$, and the blocks are just translates of the basic blocks; i.e.,

$$Na + b = \{0\} \cup \Phi(x_i) + b$$

for some $x_i \neq 0$.

PROOF.

$$Na = \{xa \mid x \in N\} = \{f_x(a) \mid x \in N\} =$$
$$\{f_x(a) \mid f_x \in \Phi \cup \{\zeta\}\} = \{0\} \cup \Phi(x_i),$$

where $a \in \Phi(x_i)$.

REMARK. This correspondence between planar near-rings and Frobenius groups is by no means bijective. In fact, different choices of the $\chi\colon I_0 \to \{0,1\}$ and/or different choices of the $e_k \in \Phi(x_k)$, $k \in K$, could lead to different but isomorphic near-rings or non-isomorphic near-rings. See [6, 2.6] for several examples illustrating this point. But, regardless of the χ and/or the e_k, the blocks are always the same, as shown by lemma 2.

2. BALANCED INCOMPLETE BLOCK DESIGNS (BIBD) FROM FROBENIUS GROUPS.

BIBD's have long played an important role in statistics with the design of experiments, and in coding theory.

Let V be a finite non-empty set whose elements are called "points" or "varieties", let $\emptyset \neq \mathcal{B}$ be a family of subsets of V whose elements are called "blocks" or "lines", and let $\in$ denote the usual set theoretic membership relation between points and blocks. Then $(V,\mathcal{B},\in)$ is an incidence structure.

DEFINITIONS. Consider an incidence structure $(V,\mathcal{B},\in)$ as above. If 1) each $x \in V$ belongs to exactly r blocks of $\mathcal{B}$, and 2) each block $B \in \mathcal{B}$ contains exactly k points, then $(V,\mathcal{B},\in)$ is a *tactical configuration*. If, in addition, 3) every pair of distinct points of V belong to exactly λ blocks of $\mathcal{B}$, then $(V,\mathcal{B},\in)$ is a BIBD.

NOTATION. Let G be a finite Frobenius group with kernel N and complement Φ. We can, and do, think of Φ as a group of fixed point free automorphisms acting on N. Assume the operation in N is written additively. For $0 \neq a \in N$, $\Phi(a) = \{\phi(a) \mid \phi \in \Phi\}$ is an orbit of Φ of cardinality the same as Φ. The union of a non-trivial orbit $\Phi(a)$ with the trivial orbit $\{0\} = \Phi(0)$ will be denoted by $\Phi_0(a)$. I.e., $\Phi_0(a) = \{0\} \cup \Phi(a)$, and is called a "basic block". Translates of the basic blocks will be called "blocks"; i.e., sets of the form

$$\Phi_0(a) + b = \{\phi(a) + b \mid \phi \in \Phi \quad \text{or} \quad \phi = 0\}$$

where $0 \neq a,\ b \in N$. Let $\mathcal{B} = \{\Phi_0(a) + b \mid 0 \neq a,\ b \in N\}$ We will soon see that $(N,\mathcal{B},\in)$ is always a tactical configuration.

The basic blocks fall into exactly one of two types. Let $\Phi_0(g_1), \cdots, \Phi_0(g_s)$ denote those basic blocks that are subgroups of $(N,+)$, and let $\Phi_0(e_1), \cdots, \Phi_0(e_t)$ denote those that are not. So the representative g_i indicates $\Phi_0(g_i)$ is a subgroup, e_j indicates $\Phi_0(e_j)$ is not a subgroup, s gives the number of basic blocks that are subgroups, and t gives the number of basic blocks that are not subgroups.

LEMMA 3. If $\Phi_0(a) + b = \Phi_0(a') + b'$, then

i) $\Phi_0(a) = \Phi_0(a')$;

ii) $\Phi_0(a)$ is a subgroup of N^+ and $b' \neq b$ if and only if $0 \neq b' - b \in \Phi_0(a)$;

iii) if $\Phi_0(a)$ is not a subgroup of N^+, then $b = b'$.

The converse is also true.

PROOF. Suppose $\Phi_0(a)$ is a subgroup of N^+. Then $\Phi_0(a) = \Phi_0(a')$ and if $b \neq b'$, we have $0 \neq b' - b \in \Phi_0(a)$. If $b \neq b'$, then from $\Phi_0(a) = \Phi_0(a') + (b' - b)$ we conclude $0 \neq b' - b \in \Phi_0(a)$ and $b - b' \in \Phi_0(a')$. Hence there are ϕ, $\phi' \in \Phi$ such that

$$\phi(a) = b' - b = \phi'(-a').$$

Hence, $\Phi_0(a') = \Phi_0(-a)$. From $\Phi_0(a) = \Phi_0(-a) + (b' - b)$ we get $\Phi_0(a) = \Phi_0(-a) + \mu(a)$ for each $\mu \in \Phi$. Hence

$$0 \notin \{\lambda(a), \tau(a)\} \subseteq \Phi_0(a)$$

implies that $\lambda(a) = \tau(-a) + \mu(a)$ for some $\mu \in \Phi$, and so $\tau(a) + \lambda(a) \in \Phi_0(a)$, and $\Phi_0(a)$ is closed with respect to $+$. Therefore $\Phi_0(a)$ is a subgroup of N^+. Hence, if $b = b'$, we have $\Phi_0(a) = \Phi_0(a')$ also.

The proof of the converse is direct.

THEOREM 4. The incidence structure $(N, \mathcal{B}, \in)$ is a tactical configuration with parameters

$$(v,b,r,k) = \left(|N|, \frac{vs}{k} + tv, s + kt, |\Phi| + 1\right)$$

where $b = |\mathcal{B}|$ and $v = |N|$.

PROOF. By definition, $|N| = v$ and $k = |\Phi| + 1$. For each $\Phi_0(g_i)$, there are $[N^+ : \Phi_0(g_i)] = \frac{v}{k}$ blocks, one for each coset. For each $\Phi_0(e_j)$, there are v blocks. These follow from lemma 3. Hence $b = \frac{vs}{k} + tv$.

For each $\Phi_0(g_i)$, $x \in N$ belongs only to the coset $\Phi_0(g_i) + x$; hence x belongs to s of these blocks. To each $\Phi_0(e_j)$, as y varies through N , x will occur in $k = |\Phi_0(e_i)|$ different $\Phi_0(e_i) + y$, each element $u \in \Phi_0(e_i)$ giving exactly one solution to the equation $u + y = x$. Hence, x belongs to kt of these blocks. Altogether $r = s + kt$.

We now turn our attention to deciding when our tactical configuration is a BIBD. Actually, we will obtain a very simple criterion: the tactical configuration $(N,\mathcal{B},\in)$ is a BIBD if and only if $0 \in \{s,t\}$.

THEOREM 5. The blocks $\mathcal{B}_g = \{\Phi_0(g_i) + x | x \in N , 1 \leqslant i \leqslant s\}$ form a BIBD $(N,\mathcal{B}_g,\in)$ if and only if $t = 0$. In this case, $\lambda = 1$, the group N^+ is elementary abelian of order p^m, p a prime, m composite, if $s > 1$. The blocks are just the lines of a finite affine space.

PROOF. Suppose $t = 0$. Then each basic block $\Phi_0(a)$ is a subgroup of N^+. Take $x, y \in N$, $x \neq y$. There is exactly one g_i such that $x - y \in \Phi_0(g_i)$. In fact,

$$x - y \in \Phi_0(x - y) .$$

Hence, $x, y \in \Phi_0(g_i) + y$ and each pair $x, y \in N$ belongs to at least one block.

Suppose $x, y \in \Phi_0(a) + c$. Then there are $\phi, \lambda \in \Phi_0 = \Phi \cup \{0\}$ such that $x = \phi(a) + c$, $y = \lambda(a) + c$, so $x - y = \phi(a) - \lambda(a) = \mu(a)$ for some $\mu \in \Phi_0(a)$, since $\Phi_0(a)$ is a subgroup. Hence $x - y \in \Phi_0(a) = \Phi_0(x - y)$. Thus the two distinct elements $x, y \in N$ belong to exactly $\lambda = 1$ block and we get a BIBD.

We will now define a multiplication for any one of the basic blocks $(\Phi_0(g_i),+)$. Define $0 \cdot x = 0$ for all x in $\Phi_0(g_i)$. Let $e_i = g_i$. For $0 \neq a \in \Phi_0(g_i)$, there is a unique $\phi_a \in \Phi$ such that $\phi_a(e_i) = a$. Define $a \cdot x = \phi_a(x)$. One can show directly, or by using the results in [6] that

$$(\Phi_0(g_i),+,\cdot)$$

is a near-field. Hence, $(\Phi_0(g_i),+)$ is an elementary abelian p-group for some prime p [21]. Thus, every non-zero element of N has additive order p. If p is odd, then $|\Phi|$ is even and so N^+ is abelian [15, Th. 10.3.1]. Hence H is an elementary abelian p-group. Since Φ acts fixed point free on N, $|\Phi| \mid |N| - 1$, so if $p^n = |\Phi_0(a)|$ and $p^m = |N|$, then $(p^n - 1)|(p^m - 1)$ and m is composite if $\Phi_0(a) \neq N$; i.e., if $s \neq 1$.

The basic blocks $\Phi_0(g_i)$ form a partition of N as defined by Baer in [2]. Applying Satz 2.3 of [2], one gets that the $\Phi_0(g_i) + c$ are lines of an affine space.

Now suppose that $(N, \mathcal{B}_g, \in)$ is a BIBD. If

$$x,\ y \in \Phi_0(a) + c$$

where $\Phi_0(a)$ is a subgroup of N^+, then $\Phi_0(a) = \Phi_0(x - y)$ as above. Hence, $\lambda = 1$ by hypothesis and lemma 3. If $t \neq 0$, then there are $x, y \in N$ such that $x - y = e_1$, and so

$$x - y \in \Phi_0(x - y) = \Phi_0(e_1)\ .$$

But if $x, y \in \Phi_0(g_i) + c$, then, as above, $x - y \in \Phi_0(g_i)$, and so $\Phi_0(g_i) = \Phi_0(e_1)$. Hence $t = 0$.

NOTE. To each elementary abelian group of order p^m, p a prime and m composite, there are Frobenius groups with kernel N of order p^m and with each basic block a subgroup of N^+. See [8].

NOTATION. For $x, y \in N$, $x \neq y$, let $[x,y]$ denote the number of blocks containing both x and y.

LEMMA 6. For $x \neq y$, $x, y \in N$, $[x,y] = [0,y - x]$.

PROOF. For $x, y \in \phi_0(a) + b$, $y = \phi(a) + b$ for some $\phi \in \Phi_0$. Hence $y - x \in \Phi_0(a) + (b - x)$. Similarly $x = \lambda(a) + b$ implies $0 = \lambda(a) + (b - x)$. Hence $0, y - x \in \Phi_0(a) + (b - x)$.

Conversely, if $0, y - x \in \Phi_0(a) + b$, then

$$x,\ y \in \Phi_0(a) + (b + x)\ .$$

It is elementary to define the desired bijection.

LEMMA 7. Assume $s = 0$; i.e., no basic block is a subgroup. Then $[0,z] = |\Phi| + 1$ for $0 \neq z \in N$.

PROOF. One calculates directly that $0,\ z$ belong to the distinct blocks $\Phi_0(z)$ and $\Phi_0(-z) + z$.

We now show the existence of a bijection between the remaining blocks containing $\{0,z\}$ and the non-identity elements $\phi \in \Phi$. Hence, there are exactly $|\Phi| - 1$ more blocks containing $\{0,z\}$, giving $|\Phi| + 1$ in all.

For $\phi \in \Phi,\ \phi \neq 1_N,\ 1_N - \phi$ is a bijective mapping, so there is a unique $a_\phi \in N$ such that $a_\phi - \phi(a_\phi) = -z$. Hence $0,\ z \in \Phi_0(a_\phi) - a_\phi$. If $\Phi_0(a_\phi) - a_\phi = \Phi_0(a_\lambda) - a_\lambda$, then $a_\phi = a_\lambda$, and from $\phi(a_\phi) - a_\phi = z = \lambda(a_\lambda) - a_\lambda$ we get $\phi(a_\phi) = \lambda(a_\lambda) = \lambda(a_\phi)$. Since Φ is fixed point free and

$$\phi,\ \lambda \in \Phi \setminus \{1_N\},$$

we have $\phi = \lambda$. Hence $\phi \to \Phi_0(a_\phi) - a_\phi$ is injective.

Similarly, $0,\ z \in \Phi_0(a) + b,\ b \notin \{0,z\}$, implies $-b = \phi(a)$ for some $\phi \in \Phi$. Hence $\Phi_0(a) = \Phi_0(-b)$ and consequently $z = \mu(-b) - (-b)$ for some $\mu \in \Phi$. If $\mu = 1_N$, then $z = 0$. Hence

$$\Phi_0(a) + b = \Phi_0(-b) + b = \Phi_0(a_\mu) - a_\mu .$$

Hence, $\phi \to \Phi_0(a_\phi) - a_\phi$ is surjective.

THEOREM 8. The blocks

$$\mathcal{B}_e = \{\Phi_0(e_j) + x | x \in N, 1 \leqslant j \leqslant t\}$$

form a BIBD $(N, \mathcal{B}_e, \in)$ if and only if $s = 0$. In this case, $\lambda = k = |\Phi| + 1$.

PROOF. If $s = 0$, then lemmas 6 and 7 apply. Together with theorem 4, we get a BIBD with $\lambda = k = |\Phi| + 1$.

Suppose $(N, \mathcal{B}_e, \in)$ is a BIBD. So there is a λ such that $x \neq y$, $x, y \in N$ implies $[x,y] = [0, x - y] = \lambda$. Suppose $\Phi_0(z)$ is a subgroup of N^+. There are $k(k-1)$ possible equations

$$\phi_i(x) = \phi_j(x) + z, \ \phi_i \neq \phi_j, \ \phi_i, \phi_j \in \Phi_0 .$$

For each one we get

$$z = \phi_j(-x) - \phi_i(-x) = \phi_k[\phi_i(-x)] - \phi_i(-x),$$

an element of $\Phi_0(\phi_i(-x)) - \phi_i(-x)$, if $\phi_i \neq 0$. If $\phi_i = 0$, then $0, z \in \Phi_0(-x)$. Also $0 \in \Phi_0(\phi_i(-x)) - \phi_i(-x)$ if $\phi_i \neq 0$. Hence, each pair $\phi_i, \phi_j \in \Phi_0$, $\phi_i \neq \phi_j$ identifies a block containing $\{0,z\}$. Conversely, if $0, z \in \Phi_0(a) + b$, then $0 = -b + b$ and so $\Phi_0(a) = \Phi_0(-b)$. Hence $z = \phi(-b) - 1_N(-b)$, $\phi \neq 1_N$. We write this as $-1_N(-b) = 1_N(b) = \phi(b) + z$, and so every block $\Phi_0(a) + b$ containing $\{0,z\}$ comes from some pair $\phi_i, \phi_j \in \Phi_0$ as described above.

We will now see that all these blocks are equal to $\Phi_0(z)$. Let $a \in \Phi_0(z)$ and recall that $\Phi_0(z)$ is a subgroup of N^+. Then $-b = -a + z \in \Phi_0(z)$ and so $b \in \Phi_0(z)$. Hence there are $\phi_i, \phi_j \in \Phi_0$, and a g_k such that $\phi_i \neq \phi_j, \phi_i(g_k) = -b$, $\phi_j(g_k) = -a$, and $\Phi_0(g_k) = \Phi_0(z)$. So $\phi_k(g_k) = \phi_j(g_k) + z$ and $\Phi_0(z)$ is the block derived from ϕ_i, ϕ_j. There are $k(k - 1)$ distinct pairs $a, b \in \Phi_0(z)$ such that $a - b = z$, each yielding distinct pairs ϕ_i, ϕ_j. Hence, all such pairs ϕ_i, ϕ_j actually yield $\Phi_0(z)$ and we get $[0,z] = 1$. Hence $\lambda = 1$. But $\Phi_0(z) \notin \mathcal{B}_e$, so $0, z$ do not belong to a single block in $\mathcal{B}_e$, and this contradicts $(N,\mathcal{B}_e,\in)$ being a BIBD. Thus $s = 0$.

COROLLARY 9. $\Phi_0(z)$ is a subgroup of N^+ if and only if $[0,z] = 1$.

PROOF. That $[0,z] = 1$ was shown in the theorem.

Assume $[0,z] = 1$ and $\Phi_0(z)$ is not a subgroup. We have $0, z \in \Phi_0(z)$, and if $\phi_i(x) = \phi_j(x) + z$, then

$$0,\ z \in \Phi_0(\phi_i(-x)) - \phi_i(-x) .$$

But $[0,z] = 1$ implies that $\Phi_0(\phi_i(-x)) - \phi_i(-x) = \Phi_0(z)$, and so $\phi_i(-x) = 0$ by lemma 3. This being true for $\phi_i \neq 0$

forces $x = 0$ and $\phi_j(x) = 0$. Hence $z = 0$, a contradiction. Therefore $\Phi_0(z)$ is a subgroup.

THEOREM 10. The tactical configuration $(N,\mathcal{B},\in)$ is a BIBD if and only if $0 \in \{s,t\}$.

PROOF. If $(N,\mathcal{B},\in)$ is a BIBD, then corollary 9 implies $0 \in \{s,t\}$. If $0 \in \{s,t\}$, then apply theorems 5 and 8.

Combining corollary 9 and lemma 3 we get the important

THEOREM 11. The following are equivalent for $0 \neq a \in N$:

1) $\Phi_0(a)$ is a subgroup of N^+;
2) $[0,a] = 1$;
3) there are $b, c \in N$, $b \neq c$, such that $\Phi_0(a) + b = \Phi_0(a) + c$;
4) there is an element $b \neq 0$ such that $\Phi_0(a) + b = \Phi_0(a)$;
5) $\Phi_0(a) = \Phi_0(-a) + a$.

3. EXAMPLES

1. For a finite abelian group $(A,+)$ with $(|A|,6) = 1$, the automorphism $\phi(x) = -x$ is fixed point free of order 2. Hence $\Phi = \{1_A,\phi\}$ is non-trivial and fixed point free on A. Since each block has exactly three elements and $3 \nmid |A|$, we have $s = 0$, and so $(A,\mathcal{B},\in)$ is a BIBD with $\lambda = k = 3$ [11].

2. For any finite field $(F,+,\cdot)$ of order p^n, p a prime, and for any divisor d, $1 < d < p^n - 1$, of $p^n - 1$, there is a unique subgroup Φ of $(F^* = F \setminus \{0\},\cdot)$ of order d. This subgroup can be considered as a fixed point free automorphism group of $(F,+)$. If $d = p^m - 1$, then each basic block is a subgroup, so $t = 0$. If $d \neq p^m - 1$, then no basic block is a subgroup, so $s = 0$. In any event, $(F^+,\mathcal{B},\in)$ is always a BIBD [8].

3. It may happen that $0 \notin \{s,t\}$. Let $N^+ = Z_9^+$, the additive group of integers modulo 9. For Φ, take $\{\overline{1},\overline{8}\}$ where $\overline{1}(x) = x$ and $\overline{8}(x) = 8x$. So Φ is fixed point free on N^+ with basic blocks

$$\Phi_0(1) = \{0,1,8\} \qquad \Phi_0(2) = \{0,2,7\}$$
$$\Phi_0(3) = \{0,3,6\} \qquad \Phi_0(4) = \{0,4,5\}$$

Only $\Phi_0(3)$ is a subgroup of N^+, so $s = 1$ and $t = 3$.

4. It may happen that Φ is not abelian, hence not cyclic. Let $(F,+,\cdot)$ be a finite near-field that is not a field, let $N^+ = F^+ \oplus F^+$, and let $\Phi \cong (F^*,\cdot)$ where $\phi_a \in \Phi$ is defined by $\phi_a(x,y) = (ax,ay)$. Then Φ is fixed point free on N^+ and Φ is not abelian.

5. The efficiency E of a BIBD is given by $E = \lambda v/rk$,

and if $\lambda = k$, as is often the case with our examples, then $E = \frac{v}{r}$. Take p to be an odd prime, and let d of 2. above be $(p^n - 1)/2$. So, if $p^n \neq 3$, we get $r = p^n + 1$ and so $E = p^n/p^n + 1$. So one can construct designs with efficiency arbitrarily close to unity.

6. For an example where N^+ is not abelian, see [8].

7. Let $\{1\} \neq \Phi$ act fixed point free on $N = N_1 \oplus \cdots \oplus N_m$, where each N_i is a Sylow subgroup of N. (Recall N is nilpotent.) The elements Φ restricted to N_i defines Φ_i isomorphic to Φ and Φ_i is fixed point free on N_i.

Conversely, let $N = N_1 \oplus \cdots \oplus N_m$ be a decomposition of a nilpotent group into its Sylow subgroups. Suppose each N_i has a non-trivial fixed point free group of automorphisms Φ_i, and that for $1 \leqslant i < m$, $f_i: \Phi_i \to \Phi_{i+1}$ is an isomorphism. For $\phi_1 \in \Phi_1$, define $\phi \in \Phi$ by

$$\phi(n_1, \cdots, n_m) =$$
$$(\phi_1(n_1), f_1(\phi_1)(n_2), f_2(\phi_2)(n_3), \cdots, f_{m-1}(\phi_{m-1})(n_m))$$

where $\phi_i = f_{i-1}(\phi_{i-1})$. Then Φ is a non-trivial fixed point free group of automorphisms on N and Φ is isomorphic to Φ_1 [3].

8. The next example provides motivation for further research, but first we need some

DEFINITIONS. Let V be a non-void finite set with $|V| = v$. Let P denote all $v(v-1)/2$ subsets of V each consisting of two elements of V. Consider a partition $A = \{A_1, \cdots, A_m\}$ of P. Then A is an *association scheme* on V if the following condition is satisfied:

given $\{x,y\} \in A_h$, the number of $z \in V$ for which $\{x,z\} \in A_i$ and $\{y,z\} \in A_j$ depends only on the indices h, i, $j \in \{1,2,\cdots,m\}$, not on x and y.

The number is denoted by $p(h,i,j)$ and notice that the definition implies $p(h,i,j) = p(h,j,i)$. The sets A_i are called the *classes* of A, and $\{x,y\} \in A_i$ is also expressed by saying that x and y are the ith *associates*, $i = 1,2,\cdots,m$. The integer m is called the *class number* of A. If $m = 1$ or $m = v(v-1)/2$, then it is uninteresting, so one usually has $1 < m < v(v-1)/2$.

A *partially balanced incomplete block design* (PBIBD) is a tactical configuration $(V,B,\in)$, together with an association scheme A on V, such that given $\{x,y\} \in A_i$, the number $[x,y]$ of blocks containing $\{x,y\}$ depends only on

$$i \in \{1,2,\cdots,m\},$$

and not on x and y. This number is denoted by λ_i.

In each association scheme, given $x \in V$, the number of

$y \in V$ such that $\{x,y\} \in A_i$ depends only on $i \in \{1,2,\cdots,m\}$, and not on x. This number is denoted by n_i.

A PBIBD has in addition to the usual parameters v, r, b, k the additional parameters $p(h,i,j)$, n_i, λ_i where h, i, $j \in \{1,2,\cdots,m\}$. For further information, consult [10].

In [15], Hall gives a powerful method for constructing PBIBD. Let G be a transitive permutation group on V with $|V| = v$, let a subgroup S of G be intransitive, let $B_1 = \{a_1,\cdots,a_k\}$ be a union of orbits of S, and let S_1, a subgroup of G, be the stabilizer of B_1. Suppose the cosets of S_1 in G are S_1X_1, $S_1X_2,\cdots,S_1X_b$ with $X_1 = 1$. Then our blocks are $\mathcal{B} = \{B_j = B_1X_j | j = 1,2,3,\cdots,b\}$. Thus $(V,\mathcal{B},\in)$ is an *orbital design* as defined by Higman [17]. Hall goes on to construct a PBIBD from $(V,\mathcal{B},\in)$ by determining an association scheme.

Let $G_1 \subset G$ be the stabilizer of a_1, and let $\{a_1\}$, Δ_1, Δ_2, $\cdots$, Δ_u, Δ_{u+1}, Δ'_{u+1}, $\cdots$, Δ_{u+v}, Δ'_{u+v} be the orbits of G_1 on V, where Δ_{u+i}, Δ'_{u+i} are "paired orbits"; i.e., $\Delta'_{u+i} = \{a_1x | x \in G$ and $a_1 \in \Delta_{u+i}x\}$, $\Delta''_{u+i} = \Delta_{u+i}$ and each Δ_j is "self-paired"; i.e., $\Delta'_j = \Delta_j$. See [16] and [20, p.44]. Now $\{a_1,a_i\} \in A_s$ if $s \leqslant u$ and $a_i \in \Delta_s$, or if $a_i \in \Delta_{u+j} \cup \Delta'_{u+j}$ and $s = u + j$. Also $\{a_f, a_h\} \in A_s$ if there is an $x \in G$ and $\{a_1,a_i\} \in A_s$ such that $a_1x = a_f$ and $a_ix = a_h$.

We now apply this method of Hall. Consider $N = V = Z_9$

and $S = \Phi = \{\overline{1},\overline{8}\}$ of 3. above. We can and do think of N as a permutation group on V, and certainly S is a permutation group on V. Let $G = [S,N] \subset S_V$ be the subgroup of the symmetric group S_V on V generated by $S \cup N$. Then G is really a Frobenius group of order 18, and S as a subgroup of G has index 9. Letting $B_1 = \{0,1,8\}$, then one gets the following blocks in $\mathcal{B}$.

$$B_1 = \{0,1,8\} \quad B_4 = \{3,4,2\} \quad B_7 = \{6,7,5\}$$
$$B_2 = \{1,2,0\} \quad B_5 = \{4,5,3\} \quad B_8 = \{7,8,6\}$$
$$B_3 = \{2,3,1\} \quad B_6 = \{5,6,4\} \quad B_9 = \{8,0,7\}\ .$$

Constructing the association scheme we use $G_1 = \Phi$ with its orbits $\{0\} = \{a_1\}$, $\{1,8\} = \Delta_1$, $\{2,7\} = \Delta_2$, $\{3,6\} = \Delta_3$, and $\{4,5\} = \Delta_4$, all self-paired. Hence

$$A_1 = \{\{\ell,\ell+1\} \mid \ell \in V\}, \qquad A_2 = \{\{\ell,\ell+2\} \mid \ell \in V\}\ ,$$
$$A_3 = \{\{\ell,\ell+3\} \mid \ell \in V\}, \qquad A_4 = \{\{\ell,\ell+4\} \mid \ell \in V\}\ .$$

One sees directly that each $n_i = 2$, $\lambda_1 = 2$, $\lambda_2 = 1$, $\lambda_3 = \lambda_4 = 0$. Certainly $v = 9$, $r = 3$, $b = 9$, and $k = 3$. We need only give the parameters $p(h,i,j)$. The especially interesting thing about this example is that they can be given as follows: let $(h,i,j) = (f(1), f(2), f(3))$ where

$$f:\{1,2,3\} \to \{1,2,3,4\}\ .$$

Then

$$p(h,i,j) = p(f\circ\pi(1), f\circ\pi(2), f\circ\pi(3))$$

for each permutation π of $\{1,2,3\}$. The f such that

$$p(f(1),f(2),f(3)) = 1$$

are given by $f_i(j)$, $1 \leqslant i \leqslant 7$, $1 \leqslant j \leqslant 3$ and the following table. All other $p(h,i,j) = 0$.

i \ j	1	2	3
1	1	1	2
2	1	2	3
3	1	3	4
4	1	4	4
5	2	2	4
6	2	3	4
7	3	3	3 .

E.g., $f_3(1) = 1$, $f_3(2) = 3$, $f_3(4) = 4$. Consequently $p(1,3,4) = p(1,4,3) = p(3,1,4) = p(3,4,1) = p(4,1,3) = p(4,3,1) = 1$.

4. FUTURE DEVELOPMENTS

In this section, we point toward specific problems and general areas of future but related research.

1. Perhaps the most fruitful direction of future research is in the direction suggested by example 8 in 3. above.

What kinds of PBIBD's does one get using Hall's method? In particular, if the B_1 is the union of orbits of a fixed point free group of automorphisms Φ acting on a group N^+, what are the resulting designs? The extra symmetry observed in example 8--i.e., that the p(h,i,j) is independent of the order h, i, j--warrants further investigation.

2. Each pair (N^+,Φ) gives a tactical configuration. What is the structure of the automorphism group of these tactical configurations? In particular, what happens if the tactical configuration is also a BIBD? The PBIBD's from (N^+,Φ), using Hall's method, also might have interesting automorphism groups.

3. It sometimes happens that two pairs $(N^+,\Phi),(N^+,\Phi_1)$ give tactical configurations, BIBD's, or PBIBD's with the same parameters. Are the resulting incidence structures ever isomorphic? Perhaps the structure of the corresponding automorphism groups will be of use here.

4. Only one example of a BIBD which has been obtained using the methods described in this paper, is also a Möbius plane. In this case $|N| = 2^2 + 1 = 5$. It was indicated in [14] that if $|N| = n^2 + 1$ and $|\Phi| = n$, then the resulting tactical configuration is, in addition to being a BIBD, also a

Möbius plane. But the additive group of the field of 37 elements has a Φ of order 6. Hence one would get a Möbius plane in which n is not a power of 2, contradicting the well known result that n must be a power of 2. See Theorem 6.2.14 of [10]. These incorrect observations in [14] are based on the incorrect main result of [5]. So there remain the following questions. What are the group pairs (N^+,Φ) such that $|N| = n^2 + 1$ and $|\Phi| = n$? Are any of the resulting BIBD's also a Möbius plane? Can one find such a pair when N is infinite? Certainly if $p = n^2 + 1$ is a prime, the cyclic group of order p serves for such as N^+, but if $p \neq 5, 17$, it is not known if the BIBD is an inverse plane. It is if $p = 5$ and is not if $p = 17$. It is an old unsolved problem to determine how many primes are of the form $p = n^2 + 1$. Regarding the first question, it is known to the authors that N^+ must be abelian of order p^α, $2 \neq p$ a prime, and α odd.

5. With each pair (N^+,Φ) there are two additional near-rings that arise quite naturally. Let D be the distributively generated near-ring obtained from Φ [19], and let M be the near-ring of zero-fixing mappings on N that commute with elements of Φ [4]. What information do these near-rings contain concerning the associated tactical configurations, BIBD's, and PBIBD's? That there is some information

is indicated by observing that if $\Phi_0(a)$ is a subgroup of N^+, then $\Phi_0(a)$ is a minimal D-subgroup of N^+, and from [4, 5.6] the number of minimal right ideals of M is equal to the number of basic blocks defined by Φ on N^+. That there is more information is indicated by examining specific examples in detail.

REFERENCES

1. M. Anshel and J. R. Clay, *Planar algebraic systems: some geometric interpretations*, J. Algebra 10 (1968), 166-173.

2. R. Baer, *Partitionen abelscher Gruppen*, Arch. Math. 14 (1963), 73-83.

3. L. Bertani, *Costruzione di disegni regolari*, to appear.

4. G. Betsch, *Some structure theorems on 2-primitive near-rings*, Proc. Colloquium Assoc. Rings, Modules, and Radicals, Keszthély (Hungary), 1971.

5. P. Biscarini, *Una caratterizzazione dei piani finiti di Möbius*, Bollettino della Unione Matematica Italiana 6 (1970), 993-997.

6. J. R. Clay, *The near-rings on groups of low order*, Math. Zeitschr. 104 (1968), 364-371.

7. ____________, *Some algebraic and geometric aspects of planarity*, Atti del convegno di geometria combinatoria e sue applicazioni, Perugia (1971), 163-172.

8. ____________, *Generating balanced incomplete block designs from planar near-rings*, J. Algebra 22 (1972), 319-331.

9. W. G. Cochran and G. M. Cox, *Experimental Designs*, 2nd Ed., Wiley and Sons, New York, 1957.

10. P. Dembowski, *Finite Geometries*, Springer-Verlag, New York, 1968.

11. G. Ferrero, *Stems planari e BIB-disegni*, Riv. Mat.Univ. Parma (2) 11 (1970), 79-96.

12. ___________, *Qualche disegno geometrico*, Le Matematiche 26 (1971), 1-12.

13 ___________, *Su una classe di nuovi disegni*, Rendiconti, Classe di Scienze (A) 106 (1972), 419-430.

14. ___________, *Su certe geometric gruppali naturali*, to appear.

15. D. Gorenstein, *Finite Groups*, Harper and Row, New York, 1968.

16. M. Hall Jr., *Designs with transitive automorphism groups*, Proceedings of Symposia in Pure Mathematics, A.M.S. 19, T. L. Motzkin, ed., (1971), 109-113.

17. D. G. Higman, *Intersection matrices for finite permutation groups*, J. Algebra 6 (1967), 22-42.

18. B. Huppert, *Endliche Gruppen I*, Springer-Verlag,Berlin, 1967.

19. R. R. Laxton, *Primitive distributively generated near-rings*, Mathematatika 8 (1961), 142-158.

20. H. Wielandt, *Finite Permutation Groups*, Academic Press, New York, 1964.

21. H. Zassenhaus, *Über endliche Fastkörper*, Abh. Math. Sem. Univ. Hamburg, 11 (1935/36), 187-220.

Mathematisches Institut der Universität
Tübingen, D.F.R. (West Germany)

University of Arizona
Tucson, Arizona

GEOMETRY OF GROUPS OF LIE TYPE

BY

BRUCE COOPERSTEIN

INTRODUCTION

The purpose of this work is to show how to associate with any group of Lie type, G, a class of incidence structures (P,L), on which G operates as a group of automorphisms, transitive on both P and L. We also report on some characterizations of these incidence structures. It is hoped that these characterizations can be applied to characterizations of the groups of type A_n, D_5, and E_6.

1. NOTATION AND DEFINITIONS

By a *graph* we mean a pair (P,Δ) consisting of a set P and a collection Δ of 2-subsets of P, $\Delta \subseteq P^{\{2\}}$. Thus (P,Δ) is undirected, without loops or multiple edges. For such a graph we let $\Delta(x)$ denote the vertices adjacent to x and $x^{\perp} = \{x\} \cup \Delta(x)$. An *incidence structure* is a pair (P,L) consisting of a set P whose elements are called points, and a set L of distinguished subsets of P called lines. (P,L) is said to be *thick* if every line has at least three points. The *point graph* of (P,L) is the graph (P,Δ) with Δ the collection of collinear pairs of points.

Let (G,P) be a transitive permutation representation of a group G in a set P and Δ be a self-paired (i.e., symmetric) orbit for the action of G on $P \times P$, Δ not the diagonal (we call Δ an *orbital*). Then if (x,y) is in Δ, so is (y,x), and so we may consider Δ to be a collection of 2-subsets of P, hence we get a graph (P,Δ). Clearly G acts as a group of automorphisms of this graph, and is transitive on points and edges.

If (G,P), Δ, are as above, then for x in P and y adjacent to x we define the *(singular) line* on x and y by

(1.1) $$xy = \bigcap_{x,y \in z^{\perp}} z^{\perp}.$$

If we set L equal to the set of all such lines, then (P,L) is an incidence structure on which G acts as a group of automorphisms, transitive on P and on L and such that (P,Δ) is the point graph of (P,L).

Properties of incidence structures (P,L) afforded by a group in this way are well known. It is easy to prove that xy is a clique (i.e., complete subgraph) and if u,v are distinct on xy then $uv = xy$. Also we can show

(1.2) If z is not on xy but is adjacent to at least two points of xy, then z is adjacent to every point of xy.

2. EXAMPLES

Often, (P,L) arising from a group G and an orbital Δ is trivial in the sense that it is not thick. This will certainly be the case if $G_x^{\Delta(x)}$ is primitive on $\Delta(x)$. Thus the rank 3 representations of HiS, McL, and Sz afford trivial incidence structures. The incidence structures arising from M_{22} acting as a rank three group on the 77 blocks of the Steiner system $S(3,6,22)$ are also trivial. In this representation, the stabilizer of a point has a system of imprimitivity for its action on one of the suborbits, so this is not a sufficient condition for the existence of thick lines.

A non-trivial example is afforded by $PSL_n(q)$ as follows. Let P be the lines of $PG(n-1,q)$, and Δ the collection of intersecting pairs of lines. The line, xy, on such a pair x,y consists of the $q+1$ lines of $PG(n-1,q)$ incident with the flag $x \cap y \subseteq \langle x,y \rangle$.

The group $PSL_n(q)$ is isomorphic to the adjoint Chevalley group of type A_{n-1} defined over $GF(q)$. The representation cited above is a parabolic representation. If other representations of $A_{n-1}(q)$ are considered, or other groups of Lie type, we can list many other examples. This suggests that the existence of lines is a "Lie" property. That this suspicion is well-founded is proved in

THEOREM 1. Suppose $G = G(K)$ is a group of Lie type defined over a field K, Π is a fundamental base for the associated root system, $\delta \subseteq \Pi$, $H = G_\delta$ is the parabolic subgroup corresponding to δ, and $P = G/H$ is the coset space of G modulo H. Then there is a set $\{\Delta_\alpha; \alpha \in \Pi - \delta\}$ of self-paired orbitals such that lines with respect to each Δ_α carry more than two points. Moreover, a line stabilizer is another parabolic subgroup.

3. Some Lie Incidence Structures, Polar and Prepolar Spaces

If (P,L) is an incidence structure, then a *subspace* is a subset X of P such that if ℓ is a line and meets X in at least two points, then ℓ is contained in X. A subspace is *singular* if it is a clique with respect to the point graph. For a subset X of P we can define the subspace spanned by X, $< X >$, to be the smallest subspace containing X, and we can show that the subspace spanned by a clique is singular if (P,L) satisfies (1.2). We can prove that the singular subspaces of the Lie incidence structures (i.e., the incidence structures asserted to exist in Theorem 1) are projective spaces, and their stabilizers are parabolic subgroups.

Many, but not all, of the remaining subspaces of Lie incidence structures have parabolic stabilizers, and we determine the different classes of parabolic subspaces.

A *prepolar space* is an incidence structure (P,L) with the "one-all" property; that is, for any non-incident point-line pair x,ℓ either x is collinear with one or all points of ℓ. The space is *non-degenerate* if no point is collinear with all the remaining points. The *rank* of (P,L) is the maximal length of a chain of singular subspaces. The thick, non-degenerate prepolar spaces are known:

THEOREM 1. (Buekenhout-Schult,[1]). If (P,L) is a thick, non-degenerate prepolar space of finite rank, then P, together with its singular subspaces is a polar space.

REMARKS. Rank two polar spaces are generalized quadrangles. In the finite case, a polar space of rank at least three essentially arises as the totally isotropic subspaces in a finite dimensional vector space with respect to a non-degenerate symmetric, alternate, or hermitian bilinear form. Hence a polar space is a geometry which is acted on by a classical group. The classical groups are groups of Lie type. The prepolar spaces of finite rank at least three arise, via Theorem 1 as Lie incidence structures where the diagram of the representation is one of

(a)  ,

or

(b) 

I have investigated many of the other incidence structures afforded by groups of Lie type, and in particular, the ones with the following diagrams:

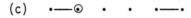

(c) 

(d) 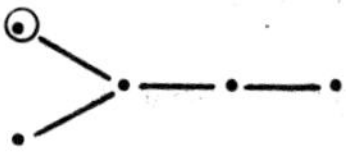

(e) 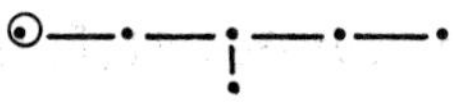

In (c) we have the groups of type A_n which, as we noted earlier, are isomorphic to the PSL_{n+1}'s. The action of this representation is the one we came across earlier, of PSL_{n+1} on the lines of projective n-space. In (d) we have a group of type D_5, which is isomorphic to Ω_{10}^{+}, the simple orthogonal group in ten variables with respect to a form with maximal Witt index. This representation corresponds to the action of the group on a class of maximal totally isotropic subspaces, two such subspaces being adjacent if they meet in exactly a space of codimension two. In (e) we are considering either of the rank three parabolic representations of a group of type E_6. We denote the associated incidence structures by $A_{n,1}(q)$, $D_{5,max}(q)$, and $E_{6,3}(q)$. Moreover, if we are considering a prepolar space afforded by a group of type X over GF(q) we will denote it by X(q).

It is not difficult to verify that the incidence structures (P,L) associated with these representations satisfy, in addition to (1.2), the following:

(3.1) The point graph (P,Δ) of (P,L) is not complete, and

(3.2) There is an integer $k \geqslant 1$ such that if w is not adjacent to x, then $\Delta(x,w) = \Delta(x) \cap \Delta(w)$, together with the lines it contains, is a thick, non-degenerate prepolar space with rank k+1. Moreover, if x,ℓ is a point-line pair with x adjacent to no point of ℓ, and we let $C(x,\ell)$ be the collection of points adjacent to x and every point of ℓ, then $C(x,\ell)$ is a singular subspace with dimension k.

We get a converse to this in the finite case.

THEOREM 2. If (P,L) is a finite incidence structure satisfying (1.2), (3.1), and (3.2), then either

(a) (P,L) is a thick, non-degenerate prepolar space of rank k+2, or

(b) k is less than four and (P,L) is $A_{n,1}(q)$, $D_{5,max}(q)$, or $E_{6,3}(q)$ for some prime power q.

4. Sketch of the Proof

Assume (P,L) satisfies the hypotheses of the theorem. For each x in P let L_x be the set of lines on x. For ℓ,m in L_x we say they are collinear or adjacent if and only if their union is a clique and define the line (of L_x) on them to be the set of lines in L_x lying in the singular subspace spanned by ℓ and m (which has the structure of a projective plane). We prove

LEMMA 3. If k is at least two then L_x also satisfies (1.2), (3.1), and (3.2) with k-1. Moreover, if L_x is a prepolar space then so is (P,L) and the rank of (P,L) is 1+ rank of L_x.

We treat the cases k = 1, 2, 3 separately.

The Case k = 1

LEMMA 4. Every line of (P,L) is contained in exactly two maximal singular subspaces with dimensions 2 and $d > 2$, d independent of the line.

Let $p(\ell)$ denote the maximal singular subspace containing ℓ which has dimension d. Let P' be the set of all $p(\ell)$ with ℓ in L. Let L' = P. Say $p(\ell)$ is incident with x if x is in $p(\ell)$. Then

LEMMA 5. (P',L') is a projective space of dimension $d+1$, and (P,Δ), the point grapn of (P,L) is isomorphic to the line graph of this projective space. Thus, (P,L) is isomorphic to $A_{d+1,1}(q)$, where there are $q+1$ points on a line of (P,L).

THE CASE $K = 2$.

By Lemma 3 we may assume L_x is isomorphic to $A_{n(x),1}(q)$ for each x in P with $n(x)$ at least four. We then prove

LEMMA 6. $n(x) = 4$ for each x in P.

In general, when (P,L) satisfies our hypotheses we can snow the subspace spanned by x, w, and $\Delta(x,w)$, $S(x,w)$, where x and w are not adjacent, is a prepolar space with rank $S(x,w) = 1 + \text{rank } \Delta(x,w)$. In the present situation $S(x,w)$ is a prepolar space $D_4(q)$. We let P' be the set of all such $S(x,w)$ and define a set L' of lines on it.

LEMMA 7. (P',L') is a prepolar space $D_5(q)$. Moreover, P can be identified with a class of maximal singular subspaces of (P',L'), and so (P,L) is isomorphic to $D_{5,max}(q)$.

THE CASE $K = 3$.

By the previous result and Lemma 3 we can assume that L_x is isomorphic to $D_{5,max}(q)$ for each x in P. Then for x and w not adjacent, $S(x,w)$ is a prepolar space of type

D_5 . Set P' equal to the set of all $S(x,w)$ with x and w not adjacent. Let $\Phi = P \cap P'$. We can snow that Φ satisfies a list of axioms due to Tits. In [2] it is shown this list of axioms characterizes the schema of R-spaces with the diagram

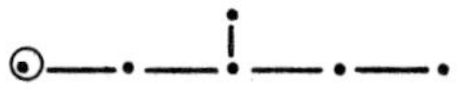 .

This suffices in the case $k = 3$. To complete the theorem we assume that (P,L) is a counterexample with k minimal. Then by Lemma 3 we must have that $k = 4$, and for each x, L_x is isomorpnic to $E_{6,3}(q)$. But by counting the number of elements in $\Delta(x)$ in two different ways we get a numerical contradiction and the proof of the theorem is complete.

REFERENCES

1. F. Buekenhout and E. E. Schult, *On the foundations of polar geometry*, Geometriae Dedicata, 3 (1974), 155-170.

2. J. Tits, *Sur la géométrie des R-espaces*, J. Math. Pures Appl., 36 (1957), 17-38.

UNIVERSITY OF MICHIGAN
ANN ARBOR, MICHIGAN

LOCALLY DUAL AFFINE GEOMETRIES

BY

MARK P. HALE JR.

Our goal is to describe the finite geometries which are "locally" dual affine; that is, in which every subplane is a dual affine plane. A geometry is a finite set of points and a collection of point sets called lines. We require that lines contain at least two points, that distinct lines have at most one point in common, and that any two points are connected by some path consisting of intersecting lines. A subgeometry is a subset which contains all of the lines passing through any two of its points; a plane is a subgeometry generated by two intersecting lines.

Two classical geometries can be made from a vector space V of dimension n over the Galois field GF(q). The points and lines of the projective geometry P(n-1,q) are the 1- and 2-dimensional subspaces of V. The points and lines of the affine geometry A(n,q) are the elements of V and the cosets of the 1-dimensional subspaces of V. A classical result on projective spaces asserts that if all planes in a non-planar geometry are projective, then the geometry is a $P(n,q)$. A similar result holds for the affine case, if lines contain at least four points [1], or parallelism is an equivalence relation on lines [3].

A slightly less classical geometry can be defined on V if V is endowed with a non-zero alternative bilinear form f. The points and lines of the symplectic geometry (relative to f) are the 1-dimensional subspaces of V which are not in the radical of f and the 2-dimensional subspaces on which f is not identically zero. This geometry is determined by the dimension n of V and the dimension k of the radical of f, so we denote it by $Sp(n,k,q)$.

A planar symplectic space $Sp(3,1,q)$ can be thought of as a projective plane $P(2,q)$ with one point (the radical) and the lines through it (the isotropic 2-spaces) removed. This is the dual of the standard construction of $A(2,q)$ from $P(2,q)$, and so we have that planar symplectic geometries are dual affine planes $DA(q)$. The local-global results on projective and affine spaces now suggest the following:

CONJECTURE: Geometries in which all planes are dual affine must be symplectic geometries.

The conjecture as stated is false for $q = 2$. We must also include the orthogonal geometries, which are refinements of the symplectic geometries, and a new family of examples constructed from the symmetric group S_n. (In the orthogonal case, our lines are the lines of the symplectic space which consist of three non-singular points; in the geometry for S_n, points are transpositions and lines are the three transpositions belonging to the symmetric groups on three element subsets of 1,...,n .) It is also possible to adjoin a radical to these geometries, in the manner of the construction given below for mixed projective and dual affine spaces. Because of these exceptions, the proof of the (corrected) conjecture for q=2 is treated separately from the general case. We employ a theorem of Shult [4] to recognize the geometries having no radical, and then reinvent the necessary radical extensions.

An obvious attack on the conjecture in the case $q > 2$ is to imitate the coordinatization of projective space. The existence of non-collinear points in the symplectic spaces causes all the new problems of the proof.

For example, in the projective case, if S and x are a subgeometry and a point contained in a projective geometry G , then the points of the subgeometry of G generated by S and x are the points lying on lines joining x to points of S. In the symplectic case, the subgeometry also contains some points not in S which are not collinear with x . These new points can be reached from x by many different two-step paths, and it must be seen that all such paths lead to the same coordinate assignment.

A further difficulty arises when two subplanes intersect in more than one point. If the points aren't collinear, we need to know that the intersection contains all the points from both planes that are not collinear with the given points (the dual-parallel classes of the points in the two planes). At the moment, this is a technical hypothesis built in to my results, which I hope to remove (i.e., that intersections of planes are either points, lines, or dual parallel classes of points).

Geometries with more varied local structure have also received some attention. Those geometries in which all subplanes are either $A(2,q)$ or $P(2,q')$ were studied by J.Hall [2] ($q=3,q'=2$) and L. Tierlinck [5]. They have shown that in such geometries, either one type or the other occurs everywhere (that is, both types cannot occur within one geometry). When we consider geometries in which each plane is either affine, projective, or dual affine, the situation changes.

There are geometries in which both projective and dual affine planes occur. To make a universal construction, consider the direct sum $V+W$ of two vector spaces over the field F. Let the points and lines of G be the projective points and lines of $V+W$ which project as points and lines on V.

(The projective points in W , which disappear, are the unseen radical of the geometry.) The proof that this construction is universal is a modification of one of the easier cases of the proof of the conjecture, blended with the coordinatization of projective space.

There are geometries in which both affine and dual affine planes occur; that is, I have constructed one, but haven't found an intelligible context for it. Naturally, it involves the standard villain, the prime 2 . It is a geometry on 18 points combining A(2,3)'s with DA(2)'s .

The existence of such a mixed breed seems slightly more reasonable if one considers the following characteristic 6 construction of DA(2) . Take as points the residues mod 6 , arc two residues if their difference is not 3. The cliques in this graph contain 3 points, are determined by the pairs within them, and regarded as lines, form the lines of DA(2) .

I close with two conjectures.

CONJECTURE 1: No geometry mixes all three types of planes.

CONJECTURE 2: Affine and dual affine planes can only be mixed in the A(2,3) , DA(2) case, as above.

NOTE: Conjecture 1 has been proved since the paper was written.

REFERENCES

1. F. Buekenhout, *Une caractérisation des espaces affins basée sur la notion de droite,* Math. Zeit. 111 (1969), 367-371.

2. J. I. Hall, *Steiner triple systems with geometric minimally generated subsystems,* Quart. J. Math. Oxford Ser. (2) 25 (1974), 41-50.

3. M. Hall Jr., *Incidence axioms for affine geometry,* J. Algebra 21 (1972), 535-547.

4. E. Shult, to appear in a work of J. J. Seidel.

5. L. Tierlinck, to appear.

UNIVERSITY OF FLORIDA
GAINESVILLE, FLORIDA

PART V

SOLVABLE GROUPS

THE HUGHES PROBLEM AND GENERALIZATIONS

BY

JOSEPH A. GALLIAN

Let G be a group, p a prime and $H_p(G) = \langle g \in G | g^p \neq 1 \rangle$. In 1957, D. R. Hughes conjectured that for any G and any p if $G > H_p(G) > 1$, then $|G:H_p(G)| = p$. A year earlier [7] he had proved this for all groups when p = 2. In 1958, Straus and Szekeres [14] showed the conjecture is correct when p = 3 and in 1959, Hughes and John Thompson [8] proved the conjecture for all finite groups which are not p-groups. Of course this result focused attention on finite p-groups and from here on, unless otherwise stated, all groups will be assumed to be finite p-groups. G. Zappa [16] in 1962 showed the conjecture is true for groups with nilpotence class at most p but in 1965 G. E. Wall [15] discovered a 5-group G of exponent 25 and with $|G:H_5(G)| = 25$. Nevertheless, the conjecture remains open for 2-generated groups and it is desirable to find conditions on G which guarantee the conjecture is valid for G. Hogan and Kappe [6] were the next contributors to the problem when in 1969 they proved that if $G'' = 1$ or G_i/G_{i+1} is cyclic for $i = 2, \ldots, c$ (where c is the class of G) then

G satisfies the Hughes conjecture. Then in 1973 Macdonald extended Zappa's theorem to the case that G has class at most $2p-2$.

The above information summarizes what was known about the Hughes conjecture when the author first encountered it and of these results only the Hogan and Kappe theorems appeared to be ripe for generalization. Thus initial efforts were focused on improving these.

The "G_i/G_{i+1} is cyclic" theorem can be generalized in a number of ways. We cite two of these.

THEOREM [1] If G_i/G_{i+1} is cyclic for $i=2,3,\ldots,p$, then G satisfies the Hughes conjecture.

THEOREM [2] If G_j/G_{j+1} is cyclic for some j and $|G_j/G_{j+1}| \geqslant |G_i/G_{i+1}|$ for $i = 2,3,5,\ldots,p-2,p$, then G satisfies the Hughes conjecture.

Analogs of the above involving the upper central factors can also be obtained (see [2] and [4]).

The "metabelian" result of Hogan and Kappe can be improved in numerous ways also. The author has shown in [2] and [4] that if $G'' \leqslant Z_{p-2}(G)$ or if G'' is Abelian and can be generated by $p-2$ elements then G satisfies the Hughes conjecture. These results are satisfactory in many cases but, in general, it is better to have theorems in which the parameter is expressed as a function of the class. One such result is the following.

THEOREM [4] Suppose G is 2-generated and has class c. If $G'' \leqslant Z_{c-2p+1}(G)$ or if G'' is Abelian and can be generated by $c-2p$ elements, then G satisfies the Hughes conjecture.

A similar theorem without the restriction that G is 2-generated is desirable.

In searching for results which break new ground the author has tried to prove that groups which possess a subgroup of maximal class satisfy the Hughes conjecture. Thus far, however, only a large number of special cases of this statement have been established. We mention two of these.

THEOREM [4] If G is 2-generated and contains a normal subgroup of maximal class then G satisfies the Hughes conjecture.

THEOREM [3] If G contains a subgroup L of maximal class with $|G:L| \leqslant p^p$ then G satisfies the Hughes conjecture.

If all the maximal subgroups of G are regular then G satisfies the Hughes conjecture and the next result motivates us to conjecture that the same conclusion holds if all of the second maximal subgroups are regular.

THEOREM [4] Suppose all of the second maximal subgroups of G are regular. If exp $(\Phi(G)) > p$ or if the minimum number of generators of G is not 3, then G satisfies the Hughes conjecture.

It is sometimes possible to show that $H_p(G)$ has the desired property indirectly. The next result gives one such method.

THEOREM [4] If $(H_p(G))'$ is Abelian and can be generated by p-2 elements then G satisfies the Hughes conjecture.

The proof of the above theorem as well as a number of others utilizes the following important lemma.

LEMMA [4] Let P be a group theoretic property which is inherited by subgroups. Suppose G is a group of minimum order with property P which does not satisfy the Hughes conjecture. Then $|G:H_p(G)| = p^2$. Furthermore, either $G' = H_p(G)$ or $|H_p(G):G'| = p$ and $\exp(G') = p$.

We conclude with some remarks on two possible generalizations of the Hughes problem. For the remainder of the paper we consider arbitrary groups (not necessarily a p-group or even finite). For a group G, a prime p and a positive integer k, define $H_{p^k}(G) = \langle g \in G \mid g^{p^k} \neq 1 \rangle$. Consider the following problems.

Weak H_{p^k}-problem: Find conditions on G and k sufficient to imply that if $G > H_{p^k}G) > 1$ then $|G:H_{p^k}(G)| = p^i$ for some $i \leqslant k$.

Strong H_{p^k}-problem: Find conditions on G and k sufficient to imply that if $G > H_{p^k}(G) > 1$ then $|G:H_{p^k}(G)| = p$.

With regard to these problems we have the following two results.

THEOREM (cf. Th. 1 in [2]). If K is a subgroup of G with $(Z_{i+p-1}(K)/Z_i(K))^{p^k} \neq 1$ for some i, then $K \leqslant H_{p^k}(G)$.

THEOREM (cf. Th. 2 in [2]). If K is a nilpotent subgroup of G and $(K_i/K_{i+p-1})^{p^k} \neq 1$ for some i, then $K \leqslant H_{p^k}(G)$.

Finite p-groups which are regular or s-minimal irregular or of class at most p ([5]) are solutions to the strong H_{p^k}-problem. On the other hand, Macdonald [12] has an example of a finite 2-group G with the properties that $H_4(G) \neq 1$ and $|G:H_4(G)| = 8$.

We ask whether the solutions of the H_p-problem (or some appropriate modification of them) mentioned in the first paragraph of this paper also provide solutions to the two H_{p^k}-problems.

Finally, mimicking Hughes and Thompson [8] we define a group H to be an H_{p^k}-group if there is some finite group G with $H = H_{p^k}(G)$ and $|G:H| = p$. Hughes and Thompson [8] have shown that H_{p^k}-groups are solvable when $k = 1$ and Kegel [10] has shown (cf. [9, p. 502]) H_{p^k}-groups are even nilpotent when $k = 1$. We ask what can be said about the structure of H_{p^k}-groups for other values of k ?

REFERENCES

1. J. A. Gallian, *The* H_p*-problem for groups with certain central factors cyclic,* Proc. Amer. Math. Soc. 42 (1974), 39-41.

2. J. A. Gallian, *On the Hughes conjecture,* J. Algebra 34 (1975), 54-63.

3. J. A. Gallian, *The Hughes conjecture and groups with absolutely regular subgroups or* ECF-*subgroups,* Proc. Amer. Math. Soc., to appear.

4. J. A. Gallian, *More on the Hughes conjecture,* to appear.

5. G. T. Hogan, *Elements of maximal order in finite* p-*groups,* Proc. Amer. Math. Soc. 32 (1972), 37-41.

6 G. T. Hogan and W. P. Kappe, *On the* H_p*-problem for finite* p-*groups,* Proc. Amer. Math. Soc. 20 (1969), 450-454.

7. D. R. Hughes, *Partial difference sets,* Amer. J. Math. 78 (1956), 650-674.

8. D. R. Hughes and J. G. Thompson, *The* H_p*-problem and the structure of* H_p*-groups,* Pacific J. Math. 9 (1959), 1097-1101.

9. B. Huppert, "*Endliche Gruppen I*," Springer-Verlag, Berlin, 1967.

10. O. H. Kegel, *Die Nilpotenz der* H_p*-Gruppen*, Math. Z. 75 (1961), 373-376.

11. I. D. Macdonald, *The Hughes problem and others*, J. Austral. Math. Soc. 10 (1969), 475-479.

12. I. D. Macdonald, Some examples in the theory of groups, in "*Mathematical Essays Dedicated to A. J. Macintyre*," Ohio University Press, Athens, Ohio, 1970.

13. I. D. Macdonald, *Solution of the Hughes problem for finite* p*-groups of class* 2p-2, Proc. Amer. Math. Soc. 27 (1971), 39-42.

14. E. G. Straus and G. Szekeres, *On a problem of D. R. Hughes*, Proc. Amer. Math. Soc. 9 (1958), 157-158.

15. G. E. Wall, On Hughes' H_p-problem, in "*Proc. Internat. Conf. Theory of Groups (Canberra, 1965)*," Gordon and Breach, New York, 1967.

16. G. Zappa, *Contributo allo studio del problem di Hughes sui gruppi*, Ann. Mat.Pura Appl. (4) 57 (1962), 211-219.

UNIVERSITY OF MINNESOTA
DULUTH, MINNESOTA

A NORMALIZER CONDITION ON FINITE p-GROUPS

BY

JOHN D. GILLAM

A subgroup H of a group G has a small normalizer provided $[N_G(H):H]$ is one or a prime. A group G is in the class SN if and only if every non-normal subgroup of G has a small normalizer.

THEOREM 1. Let G be a finite non-abelian 2-group. Then G is in the class SN if and only if G is Hamiltonian or G is a finite non-abelian homomorphic image of

$$< x,y : x^{2^{n+1}} = y^4 = 1, x^y = x^{-1+2^n} >$$

for some positive integer n.

THEOREM 2. Let p be an odd prime and G a finite non-abelian p-group. Then G is in the class SN if and only if the order of G is p^3 or $G = < x,y : x^{p^2} = y^{p^2} = 1, x^y = x^{p+1} >$.

OHIO UNIVERSITY
ATHENS, OHIO

PRODUCTS OF FORMATIONS

BY

BEN BREWSTER

If $\mathcal{X}$ and $\mathcal{Y}$ are classes of groups, the usual class product is given by:

$$\mathcal{X}\mathcal{Y} = \{G \mid G \text{ has a normal subgroup } N \in \mathcal{X} \text{ with } G/N \in \mathcal{Y}\} .$$

However, if $\mathcal{X} = QR_0(S_3)$, the formation generated by the symmetric group S_3, and $\mathcal{Y}$ = the formation of elementary abelian 2-groups, then the group $X = \langle x,y \mid x^3 = 1 = y^4 , x^y = x^{-1}\rangle$ shows that $\mathcal{X}\mathcal{Y}$ need not be a formation. Thus in Gaschütz' lecture notes [2], a formation product, $\mathcal{X} * \mathcal{Y} = \{G \mid G^{\mathcal{Y}} \in \mathcal{X}\}$, is defined for formations $\mathcal{X}$ and $\mathcal{Y}$ ($G^{\mathcal{Y}}$ denotes the $\mathcal{Y}$-residual). Products have been used as a "new from old" method of constructing formations but the class product is much more useful when determining to which product a given group belongs.

Two questions seem quite natural for formations $\mathcal{X}$ and $\mathcal{Y}$:

(I) When is $\mathcal{X}\mathcal{Y}$ a formation? and

(II) When does $\mathcal{X}\mathcal{Y} = \mathcal{X} * \mathcal{Y}$?

I have only partial results, but the test case at the conclusion indicates the progress. For convenience, consider only solvable groups. To motivate the hypothesis later, note:

(i) If $\mathcal{X}$ is S_N-closed, then $\mathcal{X}\mathcal{Y} = \mathcal{X} * \mathcal{Y}$.

(ii) If $\mathcal{Y} \subset \mathcal{Z}$ and $\mathcal{Z}$ is S_N-closed with $\mathcal{X} \cap \mathcal{Z} = 1$, then $\mathcal{X}\mathcal{Y} = \mathcal{X} * \mathcal{Y}$.

Concerning (I):

Lemma 1. Let V be an elementary abelian p-group for a prime p, $V \lhd G$ and let n be a positive integer. Then there is a group X and homomorphism ϕ from X onto G such that $\phi^{-1}(V)$ has exponent p^{n+1} and $\ker \phi = \Omega_n(\phi^{-1}(V))$ is the set of elements of order p^n in $\phi^{-1}(V)$.

There is an isomorphism $\gamma: G \to V \text{ wr } (G/V)$ and so let W be homocyclic with $\text{rank}(W) = \text{rank}(V)$ and $\exp(W) = p^{n+1}$. There is a homomorphism $\bar{\phi}$ from $W \text{ wr } (G/V)$ onto $V \text{ wr } (G/V)$ with $\ker \bar{\phi} = \Omega_n(W)$. Let $X = \bar{\phi}^{-1}(\gamma(G))$ and then $\phi = \gamma\bar{\phi}|_X$. Since $\gamma(V)$ is contained in the base group $V^{G/V}$, the conclusion follows.

Define $p\text{-exp}(\mathcal{X}) = \sup\{\exp P \mid P \in \text{Syl}_p(X), X \in \mathcal{X}\}$.

Proposition 1. Suppose that $G \in \mathcal{X}\mathcal{Y} \setminus \mathcal{X} * \mathcal{Y}$, that G has a normal subgroup K containing $G^{\mathcal{Y}}$ and that there is a set π of primes satisfying:

i) For each $p_i \in \pi$, $p_i\text{-exp}(\mathcal{X}) = p_i^{n_i}$ and $C_{p_i^{n_i}} \in \mathcal{X}$.

ii) For each $H \lhd G$ such that $H \in \mathcal{X}$ and $G^{\mathcal{Y}} \leq H$, G has a p-chief factor R/L with $R \leq H$, $L \leq K$, $RK/K \cong R/L$ and $p \in \pi$.

Then $\mathcal{X}\mathcal{Y}$ is not a formation.

The hypotheses appear intricate but, in fact, are the reality when $\mathcal{X}$ is a formation generated by a metanilpotent primitive group. The subgroup K is to give a focal point for application of Lemma 1.

To sketch the proof, let $\{H_1, H_2, \cdots, H_s\} = \{H \mid H \lhd G, H \in \mathcal{X}, G^{\mathcal{Y}} \leq H\}$ and let R_i/L_i denote a p_i-chief factor guaranteed by ii). Let $\overline{G} = G/K$ and $\overline{V}_i = R_iK/K$. By Lemma 1, for each i, there is $\phi_i : X_i \xrightarrow{\text{onto}} \overline{G}$ such that $\phi_i^{-1}(\overline{V}_i)$ has exponent $p_i^{n_i+1}$ and $\ker \phi_i = \Omega_{n_i}(\phi_i^{-1}(\overline{V}_i))$. By hypothesis $X_i \in \mathcal{X}\mathcal{Y}$. Also let $\eta : G \to \overline{G}$ be the natural homomorphism.

Let $P = G \times X_1 \times \cdots \times X_s$ and $\pi_0 : P \to G$, $\pi_i : P \to X_i$ be the projection mappings. Let

$$T = \{x \in P \mid \eta\pi_0(x) = \phi_i\pi_i(x) \text{ for } 1 \leq i \leq s\} \in R_0(\mathcal{X}\mathcal{Y}) .$$

But $T \notin \mathcal{X}\mathcal{Y}$, for if $Y \lhd T$ such that $Y \in \mathcal{X}$ and $T^{\mathcal{Y}} \leq Y$, then $\pi_0(Y) = H_i$ for some i and so $\overline{V}_i \leq \eta\pi_0(Y) = \phi_i\pi_i(Y)$. But $\phi_i^{-1}(\overline{V}) \leq \pi_i(Y)$. Thus $Y \notin \mathcal{X}$ and so $T \notin \mathcal{X}\mathcal{Y}$.

PROPOSITION 2. Let $\mathcal{X}$ be a formation such that $\pi = \{\text{primes } p \mid \text{ for some } X \in \mathcal{X}, p \,\big|\, |X/F(X)|\}$ satisfies property i) in the hypothesis of Proposition 1. Then $\mathcal{XY}$ is a formation if and only if $\mathcal{XY} = \mathcal{X} * \mathcal{Y}$.

Suppose $G \in \mathcal{XY} \setminus \mathcal{X} * \mathcal{Y}$ and $K = F(G)G^{\mathcal{Y}}$. Let $H \lhd G$ be such that $H \in \mathcal{X}$ and $G^{\mathcal{Y}} \leqslant H$. Then $H \cap K = F(H)G^{\mathcal{Y}}$ and so by Lemma 1.5 of [1], $H \neq F(H)G^{\mathcal{Y}}$. Then a chief factor $R/F(H)G^{\mathcal{Y}}$ with $R \leqslant H$ satisfies hypothesis ii) of Proposition 1 and so $\mathcal{XY}$ is not a formation. Proposition 2 then follows.

Concerning (II):

Let P_i denote a Sylow p-subgroup of S_{p^i}. Then $P_i \cong C_p$ wr P_{i-1} where the wreath product is taken with respect to the standard representation of P_{i-1}. Let $K = GF(p)$. From Mackey's Theorem, say, the following is apparent.

LEMMA 2. Suppose $P_i \lhd\lhd H$, then there is a $K(H)$-module W such that $P_{i+1} \lhd\lhd WH$.

LEMMA 3. Let V be a $K[G]$-module. Then there is an associated sequence of groups $\{X_i\}_{i=0}^{\infty}$ such that:

(1) For $i > 0$, $X_i = V(G \times X_{i-1})$, a semidirect product of V_i by $G \times X_{i-1}$.

(2) $V_i \cong V \otimes_K W_{i-1}$ for a $K[X_{i-1}]$-module W_{i-1}.

(3) $P_i \lhd\lhd X_i$.

Set $X_0 = 1$, $W_0 = K$, $X_1 = VG$ and inductively, since $P_i \lhd\lhd X_i$, there is a $K[X_i]$-module W_i such that $P_{i+1} \lhd\lhd W_i X_i$. Define $V_{i+1} = V \otimes_K W_i$ and $X_{i+1} = V_{i+1}(G \times X_i)$

where $G \times X_i$ acts on $V \otimes W_i$ by $(v \otimes w)^{(g,x)} = v^g \otimes w^x$. Then calculation verifies that induction may proceed.

To state my main result concerning (II), again it is necessary to impose some hypotheses on $\mathcal{X}$, but again these are satisfied when $\mathcal{X}$ is any formation generated by a metanilpotent primitive group.

PROPOSITION 3. Let $G \in \mathcal{Y}$, $H \lhd G$ and V a $K[G]$-module such that $VH \in \mathcal{X}$ while $V \notin \mathcal{X}$. If $\mathcal{X}\mathcal{Y} = \mathcal{X} * \mathcal{Y}$, then the sequence $\{X_i\}$ associated to G and V is a subset of $\mathcal{Y}$.

First note that $X_1 = VG \in \mathcal{X}\mathcal{Y}$ and $(VG)^{\mathcal{Y}} \leqslant V$. But V is a p-group and $V \notin \mathcal{X}$, so $VG \in \mathcal{Y}$. Inductively, $V_{i+1}(H \times 1) \in \mathcal{X}$ using the structure of a tensor product, while $X_{i+1}/V_{i+1} \in \mathcal{Y}$. Thus as above $X_{i+1} \in \mathcal{Y}$.

TEST CASE. Let $\mathcal{X} = QR_0(S_3) = \{E_3E_2$-groups with no central 3-chief factors$\}$ where E_p denotes the formation of elementary abelian p-groups. The following facts are either consequences of results presented or can be routinely verified.

Let $\mathcal{Y}$ be a formation. Then $\mathcal{X}\mathcal{Y} = \mathcal{X} * \mathcal{Y}$ if and only if $\mathcal{X}\mathcal{Y}$ is a formation.

(III) If $\mathcal{X}\mathcal{Y}$ is a formation and $2 \mid |Y|$ for some $Y \in \mathcal{Y}$, then every 3-group is a subnormal subgroup of a $\mathcal{Y}$-group.

(IV) If $\mathcal{Y} \subset S_{2'}$ = {groups of odd order} or $E_3\mathcal{Y} = \mathcal{Y}$, then $\mathcal{X}\mathcal{Y}$ is a formation.

Finally I demonstrate a method mentioned to me by John Cossey for showing that (I) and (II) do not have coincident answers, in general. Let $\mathcal{Y} = S_2$, the class of 2-groups, and $\mathcal{X} = \{G \in E_3S_2 | G$ has no central 3-chief factors$\}$. Then $\mathcal{X}\mathcal{Y} = \mathcal{X}$ while $\mathcal{X} * \mathcal{Y} = \mathcal{Y}$. Note that there are infinitely many formations F with $\mathcal{Y} \subset F \subset \mathcal{X}$.

REFERENCES

1. Bryant, Bryce and Hartley, *The formation generated by a finite group*, Bull. Austral. Math. Soc. 2 (1970) , 347-357.

2. W. Gaschütz, *Selected Topics in the Theory of Soluble Groups* (Lecture notes given in Canberra, 1969).

STATE UNIVERSITY OF NEW YORK AT BINGHAMTON
BINGHAMTON, NEW YORK

p-n GROUPS AND p-SATURATED FORMATIONS

BY

ELAYNE A. IDOWU

Let G be a finite group and p a prime number. In [1] W. E. Deskins suggested the study of the set of all maximal subgroups M of G which are "externally" related to G by the condition that $(p,[G:M]) = 1$. The intersection of this set of maximal subgroups is called the *p-Frattini subgroup* of G, $\varphi_p(G)$. $\varphi_p(G)$ is defined to be G if $G = 1$ or if p divides the index of every maximal subgroup of G.

Deskins showed in [2, Theorem 2] that $\varphi_p(G)$ is always an extension of a p-group by a nilpotent group of p'-order. Such groups are called *p-n groups*.

LEMMA 1. The following statements are equivalent:

(a) G is a p-n group

(b) G is q-nilpotent for every prime number $q \neq p$.

(c) If M is a maximal subgroup of G with $(p,[G:M]) = 1$, then $M \lhd G$.

(d) $G' \leqslant \varphi_p(G)$, where G' is the commutator subgroup of G .

(e) $G/\varphi_p(G)$ is a p-n group.

Let $\mathcal{PN}$ be the class of all p-n groups. Then $\mathcal{PN}$ is the class product of the class of p-groups by the class of nilpotent groups of p'-order. Since this product is the same as the formation product and the Fitting class product, $\mathcal{PN}$ is a Fitting formation.

A formation $\mathcal{F}$ is called *p-saturated* if $G/\varphi_p(G) \in \mathcal{F}$ implies that $G \in \mathcal{F}$. By lemma 1, $\mathcal{PN}$ is a p-saturated formation. Of course, every p-saturated formation is saturated; but $\mathcal{N}$, the formation of nilpotent groups, is a saturated formation which is not p-saturated for any prime number p .

$F_p(G)$, the *p-Fitting subgroup* of G , is the largest normal p-n subgroup of G ; i.e., $F_p(G)$ is the $\mathcal{PN}$ radical of G . If q is a prime number and $G(q)$ is the largest normal q-nilpotent subgroup of G , then

$$F_p(G) = \cap\{G(q) : q \neq p\} .$$

For every prime number t , let $f(t)$ be a formation with $f(p) \neq \emptyset$. Define a class of groups $\mathcal{F}(p)$ by $G \in \mathcal{F}(p)$ if and only if $G/F_p(G) \in f(p)$ and $G/G(t) \in f(t)$ for $t \neq p$. Then the class $\mathcal{F}(p)$ is a p-saturated formation. We call $\mathcal{F}(p)$ the formation *p-locally defined* by $\{f(t)\}$.

The class of p-n groups, the class of q-nilpotent groups $(q \neq p)$, and the class of groups with $q(\neq p)$-solvable length $\leqslant 1$ are examples of p-locally defined and hence p-saturated formations.

If we restrict our consideration to formations of solvable groups, then p-saturated is equivalent to being p-locally defined.

THEOREM A. Let F be a p-saturated formation of solvable groups. Then F can be p-locally defined.

PROOF. F is saturated, so there is a set of formations $\{f(t)\}$ which locally define F. Let $h(t) = f(t)$ if $t \neq p$ and let $h(p)$ be the smallest nonempty formation of solvable groups containing all of the $f(t)$'s. Let F^* be the formation p-locally defined by $\{h(t)\}$. F is clearly contained in F^*, and induction on the order of a solvable group G shows that $F = F^*$.

For formations of solvable groups it is also possible to prove:

THEOREM B. If F is a p-saturated formation, then every solvable group G has an F-projector K with

$$(p, [G:K]) = 1 .$$

THEOREM C. Let G be a solvable group. A subgroup K of G is a *PN*-projector of G if and only if K is a self-normalizing p-n subgroup with (p, [G:K]) = 1 .

In light of Theorem B, the only formation of solvable groups which is p-saturated for every prime number p is the class of all solvable groups.

REFERENCES

1. W. E. Deskins, *On maximal subgroups*, Proceedings of Symposia in Pure Mathematics, vol. 1 (Finite groups), Amer. Math. Soc., 1959, 100-104.

2 ____________, *A condition for the solvability of a finite group*, Ill. J. Math., vol. 2 (1961), 306-313.

UNIVERSITY OF PITTSBURGH
PITTSBURGH, PENNSYLVANIA

IRREDUCIBLE MODULES OF SOLVABLE GROUPS ARE ALGEBRAIC

BY

T. R. BERGER*

Suppose that G is a finite group, and $\underset{\sim}{k}$ is an algebraically closed field of characteristic $p > 0$. Assume that P is a p-Sylow subgroup of G. Work of Brauer [9] and Dade [10] has shown that if P is cyclic then one may find rather explicit connections between the ordinary characters of G and those of $N_G(P)$. Efforts to generalize these results so far have been unsuccessful. One of the many difficulties standing in the way of generalizations is a lack of knowledge about indecomposable modules for noncyclic p-groups at characteristic p. In the case of a cyclic p-group, the indecomposables are $|P|$ in number and are given by certain Jordan block matrices. If P is noncyclic then no such easy answer is available.

The initial observation one makes is that one need not consider all indecomposable modules. The idea is to understand

---

*Research partially supported by NSF grant GP 38879.

the indecomposable summands of $V|_P$ where V is some irreducible $\underset{\sim}{k}[G]$-module. The object of the game is then to find a suitably large class $\mathcal{J}$ of indecomposable $\underset{\sim}{k}[G]$-modules such that every "interesting" module appears in the class. Further, the class $\mathcal{J}$ should be small enough so that the "relevant" properties of the modules in the class may be known. Based upon examples, Alperin [1] has suggested two possible classes which might be interesting. These are the classes of algebraic modules and the classes of irreducibly generated modules. We shall now define these two classes of modules.

If V is a $\underset{\sim}{k}[G]$-module then let V^* be the isomorphism class of all $\underset{\sim}{k}[G]$-modules isomorphic to V. Let F be the free abelian group on the classes V^*. We let F_0 be the subgroup generated by the relations $(U \oplus V)^* - U^* - V^*$ where U is another $\underset{\sim}{k}[G]$-module. By factoring F_0 out of F we obtain a ring with addition and multiplication given by the following equations. Incidentally, we let $\{V\}$ denote the class of V in F/F_0.

(*) For $\underset{\sim}{k}[G]$-modules U and V,

(a) $\{U\} + \{V\} = \{U \oplus V\}$, and

(b) $\{U\}\{V\} = \{U \otimes_{\underset{\sim}{k}} V\}$.

The ring so defined is called the *Green ring* and is denoted by $a(G)$. The one dimensional trivial $\underset{\sim}{k}[G]$-module $V_0(G) = V_0$ gives rise to an identity $\{V_0\}$ for $a(G)$. Therefore $a(G)$

is a commutative ring with unity. A module V is called *algebraic* if there is a nonzero polynomial $p(X) \in Z[X]$ such that $p(\{V\}) = 0$. As we shall see, all irreducible modules for solvable groups are algebraic.

Another class of modules is the class of irreducibly generated ones. A module V is said to be irreducibly generated if any indecomposable summand of V is isomorphic to a direct summand of a tensor product $U_1 \otimes_{\underset{\sim}{k}} \cdots \otimes_{\underset{\sim}{k}} U_t$ where the modules U_i are all irreducible. As we shall see, for solvable groups, the irreducibly generated modules are among the algebraic modules.

These two classes of modules may or may not be interesting. At first glance, because most interesting modules of all solvable groups are included, the classifications appear to be too large to really know anything about. The problem is actually more apparent than real since the proof we give is quite constructive; that is, it gives a construction procedure for the irreducibly generated modules for solvable groups. Therefore,not only may these modules be interesting, but also, the solvable groups may constitute much too small a class of groups in which to test the meaning of "interesting".

Before proceeding to solvable groups, how do we prove that something is algebraic? The methods used here are primitive. Something is not only algebraic but also integral if it lives in an order. That is, if $\mathcal{O}$ is a commutative ring with unity which is also a finitely generated free Z-module then every element ω of $\mathcal{O}$ is *integral* in the sense that there is a *monic* polynomial $p(X) \in Z[X]$ of positive degree for which $p(\omega) = 0$. We shall focus on such rings $\mathcal{O}$ in the following way. If we want to show that a module V is algebraic then we find a finite number of indecomposable modules $V_1, \cdots, V_t$ such that the free Z-module $\mathcal{O}$ generated by $\{V_0\}, \{V_1\}, \cdots, \{V_t\}$ in a(G) is also a ring. If $\{V\} \in \mathcal{O}$ then V is algebraic.

The only other theorems which are needed are: (1) the Krull-Schmidt theorem, (2) the Mackey Decomposition theorem, and (3) the Mackey Tensor Product theorem.

Using the indecomposable direct summands of tensor powers of V we obtain the following two results.

THEOREM 1. *If* V *is algebraic then* V *is integral.*

The next result has been stated by Alperin [1].

THEOREM 2. V *is algebraic if and only if there are only finitely many isomorphism classes of modules represented among the indecomposable direct summands of tensor powers of* V.

As a corollary to Theorem 2 we obtain:

COROLLARY 3. (1) *If* V *is algebraic then any direct*

summand of V *is algebraic.*

(2) *The direct sum of algebraic modules is algebraic.*

Using the Mackey theorems it is not difficult to prove the following:

THEOREM 4. *If* H *is a subgroup of* G *and* U *is an algebraic* $\underset{\sim}{k}[H]$*-module then the induced module* $U|^G$ *is also algebraic.*

Combining Corollary 3 with this theorem we obtain:

COROLLARY 5. *If* V *is an indecomposable* $\underset{\sim}{k}[G]$*-module,* D *is a vertex for* V , *and* U *is a source for* V *upon* D *then* V *is algebraic if and only if* U *is algebraic.*

This corollary shows that the property of being algebraic is properly one belonging to p-groups. On the other hand, p- and p'-linking in G obviously has great influence upon the p-structure of G and its modules. So in passing from G to D , before we have used the p'-structure of G is to go from bad to impossible. Really, this corollary tells us that the whole problem boils down to a classification of the modules U which belong to "interesting" modules V .

Let us now limit ourselves to solvable groups G . Some of this discussion holds in general but does not clearly lead anywhere. We wish to discuss the following theorem.

THEOREM 6. *If* G *is a solvable group, and* V *is an irreducible* $\underset{\sim}{k}[G]$*-module then* V *is algebraic.*

The method of proof shows how to construct a class of modules for p-groups such that a source for V will be isomorphic to a direct summand of one of the constructed modules. In particular, the class contains only algebraic modules, and is large enough that a source for any irreducibly generated indecomposable module of a solvable group will be isomorphic to a direct summand of a module in the class.

The class is not difficult to describe. But first we must describe a module construction procedure. Let H be a subgroup of G , U a $\underset{\sim}{k}[H]$-module, and $\{x_1, \cdots, x_t\}$ a transversal for H in G. Form the modules $x_i \otimes_{\underset{\sim}{k}[H]} U = U_i$. If $y \in G$ and $yx_i = x_j h$ for $h \in H$ and $x_i \otimes u \in U_i$ then we may envision $y(x_i \otimes u)$ as being equal to $x_j \otimes (hu) \in U_j$. With this action, a trace may be defined

$$U|^G = \Sigma^{\oplus} U_i$$

which is the ordinary induced module. Viewing things this way, a norm map would be

$$\tilde{U}^G = \Pi^{\otimes_{\underset{\sim}{k}}} U_i$$

also giving a $\underset{\sim}{k}[G]$-module, which we call the *tensor induced module*. Clearly any symmetric function on the subscripts of $\{x_i, \cdots, x_t\}$ will give rise to a $\underset{\sim}{k}[G]$-module constructed from U . In fact, the functions on the subscripts need only be invariant under the permutation representation on the cosets of H in G . Our main interest will focus upon the induced and tensor induced modules for a group.

Now we may describe the class of modules. Let P be a p-group, P_0 a subgroup, and P_1/P_2 a section of P_0 . If p is odd then P_1/P_2 must be cyclic; and if $p = 2$ then P_1/P_2 must be cyclic or quaternion (including generalized quaternion). Let $\mathcal{J}(P_1/P_2)$ be the class of all indecomposable $\underset{\sim}{k}[P_1/P_2]$-modules which represent P_1/P_2 faithfully, and if P_1/P_2 is quaternion of order 2^{t+1} then we only include six possible module types of dimensions 2^t-1, 2^t-1 , 2^t+1 , 2^t+1, $2^{t+1}-1$ and $2^{t+1}-1$. These latter modules are described in a paper [11] of Dade and are exploited in [5]. We shall call these modules "Dade modules".

We may now define the class of modules $\mathcal{U}(P_0)$ as follows. For any "qualified" section P_1/P_2 of P_0 and any module $W \in \mathcal{J}(P_1/P_2)$ then place $W^{\sim P_0} \in \mathcal{U}(P_0)$. Form the class of modules $\mathcal{M}^*(P)$ by taking a subgroup P_0 of P, and taking $U_1 , \cdots , U_s \in \mathcal{U}(P_0)$, and then placing $(U_1 \otimes_{\underset{\sim}{k}} \cdots \otimes_{\underset{\sim}{k}} U_s)|^P$ in $\mathcal{M}^*(P)$. The final class $\mathcal{M}(P)$ is then the set of all modules $V_1 \otimes_{\underset{\sim}{k}} \cdots \otimes_{\underset{\sim}{k}} V_r$ for $V_i \in \mathcal{M}^*(P)$. Incidentally, the module induced by the identity upon P_0 is in $\mathcal{M}(P)$ for $P_0 = P_1 = P_2$.

We shall not actually prove Theorem 6, nor will we show exactly how the class $M(P)$ arises from the proof. But we shall describe the major steps of the proof. We shall do this by describing a four step reduction of any completely reducible $\underset{\sim}{k}[G]$-module.

Start with a completely reducible $\underset{\sim}{k}[G]$-module V.

(1) To "know" V we must "know" the irreducible summands U.

(2) To "know" U we must "know" a subgroup $H \leqslant G$ and a primitive $\underset{\sim}{k}[H]$-module W such that $W|^G \cong U$.

These first two steps hold for additive module structure. We now repeat these steps in the product module structure (unless $\dim U = 1$ in which case we quit). The module U may be decomposed into a tensor product $U \cong U_1 \otimes_{\underset{\sim}{k}} \cdots \otimes_{\underset{\sim}{k}} U_t$ of projective k[H]-modules with $\dim U_i > 1$. Further, the U_i may be taken to be "tensorially irreducible"; that is, any further decomposition of the U_i involves modules of dimension 1.

(3) To "know" the primitive k[H]-module U we must "know" the tensorially irreducible factors W of U. Each of these is a projective $\underset{\sim}{k}[H]$-module.

Lift W to a representation group H^* of H and factor out the kernel of W. This step is not necessary but it simplifies explanations. Now H^* acts upon W as a group $\overline{H}$. Further, if $\overline{F} = F(\overline{H})$ then $\overline{F}/Z(\overline{H})$ is a chief factor of $\overline{H}$

and $W|_{\overline{F}}$ is irreducible. In particular, $\overline{F}$ (ignoring most of $Z(\overline{H})$) is essentially an extraspecial r-group for some prime $r \neq p$. The next step is difficult to describe but may be distilled by the following comments. There is a subgroup $K \leqslant H$ which covers $\overline{F}$ in $\overline{H}$ and a projective $\underset{\sim}{k}[K]$-module X such that $\tilde{X}^H$ is a projective $\underset{\sim}{k}[H]$-module and, modulo a one-dimensional tensor factor, $\tilde{X}^H$ is isomorphic to W. The module X of K exhibits certain "super-primitive" qualities which we shall invoke later.

(4) To "know" the module W of (3) we must know a subgroup $K \leqslant H$ and a "super-primitive" projective $\underset{\sim}{k}[K]$-module X such that $\tilde{X}^H$ is "essentially" isomorphic to W.

This whole process may be repeated in a certain way and finally stops at a one-dimensional module for some subgroup of G. We shall not need this general decomposition. The four step procedure outlined here is described in Section 7 of [4]. One may imagine that (1), (2) and (3),(4) are exactly parallel steps; one in the additive, the other in the product module structure.

Let us now apply these steps in the proof of Theorem 6. We may proceed by induction upon $\dim V$. Since V is irreducible, step (1) is bypassed. Actually, Theorem 4 tells us that in step (2) we may assume that $V = U$ and $G = H$ so that

V is primitive. Step (3) may be taken by using a representation group $H^* = G^*$ in place of G and applying the following easy lemma.

LEMMA 7. *If* U *and* W *are algebraic* $\underset{\sim}{k}[G]$*-modules then* $U \otimes_{\underset{\sim}{k}} W$ *is algebraic.*

In particular, V is not only primitive, but also tensorially irreducible. At this point, the next lemma is useful.

LEMMA 8. *If* V *is an irreducible* $\underset{\sim}{k}[G]$*-module and* H *is a subgroup of* G *such that* (i) $V|_H$ *is irreducible, and* (ii) $[G:H]$ *is prime to* p, *then* V *is isomorphic to a direct summand of* $(V|_H)|^G$.

We are now ready to examine step (4). Of course, there is no harm in assuming that V is faithful. If $\dim_{\underset{\sim}{k}} V = 1$, clearly V is algebraic and we are through. Assume $\dim V > 1$ and let $F = F(G)$ be the Fitting subgroup of G. Recall that F is "essentially" an extraspecial r-group with $r \neq p$ and $V|_F$ irreducible. Let P be a p-Sylow subgroup of G and and set $H = PF$. Applying Lemma 8, Theorem 4, and Corollary 3 (1) we discover that $G = PF$. As we observed, the group $F/Z(G)$ is a chief factor of G . In order to take step (4) we need the following analogue to Theorem 4.

THEOREM 9. *Assume that* H *is a subgroup of* G *and that* U *is a projective* $\underset{\sim}{k}[H]$*-module. Let* H^* *and* G^* *be* *representation groups for* H *and* G *respectively. If* U *is algebraic for* H* *then* $\tilde{U}^G$ *is algebraic for* G* .

Using this theorem we may assume that G = K and V = X as in step (4). In other words, V is "super-primitive" and G = PF . Using results of [3] we now know that P is cyclic or p = 2 and P is quaternion. In this circumstance it is not difficult to show that a source for V either arises from a Jordan block matrix or is a Dade-module. Combining Corollary 5 with the following lemma completes the proof of Theorem 6.

LEMMA 10. *If* P *is a cyclic group and* V *is an indecomposable* $\underset{\sim}{k}$[P]*-module; or if* P *is a quaternion group and* V *is a Dade-module, then* V *is algebraic.*

More details in the proof of Theorem 6 may be found in [7]. Theorem 6 shows the usefulness of the method outlined in steps (1) - (4). This same mode of analysis has led to other results appearing in [2,6,8].

REFERENCES

1. J. Alperin, *The main problem of block theory*, this volume.

2. T. R. Berger, *Characters and derived length in groups of odd order*, J. of Algebra (to appear).

3. ________, *Hall-Higman Type Theorems II*, Transactions A.M.S. (to appear).

4. ________, *Hall-Higman Type Theorems V*, Pacific J. Math. (to appear).

5. ________, *Hall-Higman Type Theorems VII*, Proc. London Math. Soc. (to appear).

6. ________, *Nilpotent fixed point free automorphism groups of solvable groups*, Math. Z. 131 (1973) 305-312.

7. ________, *Solvable groups and algebraic modules*, (to appear).

8. ________ and F. Gross, *2-length and the derived length of a Sylow 2-subgroup*, (to appear).

9. R. Brauer, *On groups whose order contains a prime to the first power I*, II Amer. J. Math. 64 (1942) 401-420, 421-440.

10. E. C. Dade, *Blocks with cyclic defect groups*, Ann. of Math. (2) 84 (1966) 20-48.

11. __________, *Une extension de la théorie de Hall et Higman*, J. of Algebra 20 (1972) 570-609.

UNIVERSITY OF MINNESOTA
MINNEAPOLIS, MINNESOTA

CERTAIN FROBENIUS GROUPS ACTING FIXED-POINT-FREE ON SOLVABLE GROUPS

BY

ARNOLD D. FELDMAN

If H is a finite solvable group, let $F_1(H) = F(H)$, the Fitting subgroup of H. For $i \geqslant 1$, let $F_{i+1}(H)$ be the subgroup of H such that $F_{i+1}(H)/F_i(H) = F(H/F_i(H))$. The Fitting height of H, denoted $f(H)$, is the smallest integer i with the property that $F_i(H) = H$.

If A is a finite group, let $\ell(A)$ be the number of prime factors, counting multiplicities, in $|A|$.

Much work has been done to establish the following proposition.

CONJECTURE: Let G and A be finite solvable groups such that $(|G|,|A|) = 1$ and A acts fixed-point-freely on G; i.e., $C_G(A) = 1$. Then $f(G) \leqslant \ell(A)$.

Thompson [5] showed that if A is cyclic of prime order, then G is nilpotent; i.e., if $\ell(A) = 1$, then $f(G) = 1$.

Shult [3] showed that if $\ell(A) = 2$ and A is a Frobenius group of order pq with complement of order q and kernel of order p, then $f(G) \leqslant 2$ provided that either $|G|$ is odd or q is not a Fermat prime.

Berger [1] has shown that if A is nilpotent and Z_p wr Z_p free for all primes p, then $f(G) \leqslant \ell(A)$. His result encompasses those of many others.

It is the purpose of this paper to outline a proof of Theorem 1 below, which establishes the truth of the conjecture with certain restrictions on primes, in the case that A is a Frobenius group with cyclic kernel and a complement of prime order. The method of proof is based on that used by Shult to establish his result mentioned above.

THEOREM 1: Let $A = BQ$ be a Frobenius group with cyclic kernel B of order n and complement Q of prime order q. Suppose G is a finite solvable group such that $(|G|,|A|) = 1$, A acts fixed-point-freely on G, and the following conditions are satisfied:

(i) If q is a Fermat prime, $q = 2^m + 1$, then G has no extra-special section of order 2^{2m+1}.

(ii) If s is a prime and $\dfrac{s^m - 1}{s^{m/q} - 1}$ is a proper factor of n, then G has no elementary abelian section of order s^m.

(iii) If s is a prime and $\frac{s^m + 1}{s^{m/q} + 1}$ is a proper factor of n, then G has no extra-special section of order s^{2m+1}.

Then $f(G) \leqslant \ell(A)$.

It is easy to see that (i), (ii), and (iii) all hold if $q = 2$, yielding the following corollary.

COROLLARY: Let A be a dihedral group of order $2n$, where n is odd. If G is a finite solvable group such that $(|G|,|A|) = 1$ and A acts fixed-point-freely on G, then $f(G) \leqslant \ell(A)$.

Theorem 1 is proved by first looking at a counterexample to it such that $|A|$ and $|A| + |G|$ are minimal. In this case we show that $F(G)$ is an elementary abelian r-group for some prime r, and that $F(G) = F(GA)$, where GA is the semidirect product of G by A. Then the semidirect product $(G/F(G))A$ acts faithfully on $F(G)$; so it is sufficient to prove the following theorem, which is applied with $G/F(G)$ in place of H.

THEOREM 2: Let A be as in Theorem 1, and let H be a finite solvable group such that $(|H|,|A|) = 1$ and A acts fixed-point-freely on H. Suppose F is a finite-dimensional vector space over a finite field K of characteristic r, where $r \nmid |A|$ and $O_r(H) = 1$, and F is a faithful KHA-module. If A acts fixed-point-freely on F and conditions (i), (ii), and (iii) of Theorem 1 hold with G replaced by H, then $f(H) \leqslant \ell(A) - 1$.

Theorems 1 and 2 are proved simultaneously by induction on $|A|$. First we show that in Theorem 2 we can assume K is a splitting field for all subgroups of HA, and that F is an irreducible KHA-module. By Clifford's Theorem, $F = V_1 \oplus \cdots \oplus V_t$, where the V_i are pairwise nonisomorphic homogeneous KH-modules conjugate under A. We can index the V_i so that the stability group in HA of V_1 is HA_1, where $A_1 \leqslant A$ and, if q divides $|A_1|$, then $A_1 = B_1Q$, where $B_1 \leqslant B$. Clearly t divides $|A : A_1|$. Denoting by ρ the representation of HA_1 afforded by V_1, we show that we can assume that $\ker \rho \leqslant H$ and that the semidirect product $(H/\ker \rho)A_1$ acts faithfully and irreducibly on V_1, while A_1 acts fixed-point-freely on V_1 (so $A_1 \neq 1$). There are four possibilities for A_1:

Case I. $A_1 \leqslant B$

Case II. $A_1 = Q$

Case III. $A_1 = BQ = A$

Case IV. $A_1 = B_1Q$, where $1 < B_1 < B$.

In Case I, by a theorem of Shult [4; Theorem 4.1], A_1 is not faithful on $H/\ker \rho$, which implies A is not faithful on H, yielding $f(H) \leq \ell(A) - 1$ by induction on $|A|$ in Theorem 1.

In Case II, using the same theorem, we find that Q centralizes $H/\ker \rho$, which implies B acts fixed-point-freely on H, yielding $f(H) \leq \ell(A) - 1$ by Berger's result. Here, condition (i) is necessary in order that Shult's theorem be applicable.

We eliminate Case III by proving the following general lemma.

LEMMA: Let H be a finite solvable group such that the finite group A acts fixed-point-freely on H and $(|H|,|A|) = 1$. Let K be a splitting field for all subgroups of H. Suppose V is a finite-dimensional, irreducible, faithful KHA-module, and $\mathrm{char}(K) \nmid |A|$. Then, considered as a KH-module, V is not homogeneous.

In Case IV we apply Theorem 3 below with $H/\ker \rho$ in place of H, A_1 in place of A, and V_1 in place of V, to get that A_1 is not faithful on $H/\ker \rho$, which yields the desired result as in Case I.

THEOREM 3: Let H be a finite solvable group and $A = BQ$ a Frobenius group with $|B| = n$ and $|Q| = q$, and B cyclic as in Theorem 1. Assume $(|H|,|A|) = 1$, and V is a vector space of finite dimension over K, a splitting field for all subgroups of HA, where $\mathrm{char}(K) \nmid |A|$. Suppose A

acts fixed-point-freely on V and V is a faithful KHA-module which is homogeneous as a KH-module. If

(i) H has no elementary abelian section of order s^m, where s is prime and $\frac{s^m - 1}{s^{m/q} - 1}$ divides n

and (ii) H has no extra-special section of order s^{2m+1}, where s is prime and $\frac{s^m + 1}{s^{m/q} + 1}$ divides n,

then A is not faithful on H.

Theorem 3 is proved by induction on $\dim_K V + |H| + |A|$. We consider a maximal normal A-invariant subgroup M of H and show that if V is not homogeneous when considered as a KM-module, then B acts regularly on the elementary abelian group H/M, and also every non-trivial element of H/M is centralized by exactly one conjugate of Q in A. Setting $|H/M|=s^m$, s a prime, we find $n = \frac{s^m - 1}{s^{m/q} - 1}$, contrary to condition (i). If condition (i) is eliminated from the hypotheses of Theorem 3, counterexamples to the theorem can be constructed.

If V is homogeneous as a KM-module, we eventually reduce to the case that H is an extra-special group, and $M = Z(H)$, which is centralized by A, while A acts faithfully on $H/Z(H)$. We reduce further to the case that V is an irreducible KH-module. Then we use results of Glauberman [2; Corollary 6] to analyze the character η of the representation of HA on V . We know $\eta|_{HB}$ is the character of an irreducible representation of HB , and find that

$$\eta|_B = \varepsilon\lambda(1)1_B + \frac{1}{n}(\eta(1) - \varepsilon\lambda(1))\rho_B ,$$

where λ is an irreducible character of Z(H), 1_B is the principal character of B , ρ_B is the character of the regular representation of B , and $\varepsilon = \pm 1$.

Considering $\eta|_{HQ}$, also irreducible, we find that

$$\eta|_Q = \tilde{\varepsilon}\tilde{\lambda}(1)\tilde{\theta} + \frac{1}{q}(\eta(1) - \tilde{\varepsilon}\tilde{\lambda}(1))\rho_Q ,$$

where $\tilde{\lambda}$ is an irreducible character of $C_H(Q)$, $\tilde{\theta}$ is an irreducible character of Q , ρ_Q is the character of the regular representation of Q , and $\tilde{\varepsilon} = \pm 1$.

Finally, by studying $\eta|_A$, and using the fact that A is a Frobenius group and the principal character of A is not a constituent of $\eta|_A$, we arrive at the formula $n = \dfrac{s^m - \varepsilon}{s^{m/q} - \varepsilon}$, where $|H| = s^{2m+1}$. We then show that $\varepsilon = -1$, contrary to condition (ii). In the process of reducing to the case that H is extra-special, it may be necessary to replace A by a homomorphic image of itself that is also a Frobenius group of the same form; this is why it is necessary to prohibit $\dfrac{s^m + 1}{s^{m/q} + 1}$ from dividing n as well as prohibiting it from being equal to n . Thus Theorem 3 is proved, establishing Theorems 1 and 2.

It is also possible to use the same methods to prove a result analogous to Theorem 1, but where A is any group of order pqr, a product of 3 distinct primes. The proof uses the Lemma above and the fact that any proper subgroup of such a group is abelian or a Frobenius group with cyclic kernel and complement of prime order. The precise prime conditions necessary for the proof depend on the isomorphism type of the group A.

REFERENCES

1. T. R. Berger, *Nilpotent fixed point free automorphism groups of solvable groups,* Mathematishe Zeitschrift 131 (1973), 305-312.

2. G. Glauberman, *Correspondences of characters for relatively prime operator groups,* Canadian Journal of Mathematics 20 (1968), 1465-1488.

3. E. Shult, *Nilpotence of the commutator subgroup in groups admitting fixed point free operator groups,* Pac. Journal of Mathematics 17 (1966), 323-347.

4. ____________, *On groups admitting fixed point free abelian operator groups,* Ill. Journal of Mathematics 9 (1965), 701-720.

5. J. Thompson, *Finite groups with fixed point free automorphisms of prime order,* Proceedings of the National Academy of Sciences 45 (1959), 578-581.

RUTGERS UNIVERSITY
NEW BRUNSWICK, NEW JERSEY

UNIVERSITY OF MICHIGAN
ANN ARBOR, MICHIGAN

BOUNDING THE FITTING LENGTH OF A FINITE GROUP

BY

TREVOR HAWKES

A theorem of Dade's states that, if C is a Carter subgroup of a finite solvable group G, then the Fitting length, $\ell(G)$, is bounded by a function of the composition length of C. I discussed results from representation theory required to prove a dual of Dade's theorem; namely, that $\ell(G)$ is bounded by the number of generators of a nilpotent injector of G.

UNIVERSITY OF OREGON
EUGENE, OREGON